Environmental Issues and Waste Management Technologies in the Ceramic and Nuclear Industries VII

Related titles published by The American Ceramic Society:

Environmental Issues and Waste Management Technologies in the Ceramic and Nuclear Industries VI (Ceramic Transactions Volume 119)
Edited by Dane R. Spearing, Gary L. Smith, and Robert L. Putnam
©2001, ISBN 1-57498-116-1

Boing-Boing the Bionic Cat and the Jewel Thief
By Larry L. Hench
©2001, ISBN 1-57498-129-3

Boing-Boing the Bionic Cat
By Larry L. Hench
©2000, ISBN 1-57498-109-9

The Magic of Ceramics
By David W. Richerson
©2000, ISBN 1-57498-050-5

Environmental Issues and Waste Management Technologies in the Ceramic and Nuclear Industries V (Ceramic Transactions Volume 107)
Edited by Gregory T. Chandler and Xiangdong Feng
©2000, ISBN 1-57498-097-1

Ceramic Innovations in the 20th Century
Edited by John B. Wachtman Jr.
©1999, ISBN 1-57498-093-9

Environmental and Waste Management Technologies in the Ceramic and Nuclear Industries IV (Ceramic Transactions Volume 93)
Edited by J. C. Marra and G.T. Chandler
©1999, ISBN 1-57498-057-2

Environmental and Waste Management Technologies in the Ceramic and Nuclear Industries III (Ceramic Transactions Volume 87)
Edited by D. Peeler and J. C. Marra
©1998, ISBN 1-57498-035-1

Environmental and Waste Management Technologies in the Ceramic and Nuclear Industries II (Ceramic Transactions Volume 72)
Edited V. Jain and D. Peeler
©1996, ISBN 1-57498-023-8

Environmental and Waste Management Technologies in the Ceramic and Nuclear Industries (Ceramic Transactions Volume 61)
Edited by V. Jain and R. Palmer
©1995, ISBN 1-57498-004-1

Environmental and Waste Management Issues in the Ceramic Industry II (Ceramic Transactions Volume 45)
Edited by D. Bickford, S. Bates, V. Jain, and G. Smith
©1994, ISBN 0-944904-79-3

Environmental and Waste Management Issues in the Ceramic Industry I (Ceramic Transactions Volume 39)
Edited by G. B. Mellinger
©1994, ISBN 1-944904-71-8

For information on ordering titles published by The American Ceramic Society, or to request a publications catalog, please contact our Customer Service Department at 614-794-5890 (phone), 614-794-5892 (fax), <customersrvc@acers.org> (e-mail), or write to Customer Service Department, 735 Ceramic Place, Westerville, OH 43081, USA.

Visit our on-line book catalog at <www.ceramics.org>.

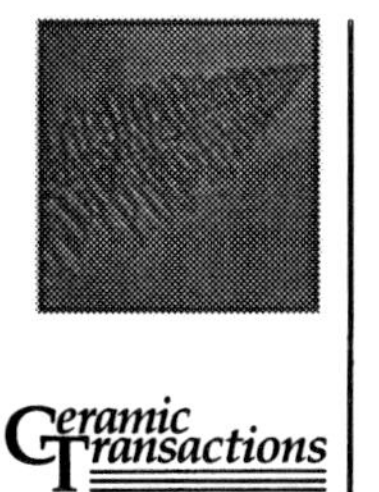

Ceramic Transactions
Volume 132

Environmental Issues and Waste Management Technologies in the Ceramic and Nuclear Industries VII

Proceedings of the Science and Technology in Addressing Environmental Issues in the Ceramic Industry symposium and the Ceramic Science and Technology for the Nuclear Industry symposium at the 103rd Annual Meeting of The American Ceramic Society, held April 22–25, 2001, in Indianapolis, Indiana, USA.

Edited by

Gary L. Smith
Pacific Northwest National Laboratory

S.K. Sundaram
Pacific Northwest National Laboratory

Dane R. Spearing
Los Alamos National Laboratory

Published by

The American Ceramic Society
735 Ceramic Place
Westerville, Ohio 43081
www.ceramics.org

Proceedings of the Science and Technology in Addressing Environmental Issues in the Ceramic Industry symposium and the Ceramic Science and Technology for the Nuclear Industry symposium at the 103rd Annual Meeting of The American Ceramic Society, held April 22–25, 2001, in Indianapolis, Indiana, USA.

Copyright 2002, The American Ceramic Society. All rights reserved.

Statements of fact and opinion are the responsibility of the authors alone and do not imply an opinion on the part of the officers, staff, or members of The American Ceramic Society. The American Ceramic Society assumes no responsibility for the statements and opinions advanced by the contributors to its publications or by the speakers at its programs. Registered names and trademarks, etc., used in this publication, even without specific indication thereof, are not to be considered unprotected by the law.

No part of this book may be reproduced, stored in a retrieval system, or transmitted in any form or by any means, electronic, mechanical, photocopying, microfilming, recording, or otherwise, without prior written permission from the publisher.

Authorization to photocopy for internal or personal use beyond the limits of Sections 107 and 108 of the U.S. Copyright Law is granted by the American Ceramic Society, ISSN 1042-1122 provided that the appropriate fee is paid directly to the Copyright Clearance Center, Inc., 222 Rosewood Drive, Danvers, MA 01923 USA, www.copyright.com. Prior to photocopying items for educational classroom use, please contact Copyright Clearance Center, Inc.

This consent does not extend to copying items for general distribution or for advertising or promotional purposes or to republishing items in whole or in part in any work in any format.

Please direct republication or special copying permission requests to the Senior Director, Publications, The American Ceramic Society, PO Box 6136, Westerville, Ohio 43086-6136, USA.

Cover photo: "Zirconium-containing crystals ...baddeleyite..." is courtesy of Pavel Hrma, and appears as figure 5 in the paper "Crystallization in High-Level Waste Glasses," which begins on page 243.

Library of Congress Cataloging-in-Publication Data

A CIP record for this book is available from the Library of Congress.

For information on ordering titles published by The American Ceramic Society, or to request a publications catalog, please call 614-794-5890.

4 3 2 1–05 04 03 02

ISSN 1042-1122

ISBN 1-57498-146-3

Contents

Vitrification and Process Technologies

Crystallization in Nuclear Waste Forms

Chemical Durability and Characterization

Alternative and Innovative Waste Forms

Preface

In 2001, The American Ceramic Society hosted several symposia focusing on five main focus areas with one of them being the impact of ceramics in energy manipulation and the environment. In today's world of increasingly stringent environmental regulations, it is critical to identify and adequately address environmental issues in the ceramic industry to ensure success. In ceramic manufacturing, companies are beginning to focus on "green ceramics," performing "life cycle analyses," and adopting "environmental stewardship" to manufacture environmentally friendly products.

In addition, ceramics and glasses play a critical role in the nuclear industry. Nuclear fuels and waste forms for low-level and high-level radioactive, mixed, and hazardous wastes are primarily either ceramic or glass. Effective and responsible environmental stewardship is becoming increasingly more important in the world.

We hope that through the symposia and subsequent proceedings, we are helping to foster continued scientific understanding, technological growth, and environmental stewardship within the field of ceramics. You hold in your hands the combined proceedings from two of the symposia, "Ceramic Science and Technology for the Nuclear Industry" and the "Science and Technology in Addressing Environmental Issues in the Ceramic Industry," from the impact of ceramics in energy manipulation focus area presented at The American Ceramic Society's 103rd Annual Meeting & Exposition held April 22–25, 2001 in Indianapolis, Indiana. This proceeding represents the thirteenth volume published by The American Ceramic Society in the areas of waste management and environmental issues in relation to ceramics. Previous proceedings on nuclear waste management and environmental issues date from 1983 and include Advances in Ceramics volumes 8 and 20 and Ceramic Transactions volumes 9, 23, 39, 45, 61, 72, 87, 93, 107, and 119.

The editors gratefully acknowledge and thank Richard A. Haber, Vijay Jain, Carol M. Jantzen, Kyei-Sing (Jasper) Kwong, Robert L. Putnam, and Camilla Warren for help in organizing the symposia; the session chairs, Robert L. Putnam, Antoine Jouan, James C. Marra, William L. Ebert, Andrew C. Buechele, John D. Vienna, and Ian L. Pegg, for their contribution in keeping the presentations running smoothly; and most importantly, the authors and reviewers, without whom a high quality proceedings volume would not be possible. Lastly, the editors thank Teresa Schott of the Pacific Northwest National Laboratory and the book publishing team at The American Ceramic Society: Mary Cassells, Sarah Godby, and Jennifer Hereth. Their support and contributions were instrumental in the publication of this volume.

Gary L. Smith

S.K. Sundaram

Dane R. Spearing

Recycling of Ceramic Byproducts

REFRACTORY RECYCLING - CONCEPT TO REALITY

James P. Bennett, Kyei-Sing Kwong
Albany Research Center - USDOE
1450 Queen Ave. SW
Albany, OR 97321
Phone: 541-967-5983
FAX: 541-967-5845
E-Mail: bennett@alrc.doe.gov

ABSTRACT

The reuse/recycling of spent refractory materials as a raw material by industry is limited by the value of the used material as a component in the process, by beneficiation costs, or by plant related factors. Economics and/or legislation continue to be the main driving forces in successful recycling programs. The lack of significant driving forces are why spent refractory materials are typically landfilled in most countries. Other factors are emerging that help to facilitate reuse/recycling, but none are strong driving forces. An evaluation will be made of successful recycling processes at a number of industries, including steel, aluminum, brass, and glass, in an effort to find similarities between them.

INTRODUCTION

Refractories removed from industrial applications have historically been landfilled in most countries of the world. The reasons for this are based on the lack of regulations affecting refractory waste disposal, the high economic costs associated with the reuse of spent refractory materials when compared to virgin raw materials, and the low perceived costs associated with landfilling spent refractory material. Refractory disposal/reuse for most companies has been and continues to be an economic decision. In contrast to the users of refractory materials, the manufacturers have long recognized advantages in the reuse of excess refractory material, off specification material, and/or floor sweepings. These materials are frequently reworked into existing formulations by refractory producers as a raw material. Benefits include reduced firing shrinkage and energy savings (because the prefired material has no water of hydration, carbonates, or other energy consuming reactions and because sintering reactions that need

To the extent authorized under the laws of the United States of America, all copyright interests in this publication are the property of The American Ceramic Society. Any duplication, reproduction, or republication of this publication or any part thereof, without the express written consent of The American Ceramic Society or fee paid to the Copyright Clearance Center, is prohibited.

energy are complete) [1]. A few companies currently process refractory waste materials as a business, but shipping distances, beneficiation costs, and/or limited demand for the processed material frequently make reuse/refractory recycling uneconomical when compared to the cost and availability of many mined virgin materials. These and other complex issues related to refractory recycling have clearly impacted material reuse considerations and applications.

According to US Census Bureau data for 1999, more than three million metric tons of refractory materials were produced in the United States with a value of 2.3 billion dollars [2]. Refractory materials produced include dense fired brick; lightweight, low density insulating brick; monolithic, lightweight fiber or castable insulation; and fused refractory materials. These refractory products can be composed of oxides (including SiO_2, Al_2O_3, MgO, CaO, ZrO_2, or mixtures of these), carbides, nitrides, carbon, metals and/or other materials. The raw materials in a refractory can be natural, sintered, or fused grains. Up to 15 different raw materials or additives may be used [3], each contributing specific traits to material performance. Bonding of the refractory raw materials is typically by sintering, chemical reactions (primarily cement, phosphate, or sol-gel), or carbon. The refractory product of today has evolved based on a goal of performance, not material recycleability.

The demands placed on a refractory product by a user vary greatly with each application. A general table listing the main application for refractories is given in table I [4].

Table I- Industrial consumption of refractories materials based on weight [4].

Industry	Percentage
Steel	70
Cement and lime	7
Ceramics	6
Chemical	4
Glass	3-4
Non-ferrous metals	2-3
Other	6

Most refractory products are used in iron and steel applications (70 pct.), followed at much lower levels by cement and lime (7 pct.), ceramics (6 pct.), and others. Refractories used in each general class are unique, with their own beneficiation problems such as impurities, quantities, and frequency of generation. Regardless of the application, the amount of refractory material removed from service is typically less than originally applied because of wear and degradation that occurs during use. The amount of refractory material eligible for reuse/recycling is further reduced because of contamination that occurs during use, removal, or storage. Contamination can be due to the process (slag, salt, or process infiltration into the refractory), from material reactions that occur during use (example - metallic antioxidant forming oxides, carbides, nitrides or sodium oxide reacting with alumina and silica to form nepheline), or from changes brought about by the process (chrome +3 going to the carcinogenic +6).

It is also known that there are differences in the amount of used refractory material that can be reclaimed based on whether the spent refractory is a brick or a monolithic. Bricks are generally considered easier to recycle versus monolithic materials, which are more prone to contamination and beneficiation difficulties. The difficulties in handling a monolithic can range from their large size to anchors contained in the refractory [5]. These and other issues complicate recycling of spent refractories. When spent refractory materials are to be reused, it is essential to keep the material as "clean" as possible [1]. This will help to minimize beneficiation costs and will help to broaden potential reuse applications.

In the United States, regulations affecting refractory waste disposal have impacted the textile and fiberglass industry the most. The regulations have at their heart the Resource Conservation and Recovery Act of 1976. The Toxicity Characteristics Leaching Procedure (TCLP) was developed because of this act and established legal limits for hazardous wastes [6]. Maximum concentrations of contaminants for some specific hazardous materials of concern to refractory users are chromium (5 mg/L) and lead (5 mg/L). This guideline has helped to encourage recycling of chrome oxide containing refractories from the textile and fiberglass industries, and has brought about a large decrease in the consumption of chrome containing refractories by the steel industry. Exemptions on regulations affecting the disposal of chrome containing refractories were granted under the 1990 National Capacity Variance Act. When the act and two extensions of it expired in May of 1994, all refractory materials became subject to strict guidelines as to material disposal if they contained hazardous material.

The high disposal cost of refractories impacted by these regulations brought about the recycling and recovery techniques for many of the chrome oxide containing refractories used in the textile and fiberglass industries. The recycling/recovery techniques currently used were developed over many years and are viewed as highly competitive among the different refractory producers. Webber [7] noted that spent refractory material removed from glass industry furnaces could be added back to the refractory during raw material fusion, but that upper limits existed as to the quantity that could be added. Once a general point is exceeded, deterioration of the physical properties occurs where impurities from the spent refractory exceed critical values.

Another example of recycling in the glass industry is a proprietary process developed and used commercially to take spent magnesia-chromia refractory material removed from glass reheat checkers and treat them by a water leach process to remove sulfates [8]. The company treating refractory materials by this process charges a fee for the processing. During the leaching process for sulfate ion removal, hexavalent chrome oxide is removed from the refractory material. The treated refractory grain obtained after leaching is sold for reuse.

A broad understanding of the issues/concerns impacting refractory reuse/recycling, of current refractory reuse and disposal practices, of potential applications, of successful recycling efforts, and of necessary steps for enacting a refractory recycling program is important if the reuse and disposal of spent refractory materials are to be widely practiced. Corporate commitment is also necessary to develop and implement a recycling program. External processing by an experienced recycler of spent refractory materials is frequently the most viable plan for a successful, long-term recycling program. When initiating a program, it is best to first recycle those materials which are the easiest to recycle, leaving a positive recycling experience in the plant. More difficult materials can be recycled later.

ISSUES/CONCERNS IMPACTING REUSE/RECYCLING

Growing concerns over legislation, the environment, and future liability have acted as driving forces for North American corporations to consider recycling [9]. Other factors; including product stewardship, pressure from refractory users, corporate policy, the high value of some refractory components (particularly fused products), and perceived marketing advantages; have also acted to encourage reuse/recycling. The high cost of land-filling material and/or limited space in landfills are also motivating factors to recycle spent refractories at a few corporations. The impact of marketplace forces for ISO 14000 certification programs (environmental management systems) may play a future role in encouraging future recycling. Material suppliers to Ford Motor Co. and General Motors, for example, are now required to have ISO 14000 certification [10].

Laws to reduce industrial wastes sent to landfills and/or the implementation of taxes on industrial wastes have encouraged the start of some refractory recycling efforts, particularly in Europe and Japan, where greater consideration is given to the reuse of spent refractory material in refractory products. A greater willingness on the part of refractory users to recycle exists because of these regulations, and because of the close proximity of refractory users and producers (helping to keep shipping costs low).

Mergers taking place within industry have consolidated and reduced the number of refractory manufactures. Although this reduction could act as a driving force for recycling, it has not. The push for recycling in the U.S. has followed the economy. In good economic times, many corporations are interested in recycling. In an economic downturn, however, industrial priorities are focused on core business activities, not recycling.

Each refractory user requires an individualized approach to reuse/refractory recycling. In the steel industry, for example, some manufacturers do not want refractories reused in "new" refractories because of concerns over decreased service life. Other steel producers do not share that concern, but don't have the storage space for spent refractory materials or don't have refractory producers or recyclers in close proximity to process spent materials. Many mills base recycling decisions strictly on economics. Does it make economic sense to recycle or to landfill spent material? Because of these and other factors, refractory recycling/reuse is very complex and must be dealt with on a case by case basis, with only broad similarities existing between plants. Some of the issues and concerns associated with recycling are listed in table II below:

Table II - Issues/concerns impacting refractory recycling.

Consumption	Location of users
Lining life	Frequency of generation
Contamination	Economics of beneficiation
Federal/state/local regulations	Health concerns
Value of components	Type of refractory
Age of material	Quantity of refractories
Mixed linings	Location of recyclers/refractory companies

It is worth noting that recycled refractory material uses can range from low to high volume applications with varying values. Recycled material consistency and quality-control over time play a large roll in its final price and in determining potential applications. The more effort given to keeping a material clean, the more potential applications exist for it. Companies involved in the recycling of spent refractories understand the importance of material cleanliness and the relationship to cost and applications. They do an excellent job of processing spent refractory material at a competitive price.

SPENT REFRACTORY RECYCLING/REUSE APPLICATIONS

Spent refractory materials may be beneficiated and reused internally, by an outside contractor, or be processed by a company specializing in recycling. Regardless of the targeted recycling/reuse application, material that is easy to process should be attempted first. More challenging materials should be processed at a later time.

Many potential internal and external applications exist or have been evaluated for spent refractory materials and are listed in table III.

Table III- Possible applications for spent refractory materials

Refractory component	Roofing granules
Slag conditioner	Tile body component
Component in cement, aggregate for concrete	Raw material for glass
	Highway road aggregate/subbase
Building component	Ferro alloy (high chrome containing materials)
Abrasive	
Fuel source (SiC, C containing materials)	Insulating powder
	Filler for bulk items
Carbon, silicon source	Landscape material
Grog in ceramic materials	Soil conditioner
Soil stabilization	Waste neutralization/treatment (acids, pathogens)

Companies that reuse/recycle spent refractory materials typically have extensive experience beneficiating spent refractory materials and have a wealth of firsthand knowledge in difficulties and pitfalls of recycling. They have established specific markets for the beneficiated spent refractory materials and have knowledge of other potential markets. Processing by these recyclers can be done on a toll basis or the spent material can be bought, processed and sold for specific applications by them. The recyclers possess beneficiation equipment such as crushers, screens, dryers, dust collectors, and bagging and storage facilities. Because of the flexibility of the recycler, they represent a viable recycling/reuse option for a refractory user not willing or able to process their materials.

SUCCESSFUL RECYCLING EFFORTS

Spent refractory materials are beneficiated by a number of techniques. Companies in the business of recycling spent refractory material typically have an operation containing all or parts of the flow sheet in figure 1. Recyclers first characterize the material to determine potential outlets for it. After a refractory is removed from an application, but before crushing, hand sorting is often necessary to segregate the different types of refractories. The different refractories occur because of zoning (the use of different types of refractory material in an application to optimize material properties), a common practice used by industry to improve refractory service life. Regardless of the origin of the refractory, the spent refractory material is reduced into a controlled particle size by crushers, then screened into different particle sizes. After screening, the processed materials should be checked to confirm chemistry. As mentioned earlier, at least one recycler of refractories removes sulfates by an aqueous process at this step.

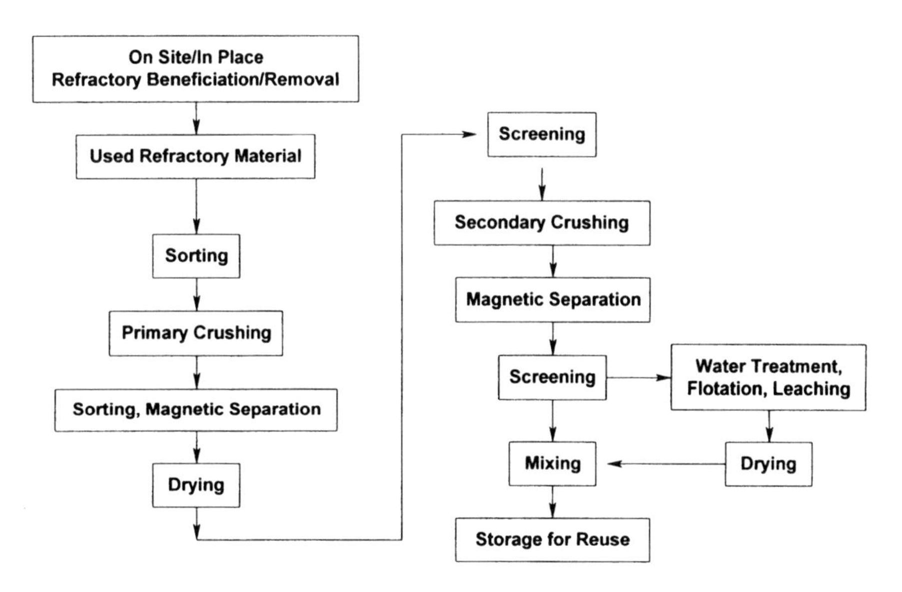

Figure 1 - General beneficiation process for refractory materials.

Flow sheets used to process spent refractory materials within an integrated steel mill (figure 2) and in an electric arc furnace shop (figure 3) illustrate some of the in-house applications for spent refractory reuse utilized by steel mills. The integrated steelmaker has more potential internal applications for spent refractory reuse than an EAF shop - if it is willing to develop these applications. The recycling flow sheets shown in figures 2 and 3 have been successful for long periods of time and have become integrated into company practice. The driving force for these applications were regulations, and the recycling effort occurred with management commitment.

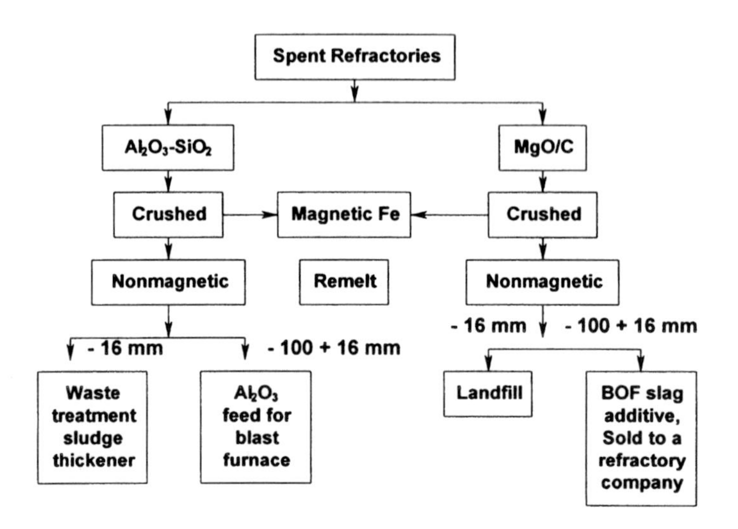

Figure 2 - Process flow sheet for spent refractory materials from an integrated steel mill.

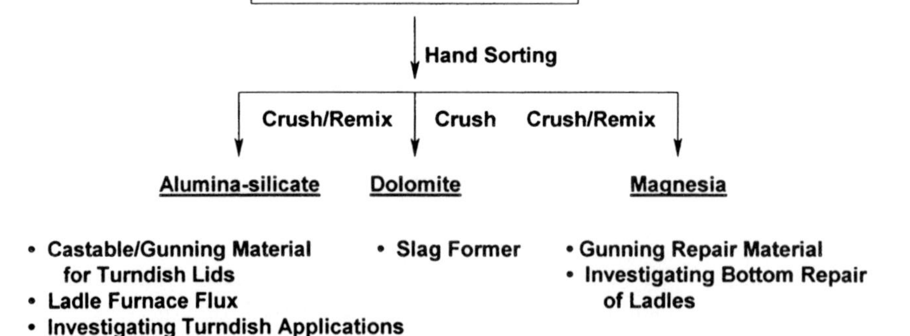

Figure 3 - Process flow sheet for refractory recycling at an EAF

Recycling of spent refractory materials in the aluminum industry has been limited primarily to carbon bake furnace brick [11]. Refractories removed from the primary and/or secondary melting furnaces are typically monolithic materials and have issues associated with impurities, material, size, bonding, and application. It may be possible to utilize the high alumina materials removed from the furnaces to satisfy alumina requirements of cement if local raw materials utilized by the cement producer are deficient in alumina. Flow sheets from two companies associated with reuse of carbon bake furnace brick are shown in figure 4.

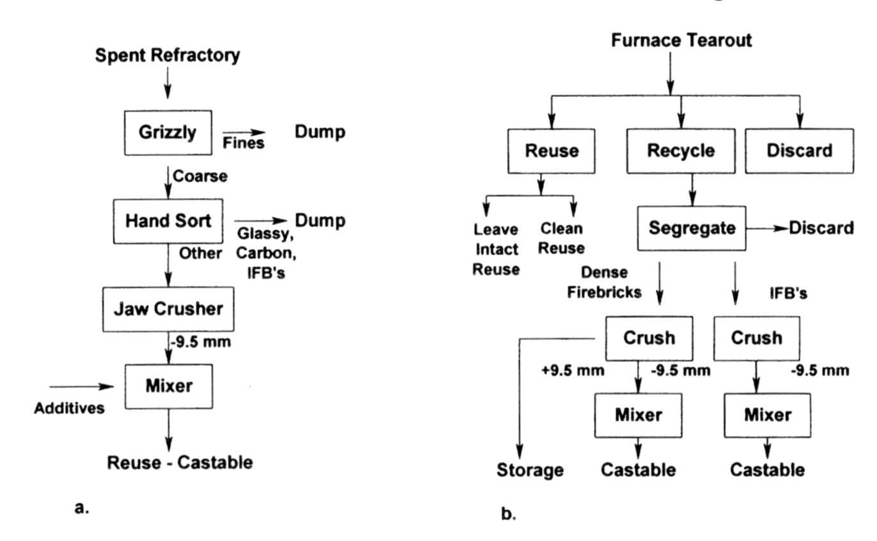

Figure 4 - Two types of process flow sheets (a. and b.) used to recycle spent refractory materials from different carbon bake furnaces of aluminum producers.

Both of the recycling programs shown in figure 4 had the support and commitment of management when enacted. The driving forces were environmental (reduced landfill space) and economic (reduced furnace rebuilding costs). It is important to note that in the flow sheets shown in figures 2-4, the refractory recycling/reuse processes involved segregation, crushing, then in house reuse of the spent refractory material. The beneficiation processes are simple, although the materials and the applications differ by plant.

ENACTING A RECYCLING PROGRAM

As mentioned earlier, when recycling/reusing spent refractory material, the commitment and backing of management is essential for a successful program.

Management should appoint one individual to head the effort and should give him strong support. An evaluation of the refractory materials coming to a plant, areas of usage, current "disposal" practices, and a cost assigned for the spent refractory handling/disposal should be made. An assessment should also be made as to possible reuse or disposal options for the materials and consideration should be given to process changes that could reduce refractory wastes being generated. This assessment should be made according to the priority listing in table IV [12].

Table IV - Priorities for spent refractory waste reuse/disposal [12].

1. Reduction in the process
2. Internal reuse
3. External reuse
4. Treatment
5. Landfill

The first priority in recycling is to reduce the amount of refractory waste being generated. Technological advances have played a significant role in reducing both the amount and the frequency of spent refractory material generation. Electric furnace design changes, for example, have reduced refractory consumption through the use of water-cooled panels, completely eliminating refractory material in these areas. Other examples are improved plant design and operation, ladle metallurgy adjustments, improved ladle and furnace maintenance and repair schedules, better slag and metallurgy control, and slag splashing [13]. Improved refractory materials; better customer service by refractory producers; hot patching; monolithic materials; zoning; gunning; subcontracted refractory installation, tearout and maintenance; the practice of the endless lining concept [14]; and refractory wear sensing have also acted to improve refractory life and reduce the amount of refractory wastes generated. Refractory wear has also been improved through the control of slag chemistry and slag foaming in the EAF [15]. Future ladle and furnace designs to reduce refractory wear may also include the use of contour linings to put the refractory in its equilibrium state from initial use. Technological advances such as these will continue to play a role in reducing the quantity and frequency of spent refractory material to be recycled. Active recycling/reuse programs, however, are necessary to reuse or recycle what refractory wastes are generated and to deal with the changing nature of refractory materials and of the process.

After considering ways to reduce refractory wastes by making improvements in the process, consideration should be given on ways the spent refractory wastes that are generated may be reused internally. An example is the potential for reuse of basic refractory materials such as MgO/C and dolomite/C as slag conditioners in electric arc furnaces or the reuse of spent MgO/C materials to freeze slag on the sidewalls of basic oxygen furnaces. Whatever the application, the economics of it should be compared with purchased materials before committing to a recycling program. Waste treatment and disposal in landfills should be a last option for refractory wastes.

CONCLUSIONS

Refractories removed from service are typically landfilled. A number of driving forces to encourage recycling have recently emerged. These include growing concerns over legislation, the environment, and future liability. Spent refractories can be reused/recycled as a raw material source in a number of products or processes ranging from low to high volume applications and with varying material values. Material cleanliness, beneficiation costs, and material consistency will play a large role in determining these applications. Barriers to material reuse are oftentimes based more on prior practices than technical issues.

A number of companies successfully process spent refractory wastes and have been doing so for a number of years. These companies have equipment and a large amount of firsthand experience they utilize to beneficiate and market the spent refractory materials. Applications for spent refractories have typically been internal rather than external. Efforts at a number of steel mills and aluminum producers have shown that successful reuse of spent refractory materials can be done with proper planning and if conditions such as storage space and company culture are conducive to recycling versus disposal. A successful recycling program must look at all costs associated with material reuse versus disposal.

REFERENCES

[1]Oxnard, R.T., "Refractory Recycling," *American Ceramic Society Bulletin*, **73**(10), Oct., 1994, pp 46-49.

[2]US Census Bureau, "Refractories 1999 - Current Industrial Reports MA327C(99)-1," 10 pp.

[3]Semler, C.E., "Overview of the U.S. Refractories Industry," Ceramic Ind., **142**(2), 1994, pp 33-38.

[4]Mosser, D.J., and G. Karhut, "Refractories at the Turn of the Millennium," Proceedings of the United International Technical Conference on Refractories 6th Biennial Worldwide Congress held in Berlin, Germany, Sept 6-9, 1999, XXV-XXX.

[5]Theriot, B.D., W.L. Greer, and K.J. Rone, "Recycling Cement Kiln Refractory Brick at Ash Grove Cement," Proceedings of the Unified International Technical Conference on Refractories 5th Biennial Worldwide Congress held in New Orleans, LA, USA, Nov. 4-7, 1997, pp 497-506.

[6]U.S. Code of Federal Regulations, Title 40--Protection of the Environment, Part 261--Identification and Listing of Hazardous Waste, July 1, 1999.
[7]Webber, R.A., "Recycling at Corhart - A 30 Year Success Story," Ceram. Eng. Sci. Pro., **16**(1), 1995, pp 214-215.
[8]Noga, J., "Refractory Recycling Developments," Ceram. Eng. Sci. Proc., **15**(2), 1994, pp. 73-77.
[9]Bennett, J.P. and K.S. Kwong, "Environmental Practices in Recycling/Reusing/Disposal of Spent Refractory Materials," Presented at the 102nd Annual Meeting of The American Ceramic Society, St. Louis, MO, May 1-2, 2000.
[10]Chalfant, R.V., "ISO 14000 and Environmental-Management Systems," July, 2000, New Steel, pp 65.
[11]Holmes, L., N.S. Schubert, A. Mooney, J. Bennett, and K.S. Kwong, "Recycling of Spent Refractory Material from Carbon Baking Furnaces," Proceedings of the United International Technical Conference on Refractories 5th Biennial Worldwide Congress held in New Orleans, LA, USA, Nov. 4-7, 1997, pp 477-486.
[12]Abrino, D.E., "Waste Minimization in Industries Using Refractory Materials," Proceedings of the United International Technical Conference on Refractories 5th Biennial Worldwide Congress held in New Orleans, LA, USA, Nov. 4-7, 1997, pp.465-471.
[13]Goodson, K.M., N. Donagly, and R.O. Russell, "Furnace Refractory Maintenance and Slag Splashing,: Iron and Steelmaker, ISS, **22**(6) 1995, pp 31-34.
[14]Alasarela, E. and W. Eitel, "How Infinite is Endless Linings of Ladles," Paper in Proceedings of the Unified International Technical Conference on Refractories, San Paulo, Brazil, Oct 31-Nov. 3, 1993, pp 1267-1278.
[15]Bennett, J.P., K.S. Kwong, R. Krabbe, C. Kerr, and E. Wilson, "Spent Refractory Reuse as a Slag Conditioning Additive in the EAF," 58th Electric Furnace Conference and 17th Process Technology Conference Proceedings, held in Orlando, FL, November 12-15, 2000, pp. 379-390.

RECYCLING OF ALUMINUM DROSS TO SIALON-BASED CERAMICS BY NITRIDING COMBUSTION

Shingo Kanehira and Yoshinari Miyamoto
Joining & Welding Research Institute,
Osaka University,
Ibaraki, Osaka 567-0047, Japan

K. Hirota and O. Yamaguchi
Department of Molecular Science & Technology,
Doshisha University,
Kyo-Tanabe, 610-0321, Japan

ABSTRACT

Sialon-based ceramics have been produced from aluminum dross by nitriding combustion mode of self-propagating high-temperature synthesis (SHS). The dross powder was blended with combustion agent of reclaimed Si. The mixture was electrically ignited in a pressurized nitrogen atmosphere below 1MPa, then, the nitriding combustion took place. The products could be sintered at 1500 C° in nitrogen atmosphere. When the sintered sialon was heated in air at 1300 C°, the mullite layer was formed at the surface with a thickness of 10-15 μm, which could improve remarkably the oxidation and corrosion resistance.

INTRODUCTION

The Self-Propagating High Temperature Synthesis (SHS) has a strong potential for recycling wasted materials because of its high energy efficiency and high reactivity with various compositions [1]. In many studies on SHS, real products are limited because of difficulties in process control and thus in product control. Moreover, the reactants such as Ti, Si, B used for the SHS reaction are costly. If industrial wastes can be used as raw materials for SHS the process cost could be reduced. We have succeeded in conversion of the silicon waste usually called "silicon sludge" that is discharged in silicon wafers production to sialon-based ceramics by the nitriding combustion mode of SHS [2].

The aluminum dross is discharged in the fusion process for casting in aluminum

To the extent authorized under the laws of the United States of America, all copyright interests in this publication are the property of The American Ceramic Society. Any duplication, reproduction, or republication of this publication or any part thereof, without the express written consent of The American Ceramic Society or fee paid to the Copyright Clearance Center, is prohibited.

industries. The surface of molten aluminum reacts with air and then Al_2O_3 and AlN slugs are formed. The mixture of such oxide, nitride, remaining aluminum, and molten salts, which is called 'aluminum dross' [3], is usually buried in the ground. However, there is a fear of generation of harmful NH_3 gas through the hydrolysis reaction of AlN.

In the present study, the Al dross could be converted to harmless sialon ceramic powder by nitriding combustion. The combustion was enhanced in a pressurized nitrogen atmosphere of 1MPa by adding the reclaimed Si as a combustion agent. It is important for the recycling process to reduce the nitrogen pressure as low as possible from an economical and safety point of view. The reclaimed Si is a byproduct of silicone production. The nitriding combustion reaction was controlled by changing the content of the combustion agent. The product was pulverized and sintered. The sintered sialon-based ceramics were evaluated in respect to mechanical, thermal and corrosion resistant properties.

EXPERIMENTAL METHOD

The chemical composition of the aluminum dross supplied by Toyo Aluminum Co. Ltd. was AlN(75 wt.%), Al(12.5 wt.%), and Al_2O_3(12.5 wt.%). The ramps of the aluminum dross were crushed into powder with a mean particle size less than 2μm using a vibration milling for 5h. It was mixed with the reclaimed Si (mean particle size ~ 8μm, purity ~93%, Fe 1.1%, Al 0.2%, Ca 0.07 %, Toho Zinc Co. Ltd.) by ball milling for 24h and provided the starting powders. These powders were reacted with nitrogen below 1MPa. The reaction temperature was measured by using a dichroismic radiation pyrometer and a W-Re thermocouple that was protected with BN spray coating against the nitridation. The product composition was identified via XRD. Product powders added 3wt.% CaO and 3wt.% Al_2O_3 as a sintering aid were compressed by CIP and the green body was sintered at 1500°C for 1h in a nitrogen atmosphere. Flexural strength, Vickers hardness, oxidation and corrosion resistances of sintered products were examined.

RESULTS AND DISCUSSION

The main products after combustion synthesis were identified as the mixture of β-sialon ($Si_{6-z}Al_zO_zN_{8-z}$: z=3) and 15R AlN-polytype ($SiAl_4O_2N_4$). The composition of starting powder is a 60/40 wt.% mixture of Al dross and reclaimed Si. The mixture was converted nearly completely to sialon phases with only a small quantity of unreacted Si.

Since the maximum combustion temperature reaches 1300–1600°C, volatile impurities in the aluminum dross are burned up and not contained in the products. The product phases suggest that the exothermic reactions of (1) and (2) could generate the following successive reaction (3).

$$3Si + 2N_2 \rightarrow Si_3N_4 - 748\ kJ/mol \quad ---(1)$$

$$Al + 1/2N_2 \rightarrow AlN - 318\ kJ/mol \quad ---(2)$$

$$Si + Al + N_2 + AlN + Al_2O_3 \rightarrow \beta\text{-sialon}(z=3) + 15R\ (SiAl_4O_2N_4) \quad ---(3)$$

Figure 1 is a SEM photograph showing the powdered product obtained from the composition of the Al dross/reclaimed Si = 60/40 wt.%. Most particles were needlelike with hexagonal facets. Mean particle size was below 10μm.

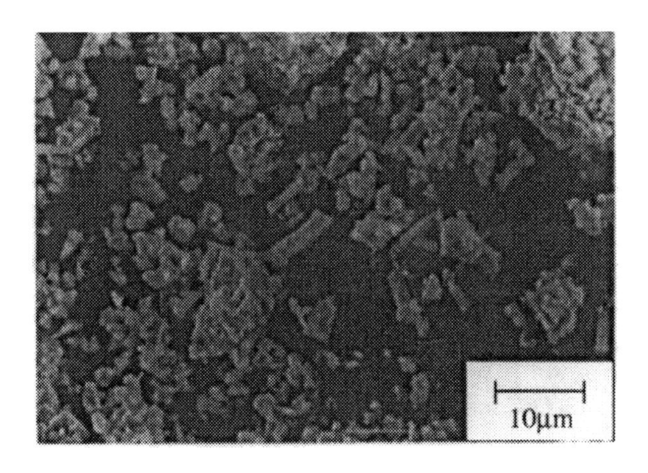

Fig. 1. SEM image of the product powder.

	Recycled samples	β
Density (g/cm^3)	2.5	3.1
Flexural strength (MPa)	110	360
Vickers hardness (GPa)	3.8	15

Table 1. Mechanical properties of sintered samples recycled from the aluminum dross mixtures comparing with the commercial β-sialon.

Mechanical properties of the recycled sintered sialon are listed in Table 1. The sintered body has a density corresponding to about 80% of theoretical. These characteristics are suitable for application to refractries.

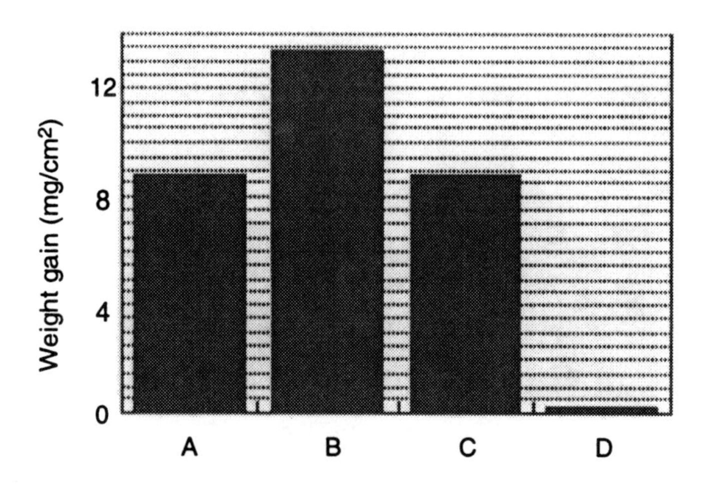

Fig. 2. Results of oxidation test for sintered recycled sialon. (A)1100°C (B)1200°C (C)1300°C (D)1300°C (Post-heated in air at 1300°C for 5h)

The results of the oxidation resistance test for recycled sialon are shown in Fig.2. The rate of weight gain increased from 1100°C to 1200°C, however, decreased at 1300°C. X-ray diffraction showed that a mullite layer was formed at the surface of a sintered body. The thickness of the mullite layer was about 15μm. The mullite layer seems to be formed through the exchange of oxygen in air and nitrogen of sialon.

In order to characterize the corrosion resistance, as-sintered and post-heated samples were subjected to corrosion tests. For the as-sintered samples, the corrosion weight in each solution increased most likely due to the reaction among the sintering aids of CaO-Al_2O_3, residual Al, and the solution at grain boundaries through pores. In contrast, the post-heated sample did not show such weight losses that can be attributed to the protective function of the mullite layer.

CONCLUSIONS

A new recycling process for industrial wastes using the nitriding combustion of SHS has been developed. It was applied to convert the aluminum dross which is discharged from aluminum industries to harmless sialon ceramics. The obtained results can be summarized as follows.

(1) The aluminum dross when blended with the reclaimed Si can be converted to sialon ceramics under less than 1.0MPa nitrogen atmosphere by the nitriding combustion.

(2) The chemical composition of the product powders was mainly sialon. The composition can be controlled with the addition of the reclaimed Si.

(3) The sialon powders could be sintered at 1500°C in nitrogen atmosphere. The mechanical properties were reduced due to a porous structure.

(4) The post-heat treatment for the sintered sialon at 1300°C in air formed the mullite layer at the surface, which could remarkably improve the oxidation and corrosion resistances of the sintered sialon.

(5) The recycled sialon is cost effective and applicable to refractories and corrosion resistant materials.

ACKNOWLEDGMENTS

The author is grateful to Toho Zinc Co. Ltd. and to Mr. T. Sakurai, Toyo Aluminum Co. Ltd., for kind offer of reclaimed silicon, reclaimed Al and aluminum dross, respectively.

REFERENCES

[1] Y. Miyamoto, Z. Li, and K. Tanihata,"Recycling processes of Si waste to advanced ceramics using SHS reaction,"*Ann. Chim. Fr.,* **20**, 197-203 (1995).

[2] Y.Miyamoto, S. Kanehira, O. Yamaguchi, and K. Kajiyama, "Recycling of Silicon-Wafers Production Wastes to Sialon Based Ceramics by Nitriding Combustion," *Environmental Issues and Waste Management Technologies in the Ceramic and Nuclear Industry V, Ceramic Transaction*, **107**, 57- 64 (2000).

[3] G. Lazzaro, M. Eltrudis, and F. Pranovi, "Recycling of Aluminum Dross in Electrolytic Pots," *Resources, Conservation and Recycling,* **10**, 153-59 (1994).

RECYCLING OF THE WASTE WATERS INTO PORCELAINIZED STONEWARE CERAMIC TILES: EFFECT ON THE RHEOLOGICAL, THERMAL AND AESTHETICAL PROPERTIES

F. Andreola, L. Barbieri, I. Lancellotti, T. Manfredini
Faculty of Engineering, University of Modena and Reggio Emilia.
Via Vignolese 905. 41100 – Modena . Italy
E-mail: andreola.fernanda@unimo.it

ABSTRACT

The importance of the porcelainized stoneware production, this product reached the 36% ($218 \cdot 10^6$ m^2) of the total Italian production in the last year, and the impossibility to reuse both the ceramic residues and the process waste waters with the similar practice developed previously for the other typologies have brought to find new solutions to applied at this kind of wastes. The aim of the work is to evaluate the possibility to recycle waste waters decanted and not, deriving from different technological ceramic cycles, into porcelainized stoneware body. The use of these waters does not change the parameters of the productive cycle, obtaining a final product with the similar quality and aesthetic characteristics required from the market and avoiding a possible spilling of the waste water after purification in sewers or in superficial river courses.

INTRODUCTION

The porcelainized stoneware is a product having very good mechanical and technical characteristics, because it shows high sintering degree, low porosity and high esthetical potentialities. This product, nowadays, is being more and more important because its production has reached the 36% ($218 \cdot 10^6$ m^2) of the total Italian production in the last year [1]. For this reason, the impossibility to reuse both the ceramic residues and the process waste waters with the similar practice previously developed for the other residues typologies have brought to find new solutions to applied at this kind of wastes. For a sustainable development it is more important the reduction of both the water discharged and water consumption, with respect to the purification process strengthening. The waste waters deriving from the productive process (preparation of bodies and glazes, glazing and cooling processes, etc.) are collected in tanks where in some cases are subjected to purification treatments with separation of sludges which contain the pollutant elements [2].

To the extent authorized under the laws of the United States of America, all copyright interests in this publication are the property of The American Ceramic Society. Any duplication, reproduction, or republication of this publication or any part thereof, without the express written consent of The American Ceramic Society or fee paid to the Copyright Clearance Center, is prohibited.

In order to realize the recycling of the waste waters derived from the different ceramic cycles inside a high quality body as the porcelainized stoneware, it was necessary a systematical study in order to highlight possible changes on the operative parameters in the different processing stages. Further it is necessary to check that the usual characteristics of the finished products remain constant. The aim of the work is to evaluate the possibility to recycle waste waters deriving from different technological ceramic cycles, inside the porcelainized stoneware body. These waters have been used for the preparation of the mixes in the wet-milling step of the ceramic cycle and to check whether the introduction makes any changes in the rheological, thermal and aesthetics characteristics of the final products. The possible aesthetics modifications on the tiles caused by the recycling have been controlled by colorimetric tests both on the base bodies and on the bodies added with four commercial pigments (pink, blue, white and yellow) in industrial percentages.

EXPERIMENTAL PROCEDURE

Materials

A typical glazed porcelainized stoneware body and waste waters deriving from different ceramic cycles (single and double firing and porcelainized stoneware), pick up from the tanks and after mixed, are used. The actual porcelainized stoneware industrial water (70% from bodies washing mill and 30% from glazes washing mill) has been taken as reference (STD). Furthermore, to verify the effect on the esthetical properties four commercial ceramic pigments (pink, blue, white and yellow) have been utilized.

Characterization

The body was dried in oven (110°C, 24hs) to eliminate the residual humidity. Different techniques have been used to characterize the dried body, such as chemical analysis by ICP (Varian Liberty 200), X-ray powder diffraction (Philips PW 3710) working with CuK_α radiation in 2θ range 5-70°, scanning rate 1°/min. Moreover the waste waters have been analyzed by ICP to determine the element concentrations and pH (pHmeter Orion 420A) and specific conductivity (Analytical Control 120) tests have been run on them. The residues obtained after the water filtration have been characterized using different techniques: EDS (EDAX PV 9900), X-ray powder diffraction (in the same conditions of the body) and particle size distribution by a laser analyzer (Analysette 22, Fritsch).

Preparation of the suspensions

Different mixtures between waste water decanted and not in different percentages (0-100% waste water not decanted) have been prepared. They have been then characterized by pH and specific conductivity tests.

The porcelainized stoneware body and the mixtures containing waste water have been used to prepare concentrated suspensions at 68 wt% solid content. This materials have been ground for 10 min in a laboratory fast ball mill (300ml) with Alubit® grinding media and deflocculated with a commercial additive (0.65 wt% dry solid). The rheological characterization has been carried out on the suspensions by a Control Rate Viscometer (VT550 Haake) with a cylinder coaxial measuring system at 30°C.

Preparation of the samples

The slurries have been dried in oven (110°C, 12hs) and successively have been moisturized (5.5 wt%) and pressed (400 Kg/cm^2) to obtain the laboratory samples. The green samples were fired in an industrial high-speed roller kiln using an industrial cycle (36 min, maximum temperature 1210°C). Using the same procedure, other laboratory samples added with commercial pigments in different percentages have been prepared: 5 wt% white ($ZrSiO_4$) and 2 wt% yellow ($(Zr.Pr)SiO_4$), blue ($(Zr.V)SiO_4$) and pink (Al_2O_3-MnO_2).

Different techniques have been used to characterize the fired samples obtained: a) water absorption (UNI EN 99/ ISO 10545-3), linear shrinkage and apparent density have been run in order to verify the sintering degree; b) colorimetric measures by a UV-Vis spectrophotometer (Perkin Elmer, Lambda19) have been made in order to verify the effect of the waste water addition on the esthetical properties of the body. The parameters CIELab L*, a* and b* were calculated by Hunter method. Differential measurements have been carried out on base body and colored bodies to obtain the ΔE* parameter using the fired samples prepared with the reference water as standard [3].

RESULTS AND DISCUSSION

Material characterization

From the X-ray analysis performed on the green porcelainized stoneware body emerges a prevalent presence of clay minerals mainly kaolinitic and in less quantity illitic, sodium and potassium feldspars and quartz. The three principal constituents (clays, feldspars, quartz) confer the plastic (unfired mechanical resistance), fluxing and inert functions respectively. The chemical analysis is showed in Table I .

The possibility of reusing waste ceramic waters in the porcelainized stoneware production cycle has required a particularized study to investigate the physical and chemical characteristic of these wastes and to evaluate their impact both on the productive cycle and on the final products. For this reason it was necessary verify that the normal characteristics of the final products does not change from the structural and esthetical point of view.

Table I. Chemical Analysis of the porcelainized stoneware body used

Oxide	wt%
SiO_2	69.45
Al_2O_3	18.99
CaO	0.80
MgO	0.35
K_2O	1.48
Na_2O	3.60
Fe_2O_3	0.59
TiO_2	1.59
L.O.I.	3.92

From the ICP, carried out on waste waters derived from the different technological cycles, and from EDS and X-ray analyses performed on the waste waters dried residues it was highlighted that these waters contain significant quantities of materials in suspension or dispersion, traceable to the clayey raw materials (Si and Al), organic substances (C and O), metals (Zn, Cu, Pb, Ca, Mg, Ba, Na and K) and not metals in solution present commonly in the composition of the glaze frits (B and chlorides). The presence of zirconium corresponds to the $ZrSiO_4$ (zircon), widely used like whitener in ceramic bodies or as white pigment for many glazes. In Table II are reported the chemical analysis of the waste and reference water used.

The particle size analysis have evidenced that the suspended dust particles are within the limit of milling sieve for the ceramic bodies (63 μm).

The pH and specific conductivity values of the waste waters are close to the reference water ($pH_{ww} = 6.89$, $CS_{ww} = 1.49$ mS/cm, $pH_{ref} = 8.21$, $CS_{ref} = 1.25$).

The proposed study foresees the decantation in tanks as only treatment of the waste waters, because the purification treatment before reusing the waste waters (as known from our previous research) renders the waters not suitable for the wet-grinding process because of the presence in solution of the coagulator compounds ($AlCl_3$, $FeCl_3$) which raise the apparent viscosity of the slurries above the industrial workability limits [4]. The waste waters recycling can not modify significantly the process parameters, therefore it is important to identify the ceramic cycle step where can happen some modifications, such as the wet grinding stage. With this aim rheological measures have been conducted in order to highlight eventual deviations of the rheological parameters (apparent viscosity, yield stress) that regulate the wet-grinding stage [5].

Table II- Chemical Analysis by ICP of the waste and reference waters.

	B	Pb	Cu	Cd	Fe	Al	Ca	Mg	Cl*
WASTE	33.75	0.00	0.00	0.00	0.00	0.02	209.50	34.35	179.02
STD	2.29	0.00	0.00	0.00	0.00	0.00	128.70	31.46	126.20

* as chlorine

Rheological Characterization

From the analysis of the flow curves obtained has been pointed out that the systems in study are non Newtonian and present a typical plastic behavior with a (τ_0) low yield stress[6]. The effect of the waste waters insertion during the preparation of the body does not change substantially the rheological behavior of the slurry. Besides, an increase rather contained of the apparent viscosity with the water introduction has been evidenced both at low shear rate (50 s^{-1}) (increase of the11%) and for high shear rate (500s^{-1}) (increase of the 5%). The analysis at two different shear rates permits to study the suspensions in two different stages of the process where the slurries have involved. Low shear rates represent critical moments as mill discharge after grinding or slurries maintenance in the tanks. High shear rates instead coincide to the grinding or the pumping stages. A good analysis of the two situations ensures a complete study of the industrial conditions.

The increase of the apparent viscosity as a function of the percentage of waste waters not decanted (Figure 1) is due to both Ca and Mg soluble composts present in the waste waters and to the increase of the solid content in the mix of waste waters.

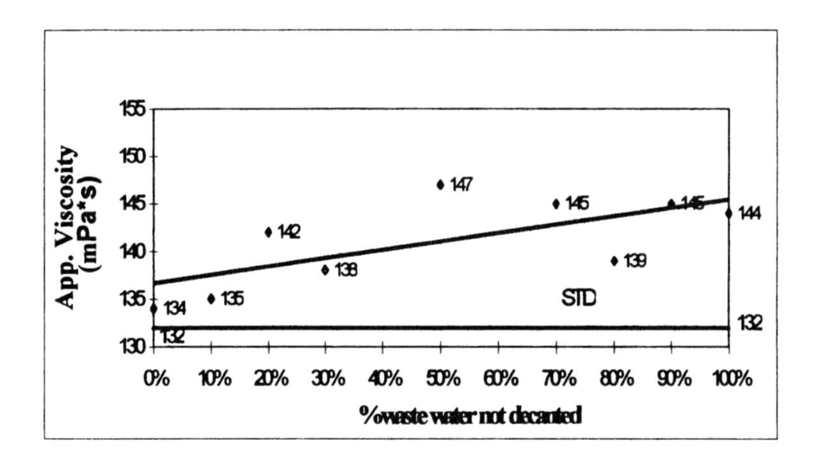

Figure 1 – Apparent Viscosity values (mPa.s) measured at the maximum shear rate (500s^{-1}) as a function of waste waters not decanted percentage.

Sintering Characterization

The sintering study, aimed to evaluate the possible changes due to the addition of the waste waters on the structural properties, was carried out by comparing water absorption, linear shrinkage and apparent density values.

The water adsorption (WA%< 0.01%) and linear shrinkage (6.64%< LS%< 6.92%) data obtained highlight that waste water reuse does not involve remarkable changes since the values do not draw significantly away from the values obtained using the industrial preparation water (AA_{STD}%< 0.01, RL_{STD}%= 6.82%), being in fully agreement with the industrial tolerances (WA%< 0.05%, LS= 6.6% - 7.0%). For what concerns the apparent density measures a slight increase of the density values as a function of the waste water percentage added (2.40<Dapp<2.41, $Dapp_{STD}$=2.38 g/cm^3) emerges, probably due to an increase in the solid content and the presence of fluxe compounds derived from the glazes contained in the waste waters. The sintering corresponding parameters measured after firing are reported in Table III.

Table III. Sintering parameters WA%. LS%. App.D for fired samples

WASTE WATER NOT DECANTED (wt%)	WATER ABSORPTION WA%	LINEAR SHRINKAGE LS%	APPARENT DENSITY (g/ cm^3)
0%	< 0.01	6.89	2.40
10%	< 0.01	6.67	2.40
20%	< 0.01	6.92	2.40
30%	< 0.01	6.83	2.40
50%	< 0.01	6.66	2.40
70%	< 0.01	6.72	2.41
80%	< 0.01	6.77	2.41
90%	< 0.01	6.79	2.41
100%	< 0.01	6.64	2.41
STD	**< 0.01**	**6.82**	**2.38**

Color Tests

In order to evaluate the effect of the waste water (not decanted) addition on the esthetical properties, fired samples have been analyzed by colorimetric tests. The measures have been carried out on both the base body and on the base body added with commercial pigments. In the ceramic industry the CIELab method is the most utilized to determine the whiteness and color of the tiles by measuring the three parameters L* (brightness) from absolute white L=100 to absolute black L=0, a* (red-green), b* (yellow-blue) elaborated from the visible spectra [3].

From the tests it emerges that the addition of the waste water not decanted into the base body provokes a decrease on L* parameter as a function of the percentage added, because of the presence of colored compounds derived from the glazes residues in the waters. However, this variation could be considered small (2.28%). It is important to note that the samples prepared with only the waste water decanted (0%) and the mixes with 10 and 20 wt% of waste water not decanted have a L* value higher than the sample prepared with the reference water (STD). Probably, this result is imputable to the minor solid content in the 10 and 20% mixes (0-1.98 wt%) with respect to the reference water (5 wt%).

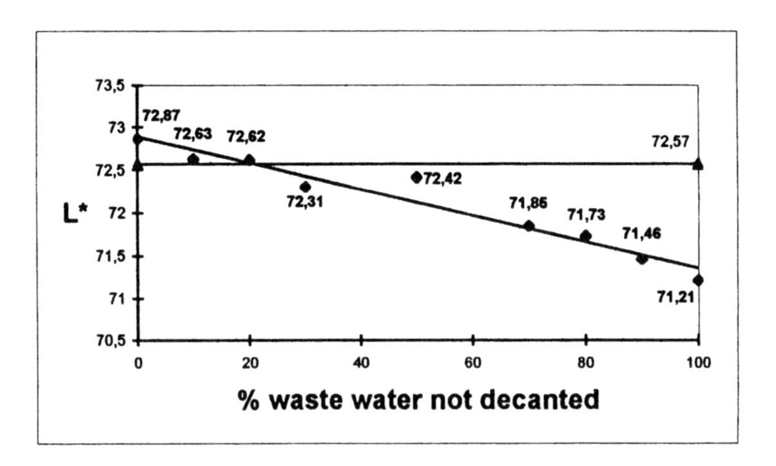

Figure 2 - L* parameter for base body as a function of waste water percentage.

As regards the base body added with the pigments, the samples have been prepared with waste water not decanted (100%) and compared with those prepared with the reference water (STD). The differences highlighted between the CIElab parameters are small. In all samples considered the waste waters not modify the desired tonality.

In order to confirm the previous considerations differential colorimetric measures have been performed. The ΔE* parameter, estimated as a function of the waste waters percentage, represents the deviation from the reference water (STD) in more satisfying way, because it takes into consideration also the changes of the a* and b* parameters on the plane of the colors [3]. The ΔE* has been calculated using the Eq 1.

$$\Delta E_i^* = \sqrt{(L_{STD}^* - L_i^*)^2 + (a_{STD}^* - a_i^*)^2 + (b_{STD}^* - b_i^*)^2} \quad (1)$$

The maximum difference between the ΔE* values obtained for the standard and for the sample prepared with only waste waters not decanted (100%) corresponds to 1.41 and the most part of the series remains (up to 70 wt%) under the industrial tolerance value around to 0.8 as observed in Figure 3. The same measures were performed on samples prepared with the base body added with the commercial pigments. From the ΔE* values obtained a more marked negative effect was observed on the samples colored with clear pigments such as white (1.05) and yellow (1.09) with respect to those added with darker pigments such as blue (0.57) and pink (0.50). Although, the ΔE* values found for the clear pigments (white and yellow colored body) are higher than the admissible tolerance value of 0.8, they are not far from it.

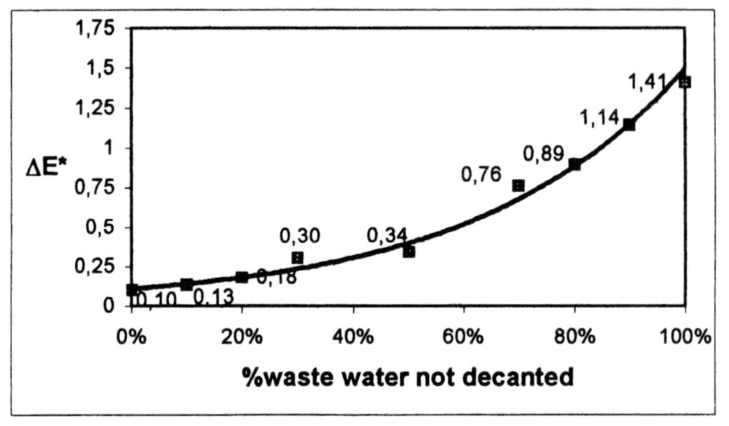

Figure 3 - ΔE* values for base body as a function of waste water not decanted percentage.

CONCLUSIONS

From this work it can be concluded that with a rational recycling it is possible to reuse waste waters derived from different productive ceramic cycles inside glazed and unglazed porcelainized stoneware bodies. This does not change the parameters of the productive cycle, obtaining a final product with the similar quality and aesthetic characteristics required from the market and avoiding a possible spilling of the waste water after purification in sewers or in superficial river courses.

REFERENCES

1. Italian national statistic study, Assopiastrelle, 1999.
2. G.Busani. and G. Timellini, "Inquinamento idrico" (Water pollution), Cap.3 in "Piastrelle Ceramiche &Ambiente" (Ceramic Tile and Enviroment) Ed. G.Busani. C. Palmonari. G. Timellini. EDICER. Sassuolo, Italy,1995.
3. P.A. Lewis, "Pigment Haandbook. Vol.1. John Wiley & Sons. 1988.
4. F.Andreola. T. Manfredini et al "Influence of waste waters. recovery muds and exhausted lime upon mixes deflocculation", *Int. Ceram. Journal..* **31**(1). 11-18 (1996).
5. T.Manfredini, G.C Pellacani, A. Corradi, P.Pozzi, " Wet grinding of raw materials in the production of traditional ceramic tiles", *Am.Cer. Bull.*70(12), 1907-13, (1991).
6. R.Tanner, "Introduction to Rheology "(Cap 1), pp2-20, in *Engineering Rheology*, rev.ed.Edited by Oxford University press, New York,1988.

MINERAL PROCESSING TECHNIQUES FOR RECYCLING INVESTMENT-CASTING SHELL

Cheryl L. Dahlin, David N. Nilsen, David C. Dahlin, Alton H. Hunt and W. Keith Collins
U.S. Department of Energy, Albany Research Center
1450 Queen Avenue, SW
Albany, OR 97321

ABSTRACT

The Albany Research Center of the U.S. Department of Energy used materials characterization and minerals beneficiation methods to separate and beneficially modify spent investment-mold components to identify recycling opportunities and minimize environmentally sensitive wastes. The physical and chemical characteristics of the shell materials were determined and used to guide bench-scale research to separate reusable components by mineral-beneficiation techniques. Successfully concentrated shell materials were evaluated for possible use in new markets.

INTRODUCTION

Investment casting, also called the lost-wax process or precision casting, is used to cast high-value and high-alloy parts that require maximum surface smoothness and rigid dimensional tolerances. The precise molds needed for this process are fabricated with a combination of zircon sand, zirconia, alumina, cristobalite, mullite, quartz, fused silica, and specialty face-coat metals or metal oxides.

Approximately 50,000 metric tons (mt) of zircon sand and 50,000 mt of silica and alumina are landfilled each year by the industry. The cost to dispose of used shell wastes is increasing because the number of landfill sites that accept such wastes is decreasing. The entire foundry industry, including the investment-casting companies, recognizes the need to recycle mold (shell) materials. Successful development of a process to recycle the investment-shell materials that are now disposed in landfills will conserve resources, lessen the importation of expensive and environmentally sensitive materials, and cause fewer hazardous materials to be landfilled.

Investment-mold materials are not currently reused in the casting process. These materials must have specific physical and chemical properties and acceptably low

To the extent authorized under the laws of the United States of America, all copyright interests in this publication are the property of The American Ceramic Society. Any duplication, reproduction, or republication of this publication or any part thereof, without the express written consent of The American Ceramic Society or fee paid to the Copyright Clearance Center, is prohibited.

levels of contaminants. To date, attempts to reuse or recycle used spent shell materials have failed because the recycled products have not met these stringent requirements. Therefore, this project's objective was to investigate the potential for using minerals beneficiation processes to produce materials that may be recycled into new markets.

This study is based on an investigation of sample waste ceramic shell assemblies from alloy castings provided by a titanium investment-casting firm. The waste material studied was intended to typify investment-casting shell waste produced throughout the industry. Processes successfully demonstrated in this investigation would have to be customized for the shell components from individual investment-casting operations. However, the basic separation technology and information presented in this study is applicable to many shell-recycling concerns of the investment-casting industry as a whole.

CHARACTERIZATION

The samples were characterized with wet chemistry (fractions), optical and scanning-electron microscopy (SEM) with both energy-dispersive and wavelength-dispersive x-ray analysis (individual grains), and x-ray diffraction. The material contained ribbons of ceramic refractory-coated, woven-wire-cloth up to approximately 60 cm (2 ft) long and 5 to 7 cm (2 in to 3 in) wide, partially- and completely-liberated woven-wire cloth ribbons, 5 cm (2 in) bolts with nuts attached, pieces of ceramic refractory up to about 15 cm (6 in) in diameter and approximately 3 cm (1 in) thick with and without fine, steel wire inclusions, and titanium-alloy splash. The facecoat appeared as a thin, dark-gray layer on exposed surfaces of the ceramic refractory shell pieces.

The facecoat represented about 1 % of the shell materials. Recycling of the facecoat material depends upon how easily the spent facecoat is liberated and separated from the other shell components and its physical and chemical characteristics. Characterization of the facecoat studied (figure 1) indicated that most of the grains whould be liberated from the alumina and zirconia grains by grinding to approximately minus 75 μm (minus 200 mesh). Investment-casting operations may use such facecoat materials as alumina, calcia, columbium (niobium), erbia, molybdenum, tantalum, tungsten, yttria, zircon, and zirconia or a composite of more than one of these materials.

The zirconia and pure-alumina grains of the material studied (figure 1) made up approximately 1 % each of the waste material. The zirconia stucco layer, which was interlocked with both the facecoat and alumina, was approximately 200 to 400 μm thick. (The zirconia stucco is applied to the partially wet facecoat, and the shell is allowed to dry before the alumina slurry is applied). The alumina layer was interlocked with both the facecoat and zirconia on the inside of the shell and the alumina/silica refractory grains of the backup stucco. The alumina layer was approximately 500 μm thick, and the tabular alumina grains range from sub-micron flour to sand nearly 1 mm in diameter. These materials are fairly typical of stuccos/slurries used in many investment-casting shells although the thickness of the materials may vary greatly from operation to operation.

The alumina/silica refractory backup stucco, shown in figure 2, constituted approximately 88 % of the waste material. In forming the shell, the alumina/silica refractory was applied as an aluminum silicate sand and flour slurry. The resultant fired ceramic refractory contains cristobalite silica that crystallizes from the ethyl silicate binder. Some other investment-casting operations use refractories that contain significantly larger percentages of higher-worth alumina, zirconia, or zircon. Alumina/zirconia/silica (AZS), in various ratios, is a high-value refractory used in high-temperature environments such as glass-making operations.

Wire mesh screen and other wires constituted approximately 7 % of the waste material. The iron wire was magnetic, but a small amount of the wire was altered to a non-magnetic, black, iron-oxide product during the process. This iron-oxide residue concentrated in gravity and high-tension conductor concentrates. Large nuts and bolts were also magnetic and represented approximately 2 % of the spent material. These materials are similar to the structural metals commonly used in other investment-casting operations.

Titanium-alloy splash scrap made up only about 1 % of the waste material. Approximately one half of the splash was visually apparent as particles from about 850 μm (20 mesh) to 9 cm (3 ½ in) in length. The titanium alloy particles are generally coated on one side with a thin layer of the alumina/silica refractory where the molten metal splashed and fused. This refractory coating is not easily removed by standard comminution or beneficiation methods, and titanium-recovery processes would have to contend with this small amount of contamination. Chemical analyses indicated that approximately half of the Ti was in the recoverable plus-850-μm material. Other investment-casting operations may produce splash scrap from super-alloys, steel, or precious metals. The value, purity, and environmental nature of the cast metals/alloys will affect the desirability of stockpiling the splash as a possible marketable product.

COMMINUTION AND SCREENING

Standard mineral-processing equipment and techniques were used to shred, crush, grind, and size the waste materials.

Materials for this experimental process were initially broken with a hammer to expose structural wires and then hand-shredded with wire cutters into pieces with a largest dimension approximately 10 cm to facilitate feeding to the jaw and roll crushers used for the experiment. A mechanical shredder is used to perform this step in some metal-recycling operations. The necessity for a shredding step depends upon

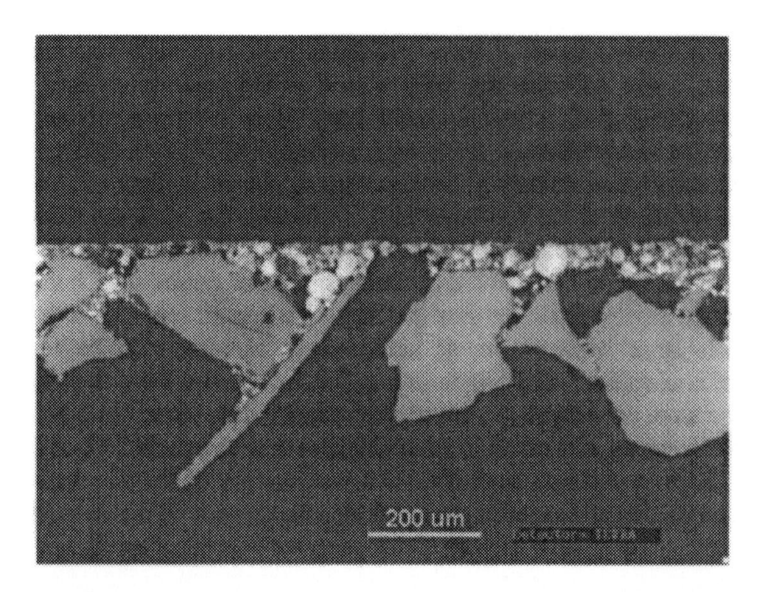

Figure 1. Backscattered-electron SEM image of near-surface, bright facecoat, light-gray zirconia, dark gray alumina, and black voids.

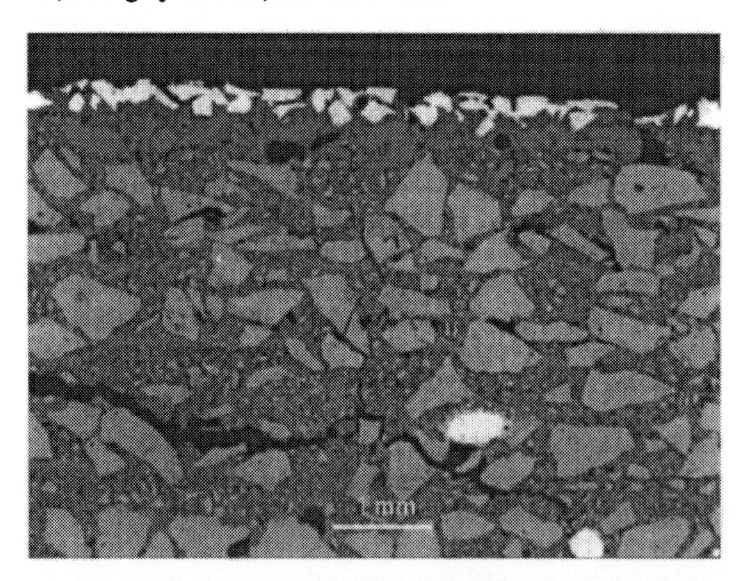

Figure 2. Backscattered-electron SEM image of bright zirconia and facecoat (top), darker gray alumina (below zirconia), and alumina/silica refractory (remaining material). Bright circular areas in the alumina/silica layer are reinforcing wires.

the size of the crushing equipment chosen for the shell-recycling process. If the material can be fed directly to crushers, then the shredding step could be eliminated from the flowsheet.

Shredded pieces of refractory with unliberated and liberated wires and small metal scrap were fed into the jaw crusher. The nuts and bolts and other large metal scrap were removed by hand prior to this step. The operation produced significant breaking and cracking of the refractories and compression of the wire cloth. The pieces were then fed into a roll crusher, and the resulting products were a relatively clean concentrate of larger wires and titanium-alloy splash, and a concentrate of smaller pieces of easily-crushed refractory material.

After crushing, the material was sized on a 0.6 cm (¼ in) screen. The oversize products included woven-wire cloth, long wires broken from the wire cloth, titanium alloy splash fragments, 5 cm (2 in) steel bolts with attached nuts, and a small amount of large pieces of refractory as shown in figure 3.

The oversize metal fragments were magnetically hand-separated into iron and titanium concentrates. Magnetic separation was also done on the minus 0.6-cm sample. The magnetic fraction consisted of small fragments of fine steel reinforcing wires with a very minor amount of attached refractory. This material is considered unrecoverable waste due to its mixed nature and insignificant volume.

The nonmagnetic, minus 0.6-cm sample was further screened on 300-μm (48 mesh), 150-μm (100 mesh), and 75-μm (200 mesh) screens. The plus 300-μm fraction was ground dry in a 18 cm x 23 cm (7 in x 9 in) rodmill in several stages to minimize fines. After the first grind, most of the brittle refractory materials passed 300 μm, with the exception of a concentrated residue of small pieces of titanium-alloy splash and remnant iron wires that were not removed during the magnetic-separation step. Seven grinding stages reduced all of the refractory material to minus 300 μm. The ground and sized samples were prepared for chemical analyses and microscopy studies. The iron and titanium concentrated in the plus 300-μm fraction, the zirconia and alumina and silica/alumina refractory in the 300 μm x 150 μm fraction, and the facecoat in the minus 75-μm fraction.

BENEFICIATION

Standard mineral beneficiation schemes may employ comminution; screening; gravity concentration; differential settling; flotation; electrostatic, high-tension, and magnetic separation; scrubbing; grease tabling; thermoadhesion; flocculation; thickening; filtering; agglomeration; and other physical processing operations. Hydrometallurgical processes may include leaching, precipitation/crystallization, biomineral processing, and solvent extraction. Thermal treatments may include drying, calcination, combustion, and roasting processes. Pyrometallurgy covers smelting processes.

Figure 3. Plain-light image of a representative split of the oversize fractions of shell components after hand-sorting, jaw- and roll-crushing, initial screening and magnetic separation. At top right are typical pieces of titanium-alloy splash. At bottom center are 5-cm (2-inch) bolt and nut sets and at bottom right is the relatively small amount of oversize pieces of refractory that remain after crushing.

The applicable operations chosen to concentrate the marketable components of the investment-casting shell studied were comminution, screening, and magnetic, gravity, and high-tension separations. The facecoat concentrate was treated by roasting, leaching, and crystallization steps to recover the facecoat material in a desired form. Details of the specific thermal and hydrometallurgical treatment of the facecoat material studied is not covered in this paper and it is doubtful that recovery of most facecoat materials would, generally, be a practical option. A conceptual flow diagram of the comminution, beneficiation, and hydrometallurgical steps developed is shown in figure 4.

The three sized fractions (300 μm x 150 μm, 150 μm x 75 μm, and minus 75 μm) were each separated on a laboratory shaking table equipped with a slime deck. Concentrate (heavy facecoat, zirconia, and alumina), middlings, coarse tailings (those that settled and banded on the table), and very-fine tailings (those that washed off the deck before they had a chance to settle) were collected; the coarse and very-fine tailings in the minus 75-μm fraction were combined for analysis. Besides tabling,

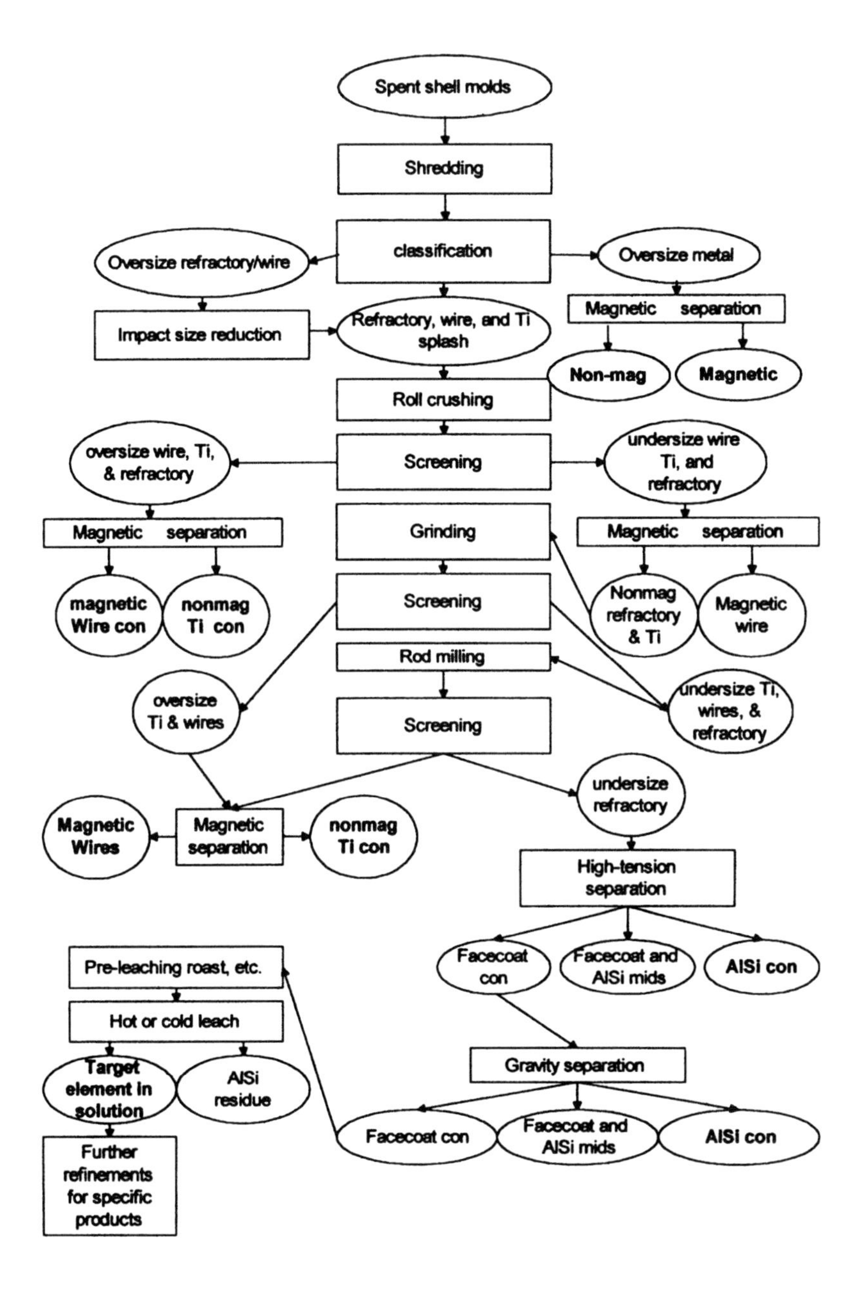

Figure 4. Conceptual beneficiation and hydrometallurgical flow diagram.

gravity separations of particles may be done on a variety of apparatuses that control settling through manipulation of gas or liquid media through gravitational or centrifugal force.

A pilot-scale electrostatic and high-tension drum separator was used for high-tension separation tests. Separation of dry granular materials may be accomplished based on differences in particle-surface conductivity. At settings of 20 kV potential and 100 rpm rotor speed, a 150-μm x 75-μm fraction of refractory mixed with facecoat was separated into conductor (facecoat concentrate, 24 % of the weight), middlings (58 %), and non-conductor (refractory concentrate, 18 %) fractions in one pass. The first-pass middlings and non-conductor fractions were removed for chemical analysis. The first-pass conductor concentrate was put through a second pass at the same settings. The resulting second-pass conductor (26 %), middlings (54 %), and non-conductor (19 %) fractions were chemically analyzed. This concentration method may be used to separate a conductive material from a non-conductive material.

A simple gravity-separation test was also performed on the conductor concentrate from high-tension separation. The facecoat material tested was readily separated from the other conducting materials by vanning. Vanning is a highly effective laboratory gravity-separation technique that does not fully predict the results of a commercial-grade gravity separator. However, considering the promising results, a gravity separation step following high-tension concentration would likely be very effective.

RECLAIMED PRODUCT SPECIFICATIONS AND POTENTIAL MARKETS

The literature suggests that reusing waste shell components in new shells is feasible[1], and attempts have been made by some of the major investment-casting companies to do so. However, due to contamination problems, there is no major reuse presently done in the industry. The emphasis of this study, therefore, was to identify potential new markets for spent shell components.

All recycling options in this study involved breaking the spent shell materials down into major, recoverable components (not necessarily single compounds or elements) and surveying markets for these components separately. The components in the material studied that were found to be physically recoverable by standard mineral beneficiation methods were concentrates of facecoat, iron (woven wire, nuts, and bolts), titanium-alloy scrap, and alumina/silica (with cristobalite) ceramic refractory. The facecoat and the alumina/silica refractory were successfully recovered by grinding, sizing, high-tension separation, and gravity concentration in concentrations high enough to be potentially recycled into other products. Other components may or may not be potentially marketable materials due to their low volume. A detailed economic analysis of a complete recycling program was beyond the scope of this study. Proceeds from the sale of the recycled products alone probably would not make the recycling circuit economical. However, recycling may be less expensive than steadily increasing landfill costs, especially if local markets for the refractories exist.

Facecoat materials may be of high value and, in some instances, may be recovered for resale either by physical beneficiation methods and/or by hydrometallurgical extraction methods.

Refined, ground, and fused alumina sells for about $575/metric ton. Although there are markets for each of these commodities when available in bulk quantities, the low volume and difficult cleaning of these materials did not warrant a market analysis or an individual separation strategy for this particular shell composition. In the conceptual flowsheet (figure 4), these materials remain part of the alumina/silica refractory ceramic grain concentrate.

Although the alumina content of the alumina/silica refractory studied was relatively modest (~40 %), the high volume produced could make the product attractive as a single-source, stable-composition feedstock. Major refractory recycling companies would be interested in purchasing such a feedstock if it came from a local supplier. Due to the bulk nature of this material, a local market would be needed to offset the relatively high cost of transport. Companies that buy materials for recycling generally require the materials to be shipped at the expense of the seller. This may involve transportation fees on the order of $0.10/mile/ton, a significant limiting cost for recycling a high-volume, low-value alumina/silica refractory but would be less of a problem for low-bulk, high-value concentrates.

Other possible markets for alumina/silica refractory materials include concrete, asphalt, mineral wool, grog, pottery, porcelain, flowable fill materials, topsoil additives, landscape paving stones, glass industry tanks, kiln and furnace fireclay linings, bricks, and cements.

Some investment-casting operations use refractories that contain significantly larger percentages of higher-valued alumina, zirconia, or zircon that may be economically shipped longer distances. An example would be materials suitable for producing AZS.

If the alumina/silica refractory is cleaned of other shell components, landfill operators may consider it a desirable cover material and the casting company may be able to negotiate lower landfill fees for a cleaned refractory material.

Recoverable wire screen and structural wires constituted approximately 7 % of the waste material. Woven-wire steel scrap is bought for $15-$25/ton by small recycling operators who compact the material into bales for resale. If the number and weight of the large nuts and bolts found in the sample is representative, then an additional 2 % of the waste was recoverable metal. In quantity, this material may also be saleable to local recyclers. The magnetic susceptibility of reinforcing metals is an important characteristic to ascertain in a recycling study and an operation considering recycling would be advised to only use magnetic structural metals.

Some titanium-alloy splash scrap was recovered from the steel wires by crushing, screening, and magnetic concentration to the non-magnetic fraction. Most of this material has a very-thin crust of alumina/silica. If stockpiled to collect marketable quantities, this type of scrap may be recyclable. There was approximately ½ % titanium alloy in the waste material, of which about half was in recoverable form.

DISCUSSION

The results of this investigation suggest that mineral beneficiation (and hydrometallurgical processes) can separate spent investment shell components into potentially marketable concentrates. Concentrates of facecoat, iron, titanium, and alumina/silica ceramic refractory were successfully produced by the experimental processes on the spent shell studied. The preliminary conceptual process flowsheet (figure 4) was developed based on the findings of the investigation.

A recycling scheme must be considered in light of a number of economic, regulatory, and environmental concerns, in addition to technological feasibility[2]. The following should be considered before a recycling operation is attempted:

- the target markets and products for each reclaimed material
- the present and projected market prices for each reclaimed material
- viable and dependable production levels
- efficiency of the reclamation system and individual processes
- equipment size, type, and level of sophistication
- capital investment, training, operating, and maintenance costs
- potential savings or costs in transportation and/or disposal
- the integration of recycling with existing or future plant processes and layout
- other businesses that may be interested in performing part or all of the recycling processes
- beneficial or detrimental impact of recycling on the physical and chemical makeup of the shell components (both products and waste)
- environmental or safety considerations of equipment or reagents used
- present and future environmental, safety, regulatory, tax, or political benefits or costs
- present and future disposal costs
- partial reclamation as a viable option

A preliminary study, as was done for this project, provides general answers to some of these questions. More detailed guidance for cost analysis of mineral-processing operations may be found in the literature[3,4] and there are some commercial concerns that have the expertise to assist in developing individualized flowsheets. A number of experiments have been done on recycling of sand-casting foundry waste that have produced materials suitable for recycling at a net benefit to the foundry[5-10]. As is true for any significant change in operations, a company must complete an analysis based on estimates of all current and projected costs and gains. If such an analysis of spent investment-shell waste indicates a net positive outcome to the operation, then recycling of waste components may be a viable option. At that point, bench-scale laboratory work can optimize beneficiation and/or hydrometallurgical processes and provide samples for marketability studies, and pilot-plant tests can determine if scaled-up processes work efficiently.

REFERENCES

[1]T.M. Peters and D.L. Twarog, "The Feasibility of Reclaiming Investment Shell Material From Investment Castings," *Illinois Hazardous Waste Research and Information Center Report* **10** 72 (1992).

[2]I.P. Murarka, *Solid Waste Disposal and Reuse in the United States*, CRC Press, Boca Raton, Florida, pp 58-82 (1987).

[3]P.G. Barnard, R.A. Ritchey, and H. Kenworth,. "Recovery of Chromite and Silica from Steel Foundry Waste Molding Sands," *U.S. Bureau of Mines Technical Progress Report* **36** p 21 (1971).

[4]F.A. Peters, "Economic Evaluation Methodology." *U.S. Bureau of Mines IC* **9147**, p 21 (1987).

[5]"Beneficial Reuse of Spent Foundry Sand." in *CWC Technology Briefs*, The Clean Washington Center, State of Washington Department of Trade and Economic Development, at url: http://www.cwc.org/briefs/industrial.html#874349822 (1996).

[6]J.P. Bennett and M. A. Maginnis, "Recycling/Disposal Issues of Refractories." *Engineering Science Proc.*,**16** [1]127-141 (1995).

[7]H. Fang, J. D. Smith, and K. D. Peaslee, "Study of Spent Refractory Waste Recycling From Metal Manufacturers in Missouri" *Resource Conservation and Recycling*, **25** 111-124 (1999).

[8]M. J. Lessiter, "Management Report: Constructing New Markets for Spent Foundry Sand." *Modern Casting*, **83** [11] 27-29 (1993).

[9]J. D. Smith, H. Fang, and K. D. Peaslee, "Characterization and Recycling of Spent Refractory Wastes From Metal Manufacturers in Missouri," *Resource Conservation and Recycling*, **25** 151-169 (1999).

[10]W.A. Stephens and T. P. Kunes, "Cutting the Cost of Disposal Through Innovative and Constructive Uses of Foundry Wastes," *AFS Transactions*, **89** 697-708 (1981).

Environmental Treatment Technology and Policy

EXPOSURE TO CRYSTALLINE SILICA IN THE ITALIAN CERAMIC TILE INDUSTRY: PRESENT STATE AND FUTURE PROSPECTS

Giorgio Timellini and Carlo Palmonari
Centro Ceramico - Bologna
Via Martelli, 26
40138 Bologna, Italy

ABSTRACT

The exposure levels to crystalline silica in the different sections of Italian industries producing ceramic tiles are presented and discussed. The areas where exposure levels are more significant are identified. The exposure levels are associated to technical, technological, organization and lay-out aspects and parameters. Possibilities of achieving a reliable reduction of the exposure to crystalline silica in the Italian ceramic tile industry are discussed. The relations among the preparation technologies of dust pressing powders (dry and wet process), the exposure levels to crystalline silica and the associated health risks are finally addressed.

INTRODUCTION

Industrial ceramics - heavy clay products, floor and wall tiles, sanitary ware, dinnerware and tableware, clay based ceramic materials for the industry, like technical/electric porcelain, clay refractories, etc. - are produced starting from raw materials substantially based on silicates (clays, feldspars, sands, etc.). The relevant bodies (mixtures of raw materials) contain significant amounts of quartz, which play indispensable functions (control of plasticity, green strength and drying and firing shrinkage) during the different phases of the manufacturing process. Hence the occurrence of airborne crystalline silica particles in the workroom of the industries listed above, and consequently the risk of exposure to this substance. The exposure conditions are usually controlled through suitable aspirations on the plants/operations which can produce particulate matter, in order to remove these particles and prevent their diffusion in the workroom environment.

To the extent authorized under the laws of the United States of America, all copyright interests in this publication are the property of The American Ceramic Society. Any duplication, reproduction, or republication of this publication or any part thereof, without the express written consent of The American Ceramic Society or fee paid to the Copyright Clearance Center, is prohibited.

It is well known that mineral dusts containing crystalline silica (quartz, cristobalite and trydimite) are highly pathogenic: prolonged inhalation of crystalline silica particles causes lung inflammation and development of acute and chronic lung disease, silicosis. Silica is also associated with the development of several autoimmune diseases[1]. In 1997 the International Agency for Research on Cancer (IARC) has classified crystalline silica as carcinogenic to humans[2]. According to IARC Working Group on silica, there is sufficient evidence in humans for the carcinogenicity of inhaled crystalline silica in the form of quartz and cristobalite from occupational sources. It is worth reminding that the words "sufficient evidence" mean that the Working Group considers that a causal relationship has been established between exposure to the agent, mixture or exposure circumstances and human cancer. That is, a positive relationship has been observed between the exposure and cancer, in studies in which chance, bias and confounding could be ruled out with reasonable confidence. The overall evaluation is the following: "Crystalline silica inhaled in the form of quartz or cristobalite from occupational sources is carcinogenic to humans. This overall evaluation is - unusually for the IARC - preceded by the statement: "... carcinogenicity in humans was not detected in all industrial circumstances studied. Carcinogenicity may be dependent on inherent characteristics of the crystalline silica or on external factors affecting its biological activity or distribution of its polimorphs ..."[2].

The reasons of concern, for the ceramics industry, associated to the above aspects (first, the close and indissoluble association between industrial ceramics and crystalline silica; second, the sufficient evidence that crystalline silica is carcinogenic if inhaled) are quite clear.

This work is focused on the ceramic tile industry, one of the industrial sectors involved in the problem under consideration. More specifically, on the Italian ceramic tile industry: which is acknowledged for its leading role worldwide as regards not only the production capacity (more than 600 millions square meters per year), but rather the technological innovation and progress. This work aims to make an updated overview of the present situation of the Italian ceramic tile industry as regards the exposure levels to crystalline silica, as a documented basis for envisaging the future trends and possible improvement programmes.

METHODOLOGICAL ASPECTS

The objective of achieving a picture of the present state and the evolution trend of the exposure levels to crystalline silica in the Italian ceramic tile industry has been pursued collecting and elaborating the results of measurement campaigns carried out in several Italian factories. These campaigns have been performed by four different laboratories - including Centro Ceramico. The measurements of the exposition level to respirable quartz have been carried out adopting personal samplers, and using, as particle separation device according to their size, a 10 mm nylon cyclone operating with a 1.7 l/min flow rate[3]. The quantitative determination of quartz has been carried out through XRD.

Two samples of production units have been selected for this study. The first sample consisted of 59 factories, suitably covering all the different types of products and production cycles. The measurement campaigns were carried out in the period 1997-2000, and provided a total of 483 results, covering all the sections of the factories. Most of these results were collected in the framework of a project including a benchmarking activity within the Italian ceramic tile industry as regards the performance and management levels in the Quality and EHS (Environment, Health and Safety) fields[4]. The results relative to this sample were included in a preceding work[5]. The second sample was smaller, and included only 10 factories, and measurement campaigns carried out in the period 2000-2001 (114 results of measures).

The description and the assessment of the exposure levels to crystalline silica in the ceramic tile factories as a whole, as well as in the different sections, have been carried out using the distribution values in specified concentration ranges. These ranges have been established with reference to: (a) the exposure limit (TLV-TWA, Threshold Limit Value - Time Weighed Average) to respirable quartz, according to ACGIH[3]: 0.05 mg/m^3, and (b) the TLW-TWA adopted until last year: 0.1 mg/m^3. In other words, the percent incidence of the measurements results lower than 0.05 mg/m3 or exceeding 0.1 mg/m3 are assumed as indicators of the present average position of the Italian ceramic tile industry with respect to the crystalline silica hazard.

RESULTS AND DISCUSSION

The distribution of the results collected, for each sector of the ceramic tile factory, in the classes specified above are reported in Table I and Table II, for the first and the second sample, respectively.

The following aspects are worth of some comments:

- no correlation has been found between the type of product (for example, porcelain stoneware, white or red single firing, etc.) and the exposure level. In other words, it is not possible to associate to any specific product a particularly high or low exposure level to crystalline silica;

- the TLV at present adopted by ACGIH - 0.05 mg/m^3 - is exceeded in several cases within the sample 1997-2000. The level 0.1 mg/m^3 is exceeded in a significant number of cases too (17.4 %, see Table I) within the same sample. This last value, in a ceramic tile sector which, as a whole, can be considered as operating in a responsible and careful way in the EHS field, represents an indicator of the technical difficulty which can be encountered in achieving a reliable compliance with very severe exposure limits like those under consideration;

Table I. Sample 1 (1997-2000). Exposure levels to respirable quartz.

Section	Number of results	Distribution (%)		
		$\leq 0.05\ mg/m^3$	0.05-0.1 mg/m^3	> 0.1 mg/m^3
Dry grinding	33	45.5	36.3	18.2
Wet grinding	51	64.7	23.5	11.8
Spray Drying	45	33.3	40.0	26.7
Pressing & Drying	123	56.1	21.9	22.0
Glazing	99	57.6	24.2	18.2
Firing	42	57.1	28.6	14.3
Others	90	76.6	13.4	10.0
All the sections	483	58.4	24.2	17.4

Table II. Sample 2 (2000-2001). Exposure levels to respirable quartz.

Section	Number of results	Distribution (%)		
		$\leq 0.05\ mg/m^3$	0.05-0.1 mg/m^3	> 0.1 mg/m^3
Dry grinding	0	-	-	-
Wet grinding	14	85.7	14.3	0
Spray Drying	16	62.5	25.0	12.5
Pressing & Drying	36	83.3	11.1	5.6
Glazing	20	80.0	20.0	0
Firing	10	80.0	20.0	0
Others	18	77.8	11.1	11.1
All the sections	114	78.9	15.8	5.3

- the second sample includes six factories which have been particularly active in the EHS field (these factories have, in particular, an environmental management system in compliance with either ISO 14001 or the requirements specified in the EMAS Scheme - see Reg. 1836/93/CE of the European Commission, at present under revision). Despite this, it is observed in Table II that, although significantly decreased, the incidence of exposure levels exceeding 0.05 mg/m^3 is still around 20 % (with only 5.3 % of the results exceeding 0.1 mg/m^3). In any case, the trend towards the reduction of the exposure to crystalline silica is documented;
- the distribution values reported in Tables I and II can be used as indicators with the purpose of identifying the sections where the crystalline silica risk is higher. The worst situation has been found, in both the first and in the second sample, in the spray drying section, where the incidence of the results respecting the ACGIH limit is the lowest, and the incidence of results exceeding 0.1 mg/m^3 is the highest. In general, it is confirmed that the sections characterized in the average by the higher exposure levels are those where the body - i.e. the powder - is prepared and processed: the sections from grinding to pressing;
- however, also in these sections, the exposure levels to respirable quartz vary within a very wide range, and this depend essentially on a combination of the following factors: i) the powder transport systems, ii) the number and the position of the aspiration points, and the overall efficiency of the aspiration systems, and iii) the floor cleaning systems adopted, and the organization of the floor cleaning operations;
- it is worth noting that also in sections where no specific source of particulate matter should exist (for example, in the kiln sections), exposure levels exceeding 0.05 or even 0.1 mg/m^3 have been measured. In general these high exposure levels can be associated to the lay-out of the factory, and to the diffusion of airborne particles from one section to a neighboring one. This diffusion effect is enhanced by all the aspiration systems, and in particular by the kilns and dryers drafts: masses of air are moving along a ceramic tile factory, and may transport pollutants - airborne quartz particles - from the section where these particles are generated to another section.

The results discussed above show a positive trend, in the Italian ceramic tile industry, towards progressively lower levels of exposure to quartz in almost all the sections. However, a reliable compliance with the more severe exposure limits (TLW-TWA for respirable quartz = 0.05 mg/m^3) recently adopted by ACGIH requires specific measures. Two types of interventions can be envisaged: i) the separation of the individual sections, through suitable screens or walls, in order to prevent circulation and transport of airborne particles from one section to a neighboring one; ii) the increase of the number of aspiration points. This last measure will entail a significant investment cost.

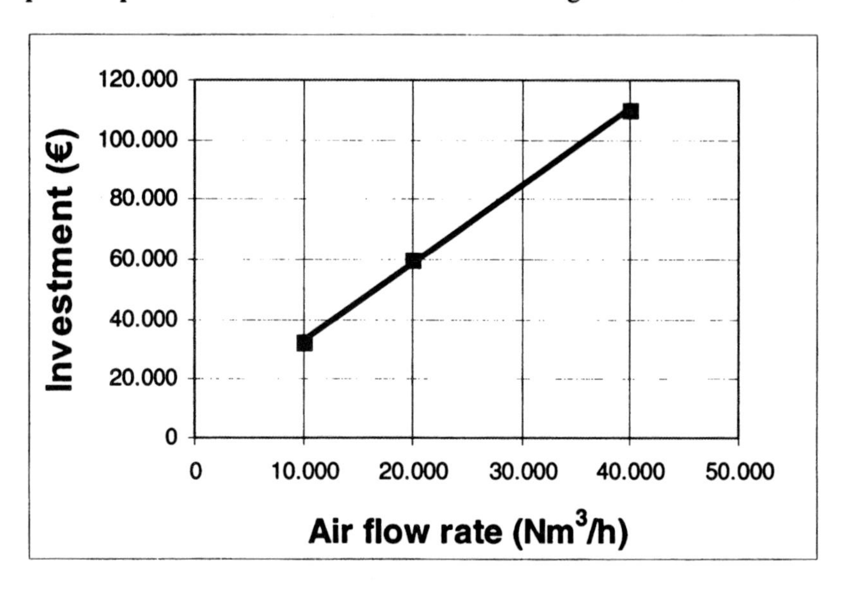

Figure 1. Fabric filters for the removal of particulate matter from gaseous emissions. Investment cost vs. air flow rate[7].

The total aspiration flow rate from any given section would increase, of course, at increasing numbers of aspiration points. In Italy (in the Sassuolo District in particular) this aspirated air must be treated in a suitable and authorized control plant, in order to ensure the compliance with rather severe emission limits. The Emilia Romagna region emission limits for particulate matter are the following: 30 mg/m^3 for emissions from grinding, spray drying and pressing operations; 10 mg/m^3 for emission from glaze preparation and application sections. The control technology which has been experienced as having suitable performances compared to the limits above is air filtration in fabric filters[6]. The relation between the investment cost and air flow rate for fabric filters (Figure 1)[7], associated to the consideration that the specific aspirated air flow rate values at present required, for each of the sections under consideration (grinding, pressing, glazing), are in the order of 100-120 Nm^3 per m^2 of tiles processed[6], gives at least a rough idea of the quite high investment cost that can be expected by each individual factory.

CONCLUSIONS

The IARC evaluation that crystalline silica inhaled in the form of quartz or cristobalite from occupational sources is carcinogenic to humans (1977) has caused a great concern in the industrial ceramics sectors.

The Italian ceramic tile industry is improving its performances, as regards the compliance with the TLV-TWA recently adopted by ACGIH for respirable quartz (0.05 mg/m^3). However, a good EHS management system may be not sufficient: considerable investments for aspiration and pollution control plants are required, in order to ensure exposure levels to respirable quartz reliably lower than the allowed value above.

In any case, it is worth devoting commitment and resources to a deep study of the inherent characteristics and the effective biological activity of the crystalline silica which is detected in airborne respirable particles in the workroom environment of ceramic factories. The results of these studies could provide a more reliable basis for an "optimal" policy against the risks from crystalline silica.

REFERENCES

[1]B.Fubini, "The role of surface chemistry in the variability of crystalline silica hazard", Proc. Int. Conf. *"Crystalline Silica and Oncogenesis"*, Bologna, Italy, Oct. 24, 2000, Ed. Centro Ceramico, Bologna (in print)

[2]IARC, "Silica, some silicates, coal dust and para-Aramid fibrils", in *Monographs on the evaluation of carcinogenic risks to humans*, Vol. 68, Ed. IARC, Lyon, 1977

[3]AIDII, "Valori limite di soglia ACGIH", *Giornale degli Igienisti Industriali*, Suppl. Vol. 26, n.1 (2001)

[4]Assopiastrelle - Snam, "Piastrelle di ceramica e refrattari. Rapporto Integrato 1998 - Ambiente, Energia, Sicurezza-Salute, Qualità", Ed. Edi.Cer, Sassuolo, 1998

[5]G.Timellini and G.F.Fregni, "The control of exposure to crystalline silica: technical and economical issues", Proc. Int. Conf. *"Crystalline Silica and Oncogenesis"*, Bologna, Italy, Oct. 24, 2000, Ed. Centro Ceramico, Bologna (in print)

[6]G.Busani, C.Palmonari and G.Timellini, "Piastrelle ceramiche & ambiente. Emissioni gassose, acque, fanghi, rumore", Ed. EDI.CER, Sassuolo (1995)
[7]Ventilazione Industriale, "Depurazione Aria. Manuale Tecnico-Pratico. Impianti", Milano, 2000

MANAGING POTENTIAL CERAMIC FIBER HEALTH CONCERNS THROUGH TECHNOLOGY AND PRODUCT STEWARDSHIP

B.K. Zoitos and J.E. Cason
Unifrax Corporation
2351 Whirlpool St.
Niagara Falls, NY 14305

ABSTRACT

Since the 1960's there has been increasing attention given to the relationship between fiber exposure and the potential for lung disease. The inhalation of some synthetic vitreous fibers poses a unique challenge to the human body, since the lung lacks a short-term mechanism to remove fibers greater than about 15 microns in length. Long-term retention of fibers in the lung increases relative dose and the possibility for adverse health effects.

In response to these issues, Unifrax Corporation, a world-wide manufacturer of refractory ceramic fiber (RCF), implemented a comprehensive product stewardship program aimed at evaluating occupational ceramic fiber exposures throughout the workplace and implementing aggressive exposure management efforts. One key aspect of this initiative included the development of new fiber chemistries capable of delivering the required high-temperature (1260°C) performance, but with the added property of being rapidly dissolved in the human lung if inhaled. In early 1999, Isofrax™ Thermal Insulation was introduced as the first fiber which is exempt from hazard classification per the European biopersistence criteria and has a continuous 1260°C use limit. Testing in furnace linings and related applications has demonstrated that Isofrax fiber is capable of surviving abusive overtemperature conditions up to 1400°C, thus allowing a wide equipment safety margin in the event of temperature overshoot. In addition, rodent inhalation testing has demonstrated that Isofrax fiber has a very short half-life in the lung.

INTRODUCTION

In response to market demand for a lung-soluble insulation fiber, Unifrax Corporation

To the extent authorized under the laws of the United States of America, all copyright interests in this publication are the property of The American Ceramic Society. Any duplication, reproduction, or republication of this publication or any part thereof, without the express written consent of The American Ceramic Society or fee paid to the Copyright Clearance Center, is prohibited.

has developed and introduced Isofrax™--a revolutionary new high-temperature insulation material. Belonging to the class of materials known as man-made vitreous fiber (MMVF), Isofrax was specifically engineered to meet three key needs: first, the new product provides superior thermal insulation at continuous use temperatures up to 1260°C (2300°F), with a wide over-temperature safety margin; in addition, it is rapidly dissolved and removed from the lung if inhaled (that is, it has low biopersistence), meeting the European regulatory criteria established in EU directive 97/69/EC for biopersistence; and finally, Isofrax can be manufactured in a wide variety of high-quality, user-friendly product forms including bulk fiber, vacuum-cast shapes, blanket and modules. Isofrax is a melt-spun fiber and is based on a patented magnesia-silica chemistry[1].

This paper describes the development of Isofrax as well as end-use applications and performance. In addition, fiber health considerations will be discussed and animal test data presented which demonstrates sharply reduced residence time of inhaled Isofrax fibers in the lung.

FIBER HEALTH CONSIDERATIONS

Since the 1960's, there has been increasing attention to the relationship between fiber exposure and lung disease. Based on both animal models and human epidemiology studies of exposed populations, this work has sought to establish dose-response relationships and has clearly determined a link between fiber lung burden over time and the potential to cause disease. Much of this work has been based on studies of exposure to highly durable fibers such as asbestos. However, in the past decade, many advances have been made in understanding the subtle details of how fiber properties may impact lung burden. In particular, these studies have demonstrated how fiber dimension and durability may affect lung burden and disease outcome.

Building on the pioneering work of Timbrell and Stanton[2-5], recent advances in fiber toxicology have examined the impact of fiber durability on lung burden and risk potential. Since long fibers in the deep lung cannot be cleared by the body's normal mechanisms, it was postulated that these fibers could be cleared if the fiber chemistry caused it to undergo corrosion and degradation in lung fluid. The degree to which this occurs and the magnitude of its impact on disease potential has been assessed in a number of long-term, chronic-exposure animal studies involving a variety of fibers and reported extensively in the scientific literature. From these studies, a clear relationship has emerged between fiber durability, lung burden and disease induction, and it has been well established that non-durable fibers are cleared from the lung much faster than more durable fibers.

Based on this information, an aggressive Product Stewardship Program (PSP) was established by the Refractory Ceramic Fiber industry which follows an integrated and proactive approach to understanding and reducing workplace exposure to RCF. Elements of this program include workplace fiber monitoring and exposure assessment, workplace controls on fiber, health effects and epidemiology studies, and open and honest communications with workers, customers and regulators. To date, these programs have been very successful in managing the RCF health issue and have helped fiber users define and implement improved handling practices when working with RCF[6].

An additional component of this effort at Unifrax has become known as the Dose-Dimension-Durability project. The DDD project has sought to decrease fiber exposure through fundamental product design. In particular, the objectives of this project are to develop fibers which (1) are of suitable dimension so as not to be respirable and (2) are non-durable or non-biopersistent if inhaled. Isofrax represents a key development in this regard and was engineered specifically to retain the high-temperature performance qualities of RCF while being soluble in the lung environment.

Fiber health concerns have also resulted in regulations regarding fiber classification. Recent legislation by the European regulatory commission DG XI established the fiber guidelines which were adopted in EU Directive 97/69/EC and its Nota Q. Based on this directive, fibers having an inhalation biopersistence half-life in the lung less than 10 days are not considered hazardous and products containing them do not require warning labels. Isofrax was specifically designed to be sufficiently non-durable to meet these criteria. Isofrax has a dissolution rate of 150 $ng/cm^2 \cdot hr$ and an inhalation biopersistence half-life of 6 days. A fiber with these characteristics would not be anticipated to cause any adverse health effects from normal occupational exposures.

ISOFRAX DESIGN CRITERIA

Isofrax was specifically designed to provide the full range of properties and product forms currently available from RCF, with the added quality of low biopersistence. To achieve this, three key performance areas were established for the product design.

Dissolution rate targets were established based on chronic animal inhalation data. In these tests, it was observed that fibers with dissolution rates greater than 95 $ng/cm^2 \cdot hr$ did not produce a disease response in chronically exposed rats[7-9]. Thus, 95 $ng/cm^2 \cdot hr$ or greater became the target for dissolution rate.

Refractoriness represents a combination of several high-temperature properties. First, a truly refractory fiber system needs to have low shrinkage at its use temperature. Based on performance of RCF, 3-4% shrinkage following exposure to 1260°C was established as the minimum necessary performance. In addition, a successful product needs to retain adequate strength at and following exposure to use temperature. For fiber, this can be measured by compression recovery, which measures the degree of "spring-back" after compression to 50% thickness. It is desirable for this parameter to approach that for RCF as closely as possible.

Finally, the product needed to be manufactured to the same dimension and quality parameters as current RCF. For a spun fiber, key components of quality are fiber diameter which ranges from 3 to 4.5 μm and fiber index (or amount of unfiberized material) which accounts for 45-60% by weight.

ISOFRAX FIBER PERFORMANCE

Refractory Aspects

Fiber refractory performance may be measured in a number of ways which ultimately determine the quality of the product in a customer application. The most direct assessment of Isofrax refractory properties may be obtained through analysis of the MgO-SiO_2 phase diagram, as shown in Figure 1. For reference, the phase diagram

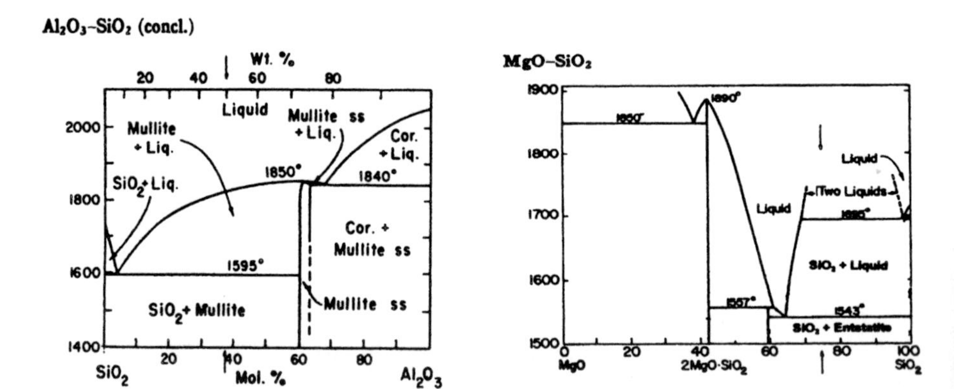

Figure 1. Phase diagrams for RCF (Al_2O_3-SiO_2) and Isofrax™ (MgO-SiO_2). Specific fiber compositions are indicated by arrows. Both fibers devitrify into two-phase systems and have similar eutectic temperatures. Adapted from Reference 10.

for the Al_2O_3-SiO_2 system (RCF) is shown as well. Both fibers are glassy as-produced, but undergo devitrification into two-phase materials on heating. In RCF, this reaction produces mullite ($3Al_2O_3 \cdot 2SiO_2$) and residual silica, while in Isofrax, enstatite ($MgO \cdot SiO_2$) and silica are formed. In addition to this similarity, RCF and Isofrax have eutectic temperatures which differ by only 52°C (1595 and 1543°C, respectively). This suggests similar thermal performance in the two products, but is not alone sufficient to guarantee equivalent refractoriness.

The eutectic point represents ultimate failure of the fiber as melting begins. In practice, however, most fibers fail at a temperature somewhat below the eutectic point due to thermal shrinkage. One of the most critical aspects of high-temperature fiber insulation is that it not shrink excessively at its use temperature. High shrinkage during use can result in insulation gaps in furnace linings or cracking of vacuum-cast parts, leading to hot-spots and loss of effective heat containment. Isofrax is designed to avoid such problems by maintaining low shrinkage up to 1260°C, with only moderate shrinkage at temperatures up to 1400°C. Indeed, Isofrax does not begin to melt until 1500°C.

These characteristics allow for a wide margin of safety when used at or below its continuous use limit of 1260°C. Figure 2 compares the shrinkage of Isofrax with RCF

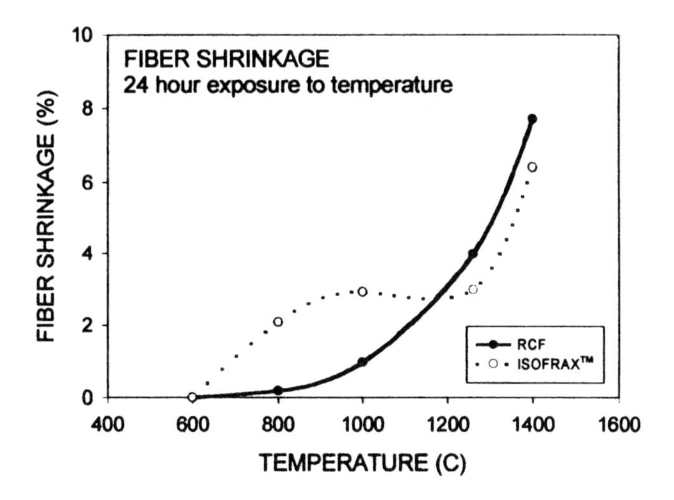

Figure 2. Comparative thermal shrinkage of Isofrax™ and RCF at temperatures from 600 to 1400°C for 24 hours.

following exposure to temperatures up to 1400°C for 24 hours. Additional tests of an Isofrax module furnace lining fired to 1400°C (140°C above the recommended use limit) for 1 week showed that even under extended over-temperature conditions, this fiber remains intact and can provide satisfactory insulation and cold-face temperature.

Shrinkage resistance alone is not fully sufficient to guarantee the suitability of a fiber for refractory applications. Devitrified fibers may undergo grain-growth if maintained at elevated temperatures. This can result in numerous surface defects as crystals grow and begin to distort the fiber surface, with the net result of the fiber becoming brittle and breaking easily. This aspect was one of the design criteria that guided the development of Isofrax. During the product development phase many chemistries were identified which were soluble and had acceptable high-temperature shrinkage; however these fibers all suffered from excessive grain growth and fragility after even short exposures to 1260°C. The unique high-silica chemistry of Isofrax allows this property to be controlled and provides the highest level of after-service fiber strength of all soluble fibers currently available. This property may be measured by assessing the 50% compression recovery value. In this test a blanket sample is subjected to use temperatures for 24 hours and then compressed to 50% of its original thickness. The amount of spring-back or recovery is then measured. This number is indicative of the "brittleness" or tendency of the fiber to break and is critical to the success of the fiber in applications where the fiber must withstand vibration or physical abrasion, such as burner impingement, opening/closing of furnace access doors, etc. Figure 3 shows compression recovery measurements for RCF, Isofrax and an experimental CMS-based fiber following exposure to 1260°C. While RCF shows the best performance, Isofrax has up to 10X better compression recovery values than CMS-based fibers and is fully suitable for most refractory applications.

ISOFRAX BIOPERSISTENCE

Isofrax was specifically developed to have low biopersistence. Fiber durability, the key driver of this property, may be measured in both laboratory and animal tests which have been devised to measure the deterioration of fibers in lung fluid. These studies have shown differences of several orders of magnitude in the dissolution rates of various fibers. Perhaps the simplest measurement method, referred to as "in vitro dissolution" can be performed in the laboratory and involves subjecting a well characterized fiber sample to simulated lung fluid (SLF) at 37°C for a period of time. As the fiber dissolves, the appearance of fiber components in the SLF is measured and used to calculate the dissolution rate of the fiber. This method[11] is rapid, simple and sufficiently accurate to distinguish small differences in durability between fibers. Results for several animal-tested fibers are shown in Table 1[11].

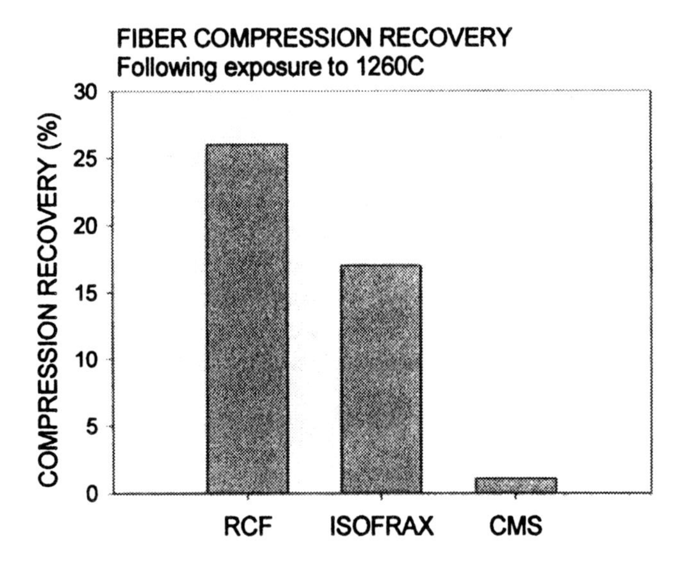

Figure 3. Fiber compression recovery following exposure to 1260°C for 24 hours. Large values indicate a flexible fiber, while smaller values indicate embrittlement has occurred. CMS represents values for a Calcia-Magnesia-Silica fiber.

Table 1. Dissolution Rates of Various Fibers Measured in Unifrax Labs at pH 7.4

Fiber	Dissolution Rate (ng/cm^2•hr)
MMVF-10 (Glasswool)	195
MMVF-11 (Glasswool)	95
MMVF-21 (Rockwool)	17
MMVF-22 (Slagwool)	220
RCF-1 (Kaolin RCF)	8
Crocidolite Asbestos	0.2
Isofrax™	150

Fiber deterioration in the lung may be directly assessed using in vivo testing. In one such test, known as the five-day inhalation-biopersistence test, rats are exposed to a fiber aerosol for a period of five days. At the conclusion of the exposure period, several rats are sacrificed and the initial lung fiber burden measured. The remaining rats are maintained and sacrificed at predetermined intervals for up to six months in order to obtain similar lung burden measurements. Based on this data, the fiber "half life" or $T_{1/2}$ is determined for the fiber. This quantity is the time required for 50% of the >20 μm fibers to disappear from the lung. A rapidly cleared fiber would therefore have a shorter half-life and reduced hazard potential. Inhalation biopersistence results are shown in Figure 4 for Isofrax. This study demonstrates the rapid clearance of Isofrax fiber and indicates an inhalation biopersistence half life of 6 days, in accord with EU directive 97/69/EC.

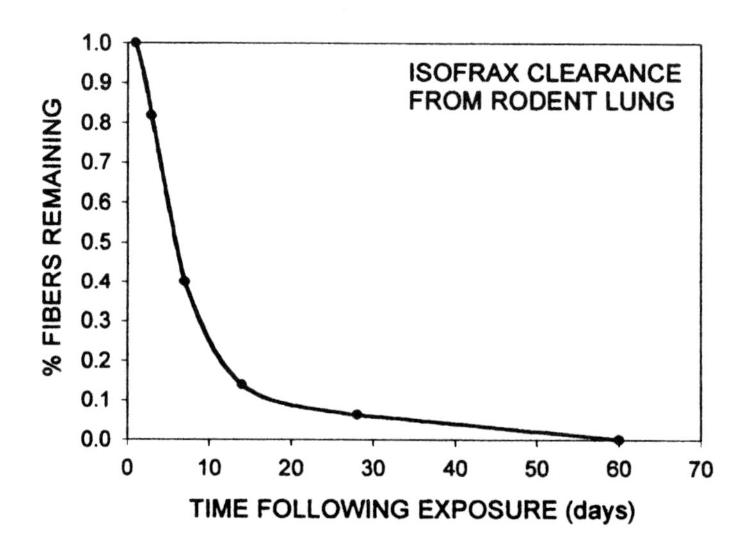

Figure 4. Isofrax clearance from rat lung following 5-day inhalation exposure. A half-life of 6 days was measured in this test.

CONCLUSIONS

Isofrax fiber, introduced in 1999 by Unifrax Corporation, represents the culmination of nearly a decade of research and development efforts to provide a non-biopersistent fiber capable of providing thermal insulation up to 1260°C. Exhaustive performance testing of the new fiber demonstrates its full suitability for use as high-temperature insulation as well as the excellent safety margin provided in the event of over-temperature situations. Isofrax has passed the five-day inhalation biopersistence protocol defined in ECB/TM26 Rev. 7 with a half-life of six days and fully meets the European criteria for a "no classification" fiber. With a reduced residence time in the lung, this fiber is anticipated to pose a lower health risk by inhalation.

REFERENCES

[1]B. Zoitos, R.E.A. Atkinson, J.R. Olson, "High Temperature Resistant Glass Fiber," US Patent #5,874,375, Feb. 23, 1999.

[2]V. Timbrell, "The Inhalation of Fibrous Dusts," *Annals of the New York Academy of Sciences*, 132: 255-273 (1965).

[3]M.F. Stanton and C. Wrench, "Mechanisms of Mesothelioma Induction with Asbestos and Fibrous Glass," *Journal of the National Cancer Institute*, 48: 797-822 (1972).

[4]M.F. Stanton, M. Layard, et al., "Carcinogenicity of Fibrous Glass: Pleural Response in the Rat in Relation to Fiber Dimension," *Journal of the National Cancer Institute*, 58(3): 587-603.

[5]M.F. Stanton, M. Layard, et al., "Relation of Particle Dimension to Carcinogenicity in Amphibole Asbestoses and other Fibrous Minerals," *Journal of the National Cancer Institute*, 67: 965-975 (1981).

[6]D. Venturin, G. Deren, M. Corn, "Qualitative Industrial Hygiene Product Life Cycle Analysis Applied to Refractory Ceramic Fiber (RCF) Consumer Product Applications," *Regulatory Toxicology and Pharmacology*, 25: 232-239 (1997).

[7]R.W. Mast, et al. "Studies on the Chronic Toxicity (Inhalation) of four types of Refractory Ceramic Fiber in Male Fischer 344 Rats," *Inhalation Toxicology*, 7: 425-67 (1995).

[8]T.W. Hesterberg, et al. "Relationship Between Lung Biopersistence and Biological Effects of Man-Made Vitreous Fibers After Chronic Inhalation in Rats," *Environmental Health Perspectives*, 102(5): 133-37 (1994).

[9]T.W. Hesterberg, et al. "Chronic Inhalation Toxicity of Size-Separated Glass Fibers in Fischer 344 Rats," *Fundamental Applied Toxicology*, 20: 464-76 (1993).

[10] E.M. Levin, C.R. Robbins, H.F. McMurdie, *Phase Diagrams for Ceramists*

Vol. I, The American Ceramic Society, Westerville, OH (1964).

[11]B.K. Zoitos, A. DeMeringo, et al., "In Vitro Measurement of Fiber Dissolution Rate Relevant to Biopersistence at Neutral pH: An Interlaboratory Round Robin," *Inhalation Toxicology*, 9: 525-540 (1997).

CHARACTERIZATION OF DEFENSE NUCLEAR WASTE USING HAZARDOUS WASTE GUIDANCE. STATUS OF THE EVOLVING PROCESS AT HANFORD.

Megan Lerchen
Pacific Northwest National Laboratory
902 Battelle Boulevard
P.O. Box 999, MSIN H6-61
Richland, WA 99352

Dr. Gertrude K Patello
Battelle Pacific Northwest Div.
902 Battelle Blvd.
P.O. Box 999, MSIN K6-24
Richland, WA 99352

David Blumenkranz
Science Applications International Corp.
3250 Port of Benton Boulevard
Richland, WA 99352

Karyn Wiemers
Holmes & Narver DMJM
3250 Port of Benton Blvd
Richland, Washington 99352

Lori Huffman
U. S. Department of Energy,
Office of River Protection
2440 Stevens Center
PO Box 450, MSIN H6-60
Richland, WA 99352

Jerry Yokel
Department of Ecology
1315 W 4th Avenue
Kennewick, WA 99336-601

ABSTRACT

Federal hazardous waste management regulations were developed with the intent of addressing industrial waste generation and disposal. These same regulations are now applicable for much of the nation's defense nuclear wastes. At the U.S. Department of Energy's Hanford Site in southeast Washington State, one of the nation's largest inventories of nuclear waste remains in storage in large underground tanks. Detailed identification of the waste's composition and the requirements driven by its designations present many challenges in planning for its acceptable treatment and disposal. An approach to resolving the interferences and limitations inherent in characterizing these high-level mixed wastes will be

To the extent authorized under the laws of the United States of America, all copyright interests in this publication are the property of The American Ceramic Society. Any duplication, reproduction, or republication of this publication or any part thereof, without the express written consent of The American Ceramic Society or fee paid to the Copyright Clearance Center, is prohibited.

discussed in the context of environmental guidance, permitting, and compliance under the hazardous waste regulations.

INTRODUCTION

The U.S. Department of Energy (DOE) is required to store, treat, and dispose of high-level waste at DOE's Hanford Site in southeast Washington. Quality data supporting the project's regulatory and engineering needs must be available.

Over the last few years, DOE has made significant progress in defining characterization requirements for treatment and final disposition of Hanford radioactive tank wastes. This effort has relied on a cooperative, teaming approach between DOE, the regulators, and the implementing contractors. As the project continues to mature, continued participation is expected to play a key role in the project's success.

Hanford Tank Waste

Hanford has 54 million gallons of high-level waste, containing 190 million curies of radioactivity, stored in 177 underground tanks. Accumulation of the waste began in 1944 with the inception of the Hanford defense production mission as part of the Manhattan Project. Current operations consist of waste receipts from activities such as deactivation and decommissioning work, analytical and processing laboratories, and ongoing tank waste management operations.

The underground tanks are within ten miles of the Columbia River, the largest river in the Pacific Northwest. Many of the tanks are past their design life, and 67 of the older tanks are known or suspected to have leaked. In addition, the newer tanks are quickly nearing their capacity. The only permanent solution is to treat and immobilize the tank waste into an inert waste form.

Tank Waste Treatment

It is planned that the dangerous waste and radioactive constituents in Hanford's high-level tank waste will be destroyed, removed, or immobilized to make durable, disposable glass waste forms through pretreatment and vitrification at the future waste treatment plant (Figure 1). The pretreatment will partition constituents between low-activity and high-level fractions to meet disposal requirements and minimize product volume. The vitrification process will combine the pretreated tank waste fractions with glass-forming materials and melt the mixture for transfer into stainless steel containers for storage and permanent disposal. The vitrified low-

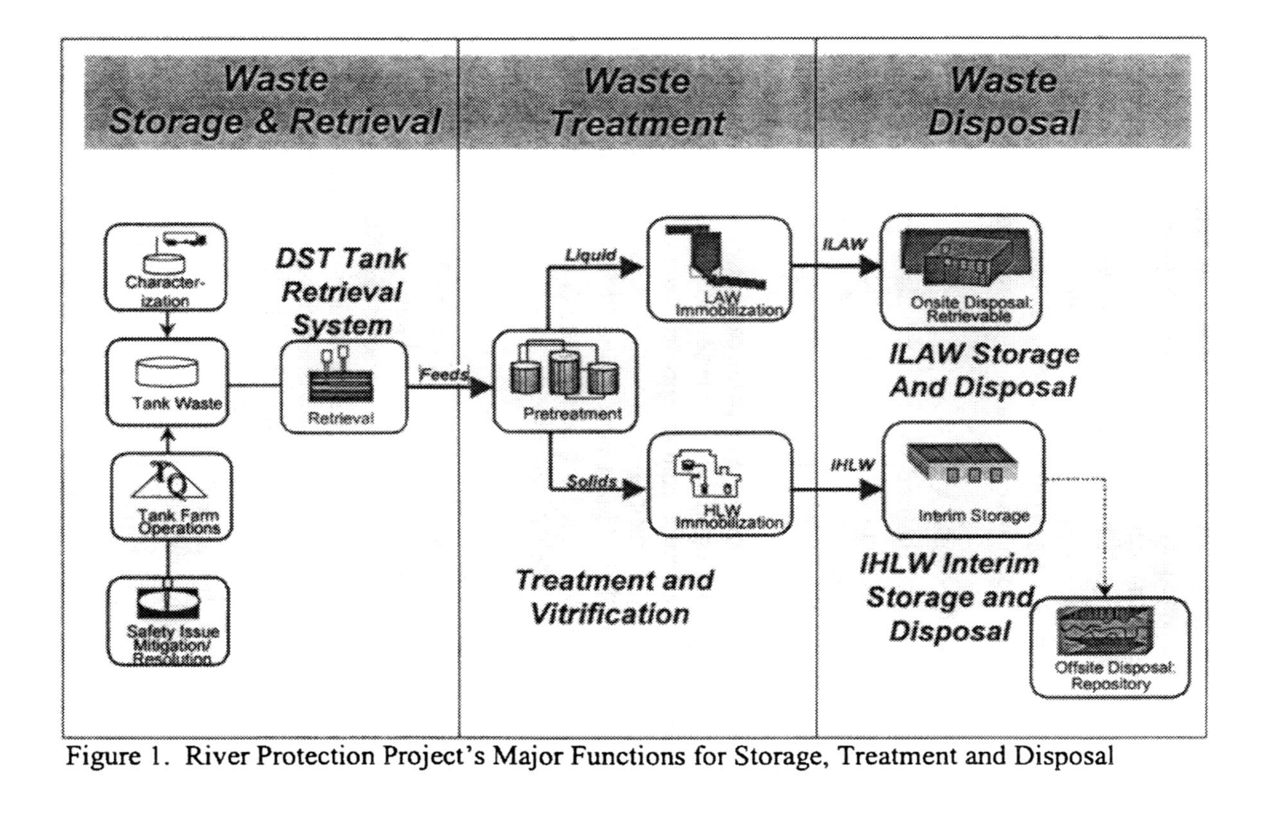

Figure 1. River Protection Project's Major Functions for Storage, Treatment and Disposal

activity waste is slated for near-surface disposal at the Hanford Site while the vitrified high-level waste will be stored for eventual disposal at the proposed federal repository. The waste treatment plant will also treat the plant off-gas to ensure that air emissions from the plants meet requirements.

A significant challenge presented by tank waste is the overall uncertainty in the detailed characterization knowledge. This adds to the difficulty in planning and designing for the tank waste treatment facilities. In lieu of certain characterization knowledge, the project has been using bounding or other conservative estimates where needed.

RCRA REGULATION OF TANK WASTE

Because of past Hanford-specific practices and its location in Washington State, Resource Conservation and Recovery Act (RCRA) requirements and their applicability to Hanford waste differ in some important aspects from other DOE sites. Under the Washington State RCRA program, the tank waste is designated for multiple RCRA waste codes. Each of these codes drives a requirement to use particular treatment technologies and/or meet particular numeric performance standards in order to meet the RCRA treatment requirements for disposal.

Tank waste characterization data will be used to support upcoming petitions to the regulators for a new treatment standard and delisting. In addition, increasingly detailed characterization data is being used to support a phased approach for RCRA permitting.

RCRA Land Disposal Restrictions

For tank waste, the applicable treatment technology under RCRA is high-level waste vitrification, or HLVIT. More waste codes are applicable to the Hanford tank waste than are covered by HLVIT. Of particular importance are those codes that trigger the numerical performance standards for the final waste form under the land disposal restrictions for constituents of concern in general industry, the underlying hazardous constituents.

In order to reduce the characterization burden in meeting the land disposal restriction requirements, the regulators will be petitioned for a new treatment standard that would be applicable to the Hanford tank waste and its waste codes. It is anticipated that the petition will need to include evidence showing that the vitrified waste form's performance would be consistent with the otherwise applicable numeric performance standards.

RCRA Listed Waste

The vitrified high-level waste will be disposed at the proposed federal geologic repository and must meet the repository's acceptance requirements, including that the waste not be regulated under RCRA. Hanford tank waste is listed under RCRA; as such it remains RCRA regulated, even after treatment, until the waste stream has been granted an exemption or exclusion from RCRA through an administrative rule-making process (otherwise known as delisting). The delisting petition, when granted, is anticipated to contain requirements for demonstrating that the conditions of the petition have been met.

RCRA Permitting

Treatment, storage, or disposal of RCRA wastes requires a permit. The RCRA Permit Application and associated risk assessment for the waste treatment plant are being developed using a phased approach to support concurrent permit application development and plant design. The results from the risk assessment will be used to set final RCRA permit conditions for waste treatment plant operations. An initial RCRA permit application has been submitted, and regulator comments are being resolved among DOE, the contractors, and the regulators.[1]

Public and stakeholder perception that the permit is protective of human health and the environment is critical in the permitting process. To this end, characterization data underpinning the permit must be objective, useful, and defensible. Without this level of data quality, it will be difficult to ensure a defensible basis for the risk assessment; one that could withstand public and stakeholder scrutiny.[2]

DATA QUALITY OBJECTIVES FOR REGULATED COMPOUNDS

DOE and the Washington State Department of Ecology (Ecology) jointly prepared a Data Quality Objectives (DQO) document to define RCRA data needs, referred to as the Regulatory DQO.[3] It is focused, in part, on land disposal restrictions and waste treatment plant permitting. At the time the DQO was conducted, the permitting and design efforts for the waste treatment plant were in their infancy. Therefore, the parties agreed that the DQO implementation would be conducted in a step-wise fashion. This allows for refining the characterization needs as data is collected and the permitting and design move forward in their own phased approach.

The outcome of the Regulatory DQO was a prioritized list of 173 compounds for analysis that was selected from an initial list of nearly 1000 regulated compounds. The selection process involved a systematic review

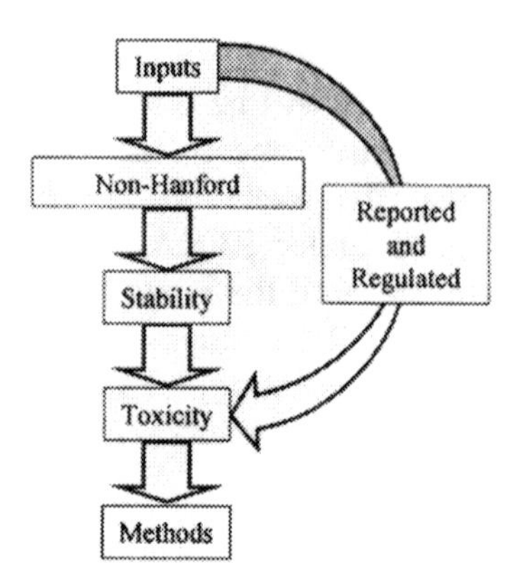

Figure 2. Logic diagram for analyte selection and prioritization process used in the Regulatory DQO.

of each compound by a team of tank waste chemistry experts, including representatives from DOE and Ecology.[4,5,6] The evaluation focused on the plausibility of the regulated compounds' existence in the tank waste matrix and a prioritization based on relative toxicity (Figure 2).[7] Target SW-846 or other environmental methods were identified for characterizing each of these prioritized compounds.[8]

RCRA CHARACTERIZATION METHODS

Characterization in support of the regulatory drivers identified above should ideally be performed using methods acceptable under the Washington State RCRA program. These are generally the methods from the Environmental Protection Agency's (EPA) SW-846 guidance document, Test Methods for Evaluating Solid Waste, Physical/Chemical Methods,[9] specific American Society of Testing and Materials (ASTM) methods, or alternative methods approved by Washington State (WAC 173-303-110).[10] Data gathered using other methods may be used to support decision making for regulatory petitions and disposal characterization, but will need to be corroborated with data from RCRA methods.

Due to the nature of the tank waste, it is likely that the methods will need to be modified or alternative methods selected. Under the Washington State RCRA Program, minor method modifications do not require formal petitioning for regulator approval (WAC 173-303-110). Conversely, the need for major method modifications or use of alternative methods require regulator approval through a formal petitioning process. If such a petition is needed, the petitioner must justify, using empirical data, why the RCRA program methods are not acceptable.

Need for Methods Modifications

RCRA characterization methods were developed for water, industrial waste, and soil samples and have, in general, not been validated for Hanford's high-level radioactive tank waste matrix. The application to high-level radioactive wastes results in unique method challenges due to characteristics of the waste that include high radioactivity, high ionic strength, pH greater than 12, solids with fine particle size, and highly insoluble solids such as mineralized forms consisting mainly of various aluminates, hydroxides, carbonates and oxides. These characterization difficulties manifest in both sample preparation and analysis.

The most frequently used EPA method modification for high-level waste is reducing sample size, often by orders of magnitude. Large quantities of sample under SW-846 should not be used when analyzing high-level waste for two main reasons. First, the larger sample size would result in an unacceptably high radiation dose to laboratory personnel. Second, the size cannot be accommodated in the environment of a shielded hot cell where all sample handling is performed remotely through use of master-slave manipulators. This is a recognized problem for radioactive waste management; thus explicit guidance for reduced sample size was issued by EPA and the US Nuclear Regulatory Commission.[11]

In addition to sample size issues, it is desirable to get a total inorganic element analysis, which requires complete digestion of a solid phase matrix. The EPA preparation methods most often use leaching or extraction for solids matrices. There are limited complete digestion methods for sample preparation in SW-846 and those that exist do not completely dissolve the more refractory phases in the tank waste matrix. Modification of the SW-846 methods or adoption of different methods from other sources such as ASTM is necessary. Some complete digestion methods have been developed for Hanford waste. However, these Hanford digestion methods have not yet been approved for use under the RCRA program. It is planned that data will be gathered to support Ecology approval of these modifications or alternative methods.

Another frequent method modification needed for the Regulatory DQO is the adoption of an analytical method to a constituent not previously identified with that method. The application of the EPA methods to new constituents is being considered a minor modification of the method, where the compound is chemically similar to the analytes that the SW-846 method was developed to target.

Approaches to Method Validation

The application of SW-846 methods to the unique radioactive waste matrix necessitated that a strategy be developed to support analytical method validation. This strategy was developed as part of the Regulatory DQO process. The initial step in this strategy has three main parts:

1. Determining method detection limits (MDLs) and estimated quantitation limits (EQLs) in water and sand to demonstrate the laboratories' ability to perform the method with standard matrices;
2. Determining MDLs and EQLs in tank waste liquids and solids to demonstrate the ability to apply the methods and establish the MDLs and EQLs for liquid and solid radioactive waste matrices; and
3. Conducting a holding time and storage condition study to understand the effect of radioactive waste sample management on the characterization data.

Parts 1 and 2 started with the preparation of an implementing plan[12] for the Regulatory DQO followed by preparation of detailed laboratory test plans for each method. Development of these test plans required the concerted efforts of chemists, statisticians, quality engineers, and regulatory analysts. This planning effort was needed to establish technical consensus on the laboratory specific methods, the data to be collected, and the data use, particularly when the Regulatory DQO or SW-846 lacked the detail needed for implementation. For example, more refined interpretive guidance for determining MDLs was adopted from 40 CFR 136, Appendix B.[13]

Additionally, technical consensus has been reached that resulted in changes to the Regulatory DQO implementation. Two cases are presented here. First, the Regulatory DQO identified hydroxide as a priority regulated compound and specified analysis by titration. The hydroxide data would be used to confirm the designation of the waste characteristic of corrosivity, which is defined by pH greater than 12.5 (WAC 173-303-090). Consensus was reached that a direct measurement using a pH probe suited to high sodium concentrations and basic conditions would satisfy the DQO intent. In the second case, ammonia was slated to be measured in the solids and liquid matrices. Consultations with tank waste chemistry experts and environmental laboratory resources confirmed that insoluble ammonia compounds would not be expected in the waste solids. Therefore, agreement was reached that ammonia would be analyzed only in the liquid matrices. The concerted, upfront test method planning by consensus has resulted in reduced sample quantity, preparation, and analysis requirements; additional opportunities for dose reduction; and more direct focus on addressing the specific DQOs.

The third part of the initial step in the Regulatory DQO is the performance of a holding time and storage condition study. Current sample retrieval and handling methods for the tank waste are not consistent with requirements of SW-846. The study will provide data on the effect of the sample handling and storage conditions and provide the data users with the information necessary to interpret and apply the analytical results. The holding time and storage condition study will start after method validation (parts 1 and 2) is completed.

Radioactive waste matrix MDLs and EQLs will be evaluated against the data needs for the permitting and petitioning needs to determine whether the method requires further refinement or replacement to achieve lower detection limits. The success of this approach is dependent on continued involvement between DOE, the regulators, and the performing contractors. This is being accomplished through monthly status meetings and the judicious use of hold points for concurrence that the overall strategy remains on track.

SUMMARY DISCUSSION

The continued refinement in our understanding of Hanford tank waste is key to supporting the ongoing efforts to permit the waste treatment facility and to gain approval for regulatory petitions. Validation of preparation and analysis methods for regulated compounds in tank waste matrices is the first step in providing technically defensible data for subsequent activities. The methods developed and data collected during implementation of the Regulatory DQO will be used in the characterization of tank waste currently in storage and preparation of the waste treatment plant permit and petitions. As these activities evolve through time, deliberate cooperative efforts among regulators, stakeholders, waste management contractors and the DOE must continue to focus on collecting quality, useful, and objective data to serve their decision-making needs.

REFERENCES

[1] BNFL. 2000a. RPP-WTP Dangerous Waste Permit Application, BNFL-5193-RCRA-01, Revision 1. BNFL, Inc., Richland, Washington, USA.

[2] C Hogue, "Federal Data Quality Soon Must Meet Quality Standards," Chemical and Engineering News, 79 [2] 7 (2001)

[3] KD Wiemers, ME Lerchen, M Miller, and K Meier, "Regulatory Data Quality Objectives Supporting Tank Waste Remediation System Privatization Project," USDOE Report PNNL-12040, Rev. 0, Pacific Northwest National Laboratory, Richland, Washington, 1998

[4] KD Wiemers, RT Hallen, H Babad, LK Jagoda, and K Meier, "A Compilation of Regulated Organic Constituents Not Associated with the Hanford Site, Richland Washington," USDOE Report PNNL-11927, Pacific Northwest National Laboratory, Richland, Washington, 1998

[5] KD Wiemers, P Daling, and K Meier, "Rationale for Selection of Pesticides, Herbicides, and Related Compounds from the Hanford SST/DST Waste Considered for Analysis in Support of the Regulatory DQO (Privatization)," USDOE Report PNNL-12039, Pacific Northwest National Laboratory, Richland, Washington, 1998

[6] KD Wiemers, H. Babad, RT Hallen, LP Jackson, and ME Lerchen, "An Assessment of the Stability and the Potential for In-Situ Synthesis of Regulated Organic Compounds in High Level Radioactive Waste Stored at Hanford, Richland, Washington," USDOE Report PNNL-11943, Pacific Northwest National Laboratory, Richland, Washington, 1998

[7] KD Wiemers, ME Lerchen, MS Miller, and NC Welliver, "Logical Selection of Analytes for Hanford TWRS Privatization Waste Feeds," 53rd Northwest Regional Meeting of the American Chemical Society, Richland, Washington, 1998

[8] KD Wiemers, ME Lerchen, and M Miller, "An Approach for the Analysis of Regulatory Analytes in High Level Radioactive Waste Stored at Hanford, Richland, Washington," USDOE Report PNNL-11942, Pacific Northwest National Laboratory, Richland, Washington, 1998

[9] EPA, "Test Methods for Evaluation Solid Waste Physical/Chemical Methods," SW-846, 3rd Edition, as amended by Updates I (July, 1992), IIA (August, 1993), IIB (January, 1995), and III, U.S. Environmental Protection Agency, Washington, D.C., 1997.

[10] Chapter 173-303 WAC. "Dangerous Waste Regulations." Washington Administrative Code, as amended, http://www.ecy.wa.gov/laws-rules/laws-etc.html

[11] Environmental Protection Agency and Nuclear Regulatory Commission, "Joint NRC/EPA Guidance on Testing Requirements for Mixed Radioactive and Hazardous Waste," 62 FR 62079, 1997

[12] GK Patello, TL Almeida, JA Campbell, OT Farmer; EW Hoppe; CZ Soderquiest, RG Swoboda, MW Urie; and JJ Wagner, "Regulatory DQO Test Plan for Determining Method Detection Limits, Estimated Quantitation Limits, and Quality Assurance Criteria for Specified Analytes," USDOE Report PNNL-13429, Pacific Northwest National Laboratory, Richland, Washington, 2001

[13] US Environmental Protection Agency, "Guidelines Establishing Test Procedures for the Analysis of Pollutants" US Code of Federal Regulations, 40 CFR 136, Appendix B, as amended

THE EUROPEAN CERAMIC TILE INDUSTRY AND THE NEW APPROACH TO ENVIRONMENTAL PROTECTION

Carlo Palmonari and Giorgio Timellini
Centro Ceramico - Bologna
Via Martelli, 26
40138 Bologna, Italy

ABSTRACT

The evolution of the environmental impact control and management activities in the last years, in Europe, for the ceramic tile sector is reviewed. The environmental impacts from ceramic industries have been studied in Italy since the early '70. Facing a legislation based on the compliance with Environmental Quality Standards, in the '70-'80 the ceramic tile industry has developed very efficient pollution control techniques. Sustainable development and integrated pollution prevention and control are the key-words of a new approach to environmental protection, which includes also environmental certification. The European ceramic industry has assumed a leading role also in this new approach.

INTRODUCTION

The prevention and reduction of pollution, in particular from industrial sources, as well as the protection of the environment, are the aim of specific national and international laws in Europe since the sixties. And in Europe, the ceramic floor and wall tile industry has been involved from the early beginning in environmental protection activities according to these laws and regulations.

The Italian ceramic floor and wall tile industry has an outstanding position in Europe: in fact, it can be acknowledged, in the last decades, as the leader at European and world level, as far as production, manufacturing techniques and technological innovation are concerned. The Italian ceramic

To the extent authorized under the laws of the United States of America, all copyright interests in this publication are the property of The American Ceramic Society. Any duplication, reproduction, or republication of this publication or any part thereof, without the express written consent of The American Ceramic Society or fee paid to the Copyright Clearance Center, is prohibited.

floor and wall tile industry has progressively extended its technological leadership also to the pollution prevention and control activities: the Italian ceramic industry, the first in Europe, has promoted studies aimed to identify all the environmental impact factors, to achieve a detailed knowledge of their intensity, correlated to product type and manufacturing technologies, as well as to develop high efficiency control techniques[1-6].

This paper aims to review the activities carried out and the main results achieved by the Italian ceramic floor and wall tile industry as regards pollution prevention and control and environmental protection. This is a very interesting case history, over an almost 30 years period, from the early seventies to now. This case history shows how an important industrial sector, through its commitment in environmental protection and a careful exploitation of the tools provided by the more recent legislation, could achieve both a reliable compliance with laws and regulations and an improved competitiveness based on environmental responsibility.

This review is subdivided in two parts, which correspond to two successive, individual and quite different phases of the European approach to environmental protection through national and international legislation. The commitment of the Italian ceramic tile industry in both these phases is outlined and characterized in detail.

THE "RECLAMATION" PHASE AND THE "COMMAND & CONTROL" APPROACH

This phase extends over the period from sixties to eighties. In most European countries, after the second world war, the industrial activities have been developed without a full awareness of the associated environmental impacts, and therefore of their possible effects on the quality of the environment. The consequences, in some industrial areas or districts, are significant contamination levels due to air, water and soil pollution, and possible hazardous effects on humans, agriculture and livestock, materials, etc. At this point, the main objective, to be achieved in the shortest time and with the highest effectiveness and efficiency, is the reclamation of the contaminated areas. The most suitable and effective tools in this context are laws, specific for each environmental aspect involved: a specific law for air pollution and gaseous emissions, a specific law for water quality and waste water, a specific law for solid wastes, etc. Each of these laws is focused on the relevant outputs (gaseous emissions, waste water, waste materials, etc., respectively) from industrial sources. An inventory of all these outputs is required to each production unit. Environmental Quality Standards (EQS) are established in each law (in terms, for example, of maximum exposure levels to airborne pollutants in the environment). Subsequently, permit conditions are established for each production unit. The permit includes output/emission limits (for example, maximum concentration or maximum mass flow rate of pollutants in gaseous emissions), suitable to ensure the compliance with the EQSs in the area where the production unit is operating, that is at local scale.

In this phase the attitude of industry to the environment protection needs is merely reactive to a "command", consisting in permit conditions that, being based on the compliance with EQSs, are the most severe and restrictive, the highest is the "pressure" of industrial sources on the local environment.

The ceramic sector that, at European level, is characterized by the most critical situation from the local environment point of view is the Italian ceramic tile industry.

Few figures can give an idea of the critical situation above: the Italian national production of ceramic tiles is now of more than 600 millions square meter per year. This production is associated to around 350 production units and 31,000 employees. Around 70 % of the Italian tile production is exported all over the world: competitiveness is therefore a primary and vital aspect. Around 80 % of the national tile production (more than 480 millions m^2/year and 280 production units) is "concentrated" in a rather small area, in the northern part of Italy: the Sassuolo ceramic District.

High concentration of ceramic factories implies high concentrations of pollution sources, high mass flow rate of pollutants in the area, and therefore a high risk of exceeding the EQSs. In summary, a very high "pressure" on the environment, and a contaminated area to be reclaimed and managed through very severe regulations and quite low emission limits.

The environmental commitment of the Italian ceramic tile industry is, in this first phase, the reaction to regulations that, for the reasons above, are more severe compared with those adopted in other countries or for other industries without the concentration problems discussed above. This reaction led to important results:

- knowledge
- development and adoption of high efficiency control technologies
- technological innovation, toward cleaner manufacturing technologies.

Few significant data on these aspects.

Knowledge: all the environmental impact factors associated to ceramic tile manufacture have been characterized, over the last decades, through extensive surveys and studies involving almost all the Italian tile factories. The results are data bases on pollutant emission factors and other impact indicators[1, 5], or reports from benchmarking activities[7]. A good knowledge is a pre-requisite for good regulations[5].

High efficiency control technologies: very high performances were required in order to ensure, also in the quite heavy conditions of the District, the compliance with EQSs. The control of gaseous emissions is a suitable example of these results[5]. As reported in Table I, the extensive adoption of gaseous emission control plants in all the ceramic tile factories of the Sassuolo District - a control level reached and maintained since the early eighties - has dramatically reduced the annual mass flow rate of the main pollutants (fluorine and lead compounds, particulate matter), compared to the "potential" emission levels, which would occur without - or before - the adoption of the control techniques under consideration. With these performance levels, acceptable air quality levels have been achieved and maintained, despite the high concentration of pollution sources.

Table I - Annual pollutant emissions in the Sassuolo District through gaseous emissions[8]

Pollutant	Potential Emission (t/year)	Present Emission (t/year)
Fluorine compounds	2,200	180
Lead compounds	$280*10^3$	$1.7*10^3$
Particulate matter	390	45

Low impact technologies: the ceramic tile manufacture technologies have been subjected to an important innovation process from the late seventies to now; and there is no doubt that the need of reducing environmental impact has been one - non the only - driving force in this process. Two emblematic results: from the innovation of the firing process (from double firing to single firing, and from traditional tunnel kilns to single layer fast firing roller kilns) to a substantial reduction of energy consumption. Now, as documented in Figure 1, the specific energy consumption in ceramic tile manufacture is around 50 % of the consumption levels before 1980[9]. A further example: waste water recycling and reuse techniques have been developed in the last decade. Now most ceramic tile factories do not release any waste water, but use waste water in order to cover a part of the water demand. The results is that, at present, in the Ceramic District of Sassuolo, against an annual total water demand of around 8 millions cubic meters, the effective water consumption - from natural water supplies - is around 3.7 millions cubic meters. This means that, in the Ceramic District, recycled waste water covers more than 50 % of the water demand for the manufacture of ceramic tiles[8].

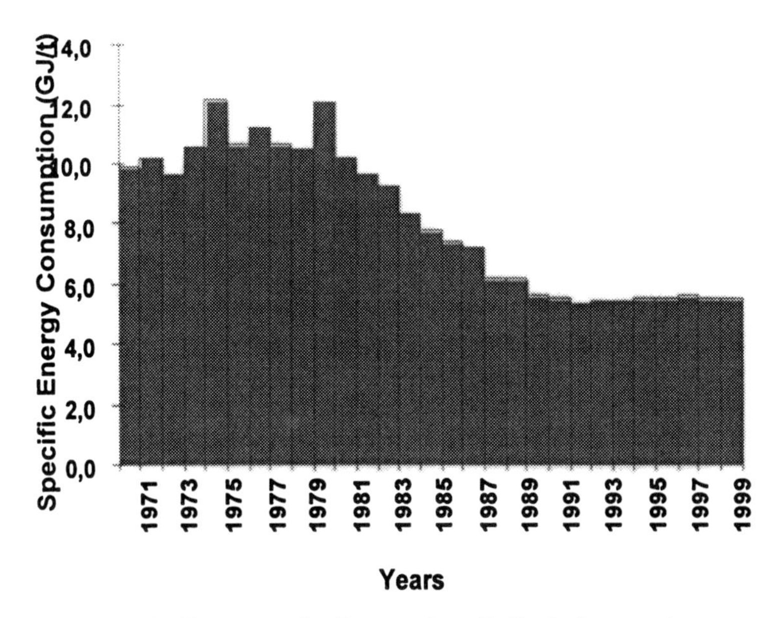

Figure 1 - Italian ceramic floor and wall tile industry. Average specific energy consumption in the period 1971-1999[9].

Of course, the last interventions on manufacturing technologies are efficient from both environmental protection and cost points of view. On the contrary, the end-of-pipe techniques - like gaseous emission control plants - imply a relevant additive cost. The costs associated to the installation and operation of these control plants have been assessed for a sample ceramic tile factory[5]: the investment cost for these plants represents around 4.5 % of the total investment for the production plants and equipment, while operative costs have been estimated as high as 3% of direct manufacturing costs. It is worth emphasizing that these control techniques have been adopted, in the last two decades, only by the Italian ceramic tile industry (not at all, or to a quite lower extent, in the other countries).

In conclusion, looking at the resources devoted and the results achieved in this first phase of activity for environmental protection, in comparison with the ceramic tile industries of other countries, the Italian ceramic tile industry can be acknowledged for the best environmental performances, but also for the highest costs associated to the protection of the environment.

THE "SUSTAINABLE DEVELOPMENT" PHASE AND THE MANAGEMENT APPROACH

In the early '90 the environmental legislation in the European Union has changed the approach to the environmental protection: the 5th Environmental Action Program (1993), containing the Community program of policy and action in relation to the environment and sustainable development, has accorded priority to integrated pollution control as an important part of the move towards a more sustainable balance between human activity and socio-economic development, on the one hand, and the resources and regenerative capacity of nature, on the other. Assuming that substantially the "reclamation" objective has already been achieved, the new objective of the environmental protection activities can be expressed as "sustainable development".

The Directive 96/61/CE (the so-called IPPC Directive on Integrated Pollution Prevention and Control) has introduced the concept of Best Available Techniques as reference - in place of the above mentioned EQSs - for the design of the measures to prevent or reduce the environmental pollution. Definitions of Best Available Techniques are reported in Table II.

The main novelty is that the IPPC Directive requires that permit conditions for the operation of industrial installations must be based on the BAT. The compliance with EQSs is obviously still required, but it is no longer sufficient: every industry is required to make any possible effort - within the limits of the "economical and technical viability" (see Table II) - in order to prevent or reduce pollutant emissions, according to the environmental performance levels associated to the BAT.

It is worth emphasizing that in this new integrated approach, the mere compliance with EQSs for the protection of the environment at local scale is overcome and replaced by a motivated interest towards the continuous improvement of the environmental performances, following the evolution of technology: with the purpose of ensuring a reliable protection of the "whole" environment (which does not accept measures that only transfer pollution to different media, or to a different place).

"Continuous improvement of environmental performances": this is a basic objective of the phase under consideration. For some aspects pursuing this objective is mandatory, in the framework of permit conditions according to the IPPC Directive discussed above. But for other aspects this objective can be adopted by industries on a voluntary basis. In effect, the environmental legislation in Europe in the last decade aims to support, in the industries, a pro-active attitude toward the protection of the environment: which is no longer only a "duty" imposed by law, to be carried out through "end-of-pipe" interventions.

Table II - Definitions of BAT (IPPC Directive, art. 2)

Best Available Techniques	"... the most effective and advanced stage in the development of activities and their methods of operation which indicate the practical suitability of particular techniques for providing in principle the basis for emission limit values designed to prevent and, where that is not practicable, generally to reduce emissions and the impact on the environment as a whole ..."
Techniques	"... includes both the technology used and the way in which the installation is designed, built, maintained, operated and decommissioned ..."
Available	"... techniques ... developed on a scale which allows implementation in the relevant industrial sector, under economically and technically viable conditions, taking into consideration the costs and advantages ..."
Best	"... most effective in achieving a high general level of protection of the environmental as a whole ..."

The protection of the environment is now an important item of the policy of the firm, as well as an objective to be taken into account starting from the design stage of the manufacturing cycle and the research and development activities of the product. The firm is induced to perceive the protection of the environment as a competitiveness factor: a factor affecting positively both the relationships among the factory and the public, and the "success", the acceptation of the product by the customers. Hence the "continuous improvement of environmental performances" included in the objectives of industrial development of the firm. Hence also specific tools designed in order to enable industries to communicate - in a credible, transparent and documented way - their commitment toward the continuous improvement of environmental performaces both in the management of the process and in the design, development and manufacture of products.

Besides the ISO "environmental" standards of the series 14000, the European Commission has developed other voluntary tools, designed to support industries in developing environmental management systems and "environmentally friendly" products, according to a policy based on the continuous improvement of the environmental performances. These tools are, respectively, the Eco-Management and Auditing Scheme (EMAS, Regulation 1836/93/CE, at present under revision), and the Ecolabel Mark (Regulation 880/92/CE, updated as 1980/2000/CE) (Figure 2). EMAS refers to processes and systems (to the production unit), while Ecolabel is a product mark. Both are intended to support the development of systems and products with lower environmental impact, and to inform the public and the consumers about the respective environmental impact and performances. Through these tools the industry communicates with the public and the stakeholders, that are more and more aware and interested to the protection of the environment. These tools, therefore, help industries in promoting and reinforcing their competitiveness.

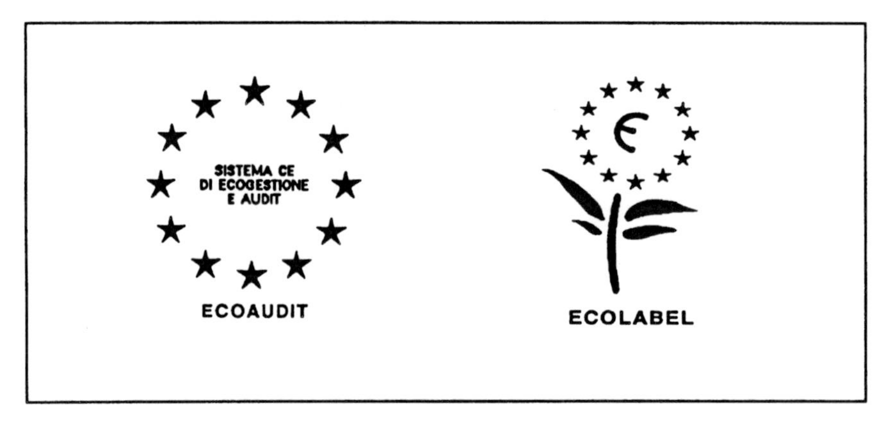

Figure 2 - The EU ecological marks.

The commitment of the Italian ceramic tile industry in this phase is very intense and motivated: the Italian industry looks at the laws and tools discussed above as a mean to recover competitiveness, exploiting the excellent environmental performances achieved. In detail:

- Italy is actively involved in the exchange of information on the BAT, according to the IPPC Directive. The Italian ceramic tile industry supports the adoption of most of the Italian manufacturing and control techniques as the Best Available Techniques. LCA (Life Cycle Assessment) methods have been used to demonstrate that this statement has a sound technical base[10]. The application of the IPPC Directive should have negligible consequences on the permit conditions in use in Italy. The situation may be quite different - more problematic - in other countries;
- more and more Italian ceramic tile industries are motivated towards certification of the environmental management system and adhesion to the EMAS scheme: around 10 Italian firms, representing some of the most important groups, has achieved this result in the last three years, and their example is being imitated by other firms;
- the European Commission is experiencing procedures for the adhesion also of industrial districts to the EMAS scheme, and the Ceramic District of Sassuolo is carrying out a pilot experience in this framework;
- a Working Group for the attribution of the Ecolabel mark to hard floor coverings, which include ceramic tiles, has been activated by the European Commission in September 2000. This group is coordinated by the Italian Environment Protection Agency (ANPA).

CONCLUSIONS

The Italian ceramic tile industry has achieved outstanding results in the protection of the environment. After an early phase (1970-1990) in which the pollution control activity has been carried out under the pressure of the environmental laws, in the last decade the efforts have been focused on the exploitation of the resources devoted to environmental protection and the high environmental performance levels achieved as competitiveness factors.

It is worth mentioning that these quite good results, in an industrial sector consisting, as discussed above, of a large number of small and medium size enterprises, have been made possible by a close and effective cooperation among the industrial sector itself, the research institutions and the public authorities. This cooperation can be acknowledged as the key success factor of the environmental protection activities. The Italian Ceramic Center - the Italian national research and experimentation center for the ceramic industry - which is held by a University Consortium including the University of Bologna, the Italian Association of ceramic products manufacturers and the regional authorities, is the main tool of this cooperation. The Italian Ceramic Center, founded in 1976, has played an important role of scientific and technical support to the environmental protection activities presented in this paper.

REFERENCES

[1]C.Palmonari, G.Timellini, B.Bacchilega et al, "Inquinamento atmosferico da industrie ceramiche. Studio di un Comprensorio: Sassuolo", Ed. Centro Ceramico, Bologna, 1978 - Volume 304 pages

[2]C.Palmonari, G.Timellini, "Pollutant emission factors for the ceramic floor and wall tile industry", *APCA Journal,* XXXII, 10, 1095-1100, (1982)

[3]C.Palmonari, F.Cremonini, A.Tenaglia, G.Timellini, "Water pollution from ceramic industries. Disposal and re-use of waste sludges. Part 1. Characterization of waste water and sludges", *Interceram*, XXXII, N. 1, 40-42, and N. 2, 48-49, (1983)

[4]G.TimelliniI, A.Tenaglia, C.Palmonari, "Water pollution from ceramic industries. Disposal and re-use of waste sludges. Part 2. Technologies for the disposal and re-use of ceramic sludges", *Interceram,* XXXII, N. 4, 25-29, (1983)

[5]G.Busani, C.Palmonari, G.Timellini, "Piastrelle ceramiche & ambiente. Emissioni gassose, acque, fanghi, rumore", Ed. EDI.CER, Sassuolo, 1995 - Volume 428 pages

[6]C.Palmonari, G.Timellini, "Air pollution from the Ceramic Industry: Control Experiences in the Italian Ceramic Tile Industry", *Ceram.Bull.,* 68, n.8, 1464-1469 (1989)

[7]Assopiastrelle-SNAM, "Piastrelle di ceramica e refrattari. Rapporto integrato 1998 Ambiente, Energia, Sicurezza-Salute, Qualità", Ed. EDI.CER, Sassuolo, 1998 - Volume 66 pages

[8]G.Busani, F.Capuano, "Quality and environmental management systems in homogeneous manufacturing areas. Environmental impact of the ceramic industry in its geographic context", *Proc. Qualicer 2000,* Ed. Camara Oficial de Comercio, Castellon, 2000 - Vol. 1, Con-77-92

[9]G.Busani, G.Timellini, "European proposals and directives on energy tax and environmental product and process certification: application and perspectives for the ceramic tile industry", *Proc. 4th Euro-Ceramics,* Vol. 13, Faenza Ed. (1995), 59p

[10]C.Palmonari,G.Timellini, "The environmental impact of the ceramic tile industry. New approaches to the management in Europe", *J. Aust. Ceram. Soc.,* 36, n.2, 23-33 (2000)

Vitrification and Process Technologies

WEST VALLEY DEMONSTRATION PROJECT: VITRIFICATION CAMPAIGN SUMMARY

R. A. Palmer and S. M. Barnes
West Valley Nuclear Services Co.
10282 Rock Springs Road
West Valley, NY 14171-9799

ABSTRACT

The vitrification campaign at the West Valley Demonstration Project (WVDP) is coming to a close. Plans are being made to flush waste-handling equipment and shut down the melting process. This paper describes the waste acceptance system used to produce high-level waste (HLW) forms acceptable for deep geologic disposal. Activities to conclude the vitrification campaign also are discussed. Additionally, incidental successes and challenges experienced during the vitrification campaign are presented.

Waste Acceptance Performance Specifications (WAPS) were used as the standards for vitrification processing of high-level waste (HLW) at the WVDP. Production records for more than 250 canisters of vitrified HLW produced provide the data needed to verify that these canisters are suitable for deep geologic disposal at the proposed federal repository. Data collected and recorded during the vitrification campaign includes: waste form chemistry, radiochemistry and durability, canister fill-height, and the results of smear testing to detect contamination on a canister. The WVDP strategy for meeting each specification called out in the WAPS is presented in this paper. Process performance over more than four years of production also is described.

CAMPAIGN SUMMARY

The West Valley Demonstration Project (Project) has produced 255 canisters of vitrified high-level waste (HLW) suitable for deep geologic disposal (as of April 2001). The first phase of the vitrification campaign was completed on June 10, 1998. Canister production rates have slowed since then because the remaining HLW is more dilute. This means that more HLW transfers are needed

To the extent authorized under the laws of the United States of America, all copyright interests in this publication are the property of The American Ceramic Society. Any duplication, reproduction, or republication of this publication or any part thereof, without the express written consent of The American Ceramic Society or fee paid to the Copyright Clearance Center, is prohibited.

to retrieve enough solid material for a complete glass batch. With more than 99.9% of the long-lived transuranic isotopes removed, the Project is now evaluating state-of-the-art equipment for the final clean-out of HLW tanks at the WVDP site. It is estimated that 275 canisters will have been filled at the completion of the vitrification campaign.

MEETING WASTE ACCEPTANCE PRODUCT SPECIFICATIONS

The Waste Acceptance Product Specifications[1] (WAPS) provide the guidelines for producing an acceptable waste form for disposal in the proposed federal repository. Using these specifications, a processing strategy was devised to demonstrate that control of critical parameters provides adequate assurance of product quality with a minimum of sampling and testing of the final product. This strategy was presented in the Waste Form Compliance Plan[2] (WCP), the document that describes the plan for meeting the WAPS. The results of work done in support of the WCP is provided in the Waste Form Qualification Report[3] (WQR).

Glass Specifications

Specification 1.1: This is the requirement for reporting the chemical composition of the glass waste form. All oxides present at greater than 0.5 weight percent must be reported. The expected range of oxides is presented in Table I. The target composition, which is comprised of 34 different oxides, is presented in Table II.

TABLE I. WVDP Glass: Expected Variations

	Weight Percent	
Oxide	Lower Bound	Upper Bound
Al_2O_3	5.43	6.57
B_2O_3	10.96	14.82
CaO	0.36	0.55
Fe_2O_3	10.22	13.82
K_2O	4.37	5.63
Li_2O	3.25	4.17
MgO	0.76	1.02
MnO	0.70	0.94
Na_2O	7.00	9.00
P_2O_5	1.02	1.38
SiO_2	38.73	43.23
ThO_2	2.67	4.09
TiO_2	0.68	0.92
UO_3	0.47	0.72
$Zr0_2$	1.12	1.52

TABLE II. WVDP Glass: Target Composition

Oxide	Weight Percent	Oxide	Weight Percent
Al_2O_3	6.00	Nd_2O_3	0.14
B_2O_3	12.89	NiO	0.25
BaO	0.16	P_2O_5	1.20
CaO	0.48	PdO	0.03
Ce_2O_3	0.31	Pr_6O_{11}	0.04
CoO	0.02	Rh_2O_3	0.02
Cr_2O_3	0.14	RuO_2	0.08
Cs_2O	0.08	SO_3	0.23
CuO	0.03	SiO_2	40.98
Fe_2O_3	12.02	Sm_2O_3	0.03
K_2O	5.00	SrO	0.02
La_2O_3	0.04	ThO_2	3.56
Li_2O	3.17	TiO_2	0.80
MgO	0.89	UO_3	0.63
MnO	0.82	Y_2O_3	0.02
MoO_3	0.04	ZnO	0.02
Na_2O	8.00	ZrO_2	1.32

An acceptable glass composition is made by first transferring HLW slurry from the primary HLW tank at the WVDP site, Tank 8D-2, to the building that houses systems and components used to vitrify the waste, the Vitrification Facility (VF). The slurry is combined with the heel from the previous batch of melter feed and vitrification process recycle streams (e.g., off-gas condensate) in the primary process vessel used to prepare melter feed, the Concentrator Feed Make-up Tank (CFMT). The slurry mixture is then sampled and concentrated to remove excess water. Following concentration, slurry samples are taken from a recirculating sampling loop. These slurry samples are then sent to the Analytical and Process Chemistry Laboratory (A&PC Lab) to be analyzed for pH, density, percent solids, anions, cations, and selected radionuclides. The 15 oxides listed in Table I are the materials closely monitored and will be reported in the Production Records.

Results from the first round of slurry sampling are used to determine what chemical additions need to be made to meet the target feed composition. The composition of the chemical batches is confirmed by sampling and analysis before the batch is pumped as a cold (radiologically) chemical slurry to be mixed with the HLW.

After the cold chemical slurry has been mixed with the HLW slurry, additional samples are taken and analyzed to confirm that both the desired slurry formulation for making the glass waste form and mass balance closure have been achieved. Verification of the slurry composition is achieved by showing that the prediction of glass leach resistance, which is based on the feed composition, exceeds the WAPS requirements. (Specific WAPS requirements are described in detail in the discussion of Specification 1.3.) If the combined waste and glass-former slurry samples fail this test, additional sampling or chemical addition operations is done and the acceptance testing repeated. The slurry batch is transferred to the process vessel used to feed slurry to the glass melter, the Melter Feed Hold Tank (MFHT), only after it has been demonstrated that the combined waste and glass former slurry sample meets the durability requirement.

The glass composition determined by analysis of the feed slurry has been shown to reliably predict the final glass composition. Approximately 10% of the HLW canisters have been sampled and analyzed for the 15 oxides present in amount greater than 0.5 weight percent. Glass composition measurements compared with feed slurry measurements are shown in Figure 1 (for selected components). At the conclusion of the vitrification campaign, an average glass composition will be calculated over the entire population and used as the value to meet Specification 1.1.

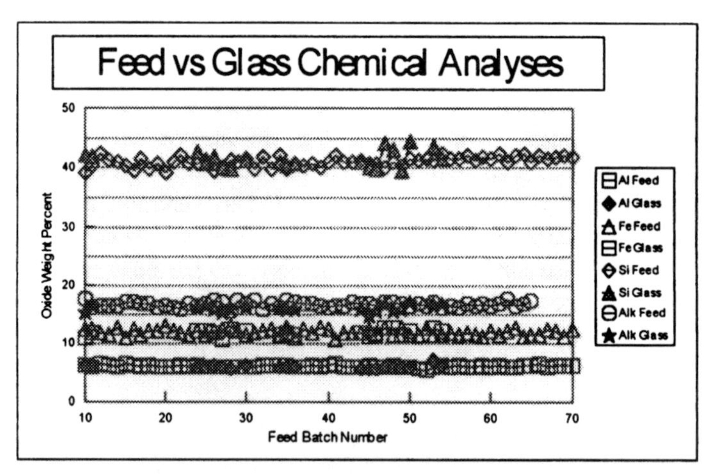

Figure 1. Feed vs. Glass Chemical Analysis

Specification 1.2: This is the Radionuclide Inventory Specification. As with Specification 1.1, this specification will be demonstrated by sampling and analyzing approximately 10% of the canisters. An average will be reported for the population. Indexed to the years 2015 and 3115, all radionuclides that have half-lives longer than 10 years and that are, or will be, present in concentrations greater than 0.05 percent of the total radioactive inventory must be reported. Twenty-nine radionuclides that will be reported are listed in Table III. To meet Specification

1.2, only Cs-137 and Sr-90 will actually be measured on the waste form itself. Other radionuclides will be calculated based on the known ratio to these more predominant species.

TABLE III. Reportable Radionuclides

C-14	Cs-135	Pu-238
Ni-59	Cs-137	Pu-239
Ni-63	Sm-151	Pu-240
Se-79	Ac-227	Am-241
Sr-90	Pa-231	Pu-241
Zr-93	Th-232	Pu-242
Nb-93m	U-233	Am-242m
Tc-99	U-233	Am-243
Pd-107	Np-236	Cm-244
Sn-126	Np-237	

Specification 1.3 : This specification is the waste form durability requirement. The glass must be shown to be more durable than a standard glass as measured by the Product Consistency Test (PCT). The WVDP will predict PCT results based on the chemical analysis of the production glass samples described for Specifications 1.1 and 1.2. Predictions will be compared to measured values obtained on a standard glass composition. (The standard glass is known as the Environmental Assessment, or EA, glass. A detailed description of the glass may be found in the WAPS[1].) The PCT predictions will be based on a DOE-accepted regression model correlating measured PCT results to glass composition. The predicted normalized PCT release for B, Li, and Na will be compared to measurements from the benchmark EA glass to demonstrate compliance with Specification 1.3. Data obtained to date is presented in Figure 2.

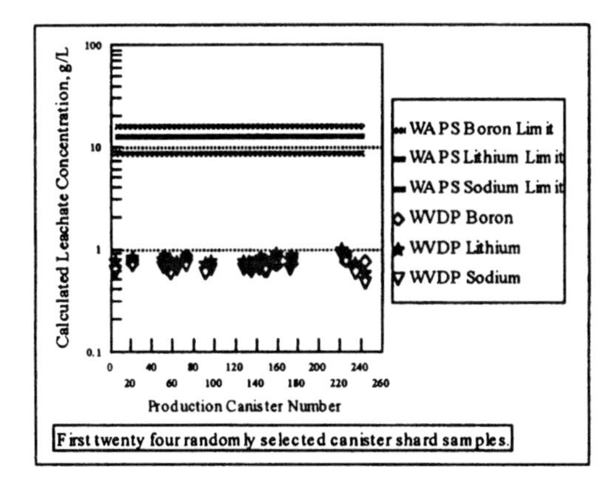

Figure 2. Calculated PCT Response of WVDP Canistered Waste Form

Specification 1.4: This specification is the requirement for Phase Stability. The glass transition temperature, T_g , crystallization behavior and tendency of the glass to form crystals, and storage temperature conditions must be reported. The T_g for target glass composition and several others in the same composition range, was measured and found to be in the range of 450 to 465°C. The effect of redox and thermal history on the glass transition temperature was found to be not significant. The crystallization behavior was reported in the WQR[3] using time-temperature-transformation (TTT) diagrams. The WVDP canister storage facility was designed to maintain the maximum glass temperatures well below 400°C. This has been maintained throughout the campaign.

Specification 1.5: This requirement is the Hazardous Waste Specification. The WVDP glass waste form is not a "listed" hazardous waste. To assess whether the vitrified glass product is a "characteristic" waste, prototypical WVDP glass compositions containing three times the expected amounts of silver, arsenic, barium, cadmium, chromium, lead, and selenium were prepared and evaluated using the Toxicity Characteristic Leaching Procedure (TCLP). In all cases, the metal extraction levels for hazardous metals were significantly below the regulatory limit.

Specification 1.6: This specification describes the International Atomic Energy Agency (IAEA) safeguards regarding the uranium and plutonium content of each canister. The total amount (in grams) as well as concentration (in grams per cubic meter) of ten of the isotopes of these elements will be reported with the radionuclide content as described in Specification 1.2.

Canister Specifications

Specifications 2.1, 2.2, 2.3, and 2.4: These requirements are for canister material, fabrication and closure, identification and labeling, and length and diameter. The canisters were fabricated from austenitic stainless steel Type 304L. The canister heads and barrel are made of ASTM A240 plate Type 304L stainless steel. The flange is 304L stainless steel pipe per ASTM A312, plate per ASTM A240, or forging per ASTM 182. The composition of ASME SFA5.9 ER308L austenitic stainless steel, the weld filler metal, was used to assemble the canister from its component parts. This 308L alloy also was used for the weld-beaded canister identification labels.

Canister integrity has been ensured by specifications of the components, specification of the method of fabrication, and a rigorous program of inspection and verification. One major component of the canister receipt inspection is a test of the leak-tightness of the as-manufactured, unfilled canister. Final, leak-tight (1 x 10^{-4} atm-cc/sec helium, as defined in Specification 2.2) weld closure of the canisters has been performed as soon as practical after filling. Close control of the autogenous, pulsed gas tungsten arc welding (GTAW) welding process and visual weld inspection has assured a leak-tight weld on each production canister.

The code used for identifying the canistered waste forms is a five character alphanumeric code of two letters and three numbers. The reference labeling technique is a bead-welding of the alphanumeric characters directly onto the canister surface using Type 308L weld metal. This labeling technique was shown to be suitable by fabricating full-sized, weld-bead-labeled canisters, handling and decontaminating the labeled canister in a manner similar to that used in the WVDP process, and establishing that the labels are still readily legible and not subject to preferential obliteration.

The specified maximum and minimum length and diameter of the unfilled WVDP canister has been designed to be safely within those required by this specification. As-built canister lengths and diameters have been provided for all production canisters.

Canistered Waste Form Specifications

Specifications 3.1, 3.2, 3.3, and 3.4: These are the requirements for exclusion of materials other than glass from the closed canisters. The precluded materials include free liquids, extraneous gases, explosive, pyrophoric, combustible, and organic materials. The vitrification process has been shown to evaporate free liquids and destroy most of the compounds in question. The primary task during vitrification operations has been to exclude all of these foreign materials from the facility (and therefore the canister) to the greatest extent possible. This has been done though administrative control and timely permanent weld closure of the canister.

The canisters were inspected before entry into the VF, after filling and just before closure to ensure they contain no prohibited materials.

Specification 3.5: This specification requires that the glass and canister material are chemically compatible. Testing had been done with nonradioactive glasses to demonstrate that there are no adverse reactions between the glass composition and the 304L stainless steel at temperatures up to and beyond 500°C which would degrade the canister integrity.

Specification 3.6: This specification requires that the canisters be at least 80 percent full. The WVDP planned that canisters be at least 85 percent full. The actual average production value is more than 91 percent full. The fill height of each production canister is determined by using a measuring device that physically probes the height of the glass in several places after the canister has cooled and been removed from the canister loading turntable.

Specification 3.7: This is the specification for removing radioactive contamination on the outside surface of the canister. The canistered waste form is decontaminated with a nitric acid and Ce^{+4} solution before being to transferred to the interim HLW canister storage facility. External surfaces are smeared according to 10 CFR 71.87(i) before transfer. The limits are 22,000 dpm/100 cm^2 for beta and gamma emitters, and 2,200 dpm/100 cm^2 for alpha emitters.

The external surfaces of the canistered waste forms also are visually inspected for visible glass. The smear surveys have resulted in only three of 255 canisters being decontaminated a second time. No visible glass has ever been detected on a canister. A second decontamination process is planned for use just before the canisters are packaged for final shipment.

The Project must provide an estimate of the material removed during the decontamination process. It also needs to provide an assessment of the unfilled canister wall thickness. The decontamination process is estimated to remove between ten and fifteen micrometers of 304L stainless steel from the canister. Ultrasonic wall thickness measurements on the as-manufactured canisters have been made and found to be within the specifications required of the fabricator. The individual results for each canister will be reported in the Production Records in compliance with Specification 3.11.

Specification 3.8: This specification limits the heat generation of the individual canisters to 1500 watts. The heat generation rate in a canister containing HLW was calculated using the Standardized Computer Analyses for Licensing Evaluation (SCALE) computer codes. The heat generation rate depends on the amount and type of radionuclides contained in the canister and decreases with time as a result of radioactive decay. Data to compute heat generation rate in the SCALE code is derived from the concentration of radionuclides in a canister. The source of this data is in WQR Section 1.2.1. Radionuclide concentrations have been established in the WQR. The corresponding heating rates are computed with the SCALE system. The maximum amount of heat generated expected is about 362 watts, as measured by 1996 radionuclide decay. The maximum values predicted for 2015 and 3115 are 238 and 3.5 watts, respectively.

Specification 3.9: This specification requires an estimate of the maximum gamma and neutron dose rates indexed to the year 2015 and the time of shipment. The limits are 10^5 rem/h for gamma and 10 rem/h for neutrons. Projections of gamma dose rates at the surface of high-level waste (HLW) canisters were made using the Standardized Computer Analyses for Licensing Evaluation (SCALE) system computer codes. The estimate of the radionuclide inventory is described in WQR Section 1.2. Results of the calculations are reported in WQR Section 3.9 where the maximum values are given as 6.4×10^3 rem/h gamma and 8.8×10^{-2} rem/h for neutron.

Specification 3.10: This is the requirement for Subcriticality. The calculated effective neutron multiplication factor, K_{eff}, must be shown to be much less than or equal to 0.90 for the canistered waste form. The value for the WVDP glass was calculated using the KENO computer code. The maximum value of K_{eff}, conservatively calculated using twice the anticipated amount of fissionable material, was found to be 4.89×10^{-3}.

Specification 3.11: This specification provides for the reporting the mass of the overall canistered waste form and its dimensions. The maximum allowed mass is 2500 kg and the canister must fit into a right-circular cylindrical cavity, 64.0 cm in diameter and 3.01 m in length. The design of the canister and the Quality Assurance provisions in place will ensure that these criteria can be met. These parameters will be checked before shipout and the results of the tests will be reported in the Storage and Shipping Records.

Specification 3.12: This specification gives the requirements for the Drop Test. A filled canister must withstand a drop of seven meters onto a flat, essentially unyielding surface without breaching. The WVDP strategy for compliance with this specification consisted of two approaches: 1) using engineering calculations to form a basis for the conclusion that the reference canister is capable of surviving a 7-meter drop; and 2) dropping nonradioactive glass-filled canisters to confirm their ability to withstand the required drop. Both methods demonstrated that the canistered waste form will survive such a test.

Specification 3.13: This specification gives the requirements for Handling Features. A flange geometry for the WVDP canister and a grapple that couples with this flange and complies with this specification was designed. The grapple has a rated capacity nearly two times that of a completely (100%) full canister and a 5,000-lift cycle lifetime.

Specification 3.14: This specification provides a requirement for reporting the concentration of plutonium in each canistered waste form. The WVDP plans to comply with this specification by taking shards removed from the top of the canistered glass, measuring the quantity of Sr-90 in the shards, and relating this value to the quantity of plutonium in the canistered waste form using scaling factors from the waste characterization program. The plutonium value will then be divided by the quantity of glass in the canistered waste form to generate the plutonium concentration value. This calculated plutonium concentration will be listed in the Production Records at the end of the vitrification campaign.

SUMMARY

The West Valley Demonstration Project has vitrified nearly all the HLW at the Project site. The strategy developed for meeting the WAPS has been successful in producing 100% acceptable canistered high-level waste. Implementation and refinement of this strategy should allow future projects to follow the same routine to obtain similar success.

REFERENCES

[1]U.S. Department of Energy, Office of Environmental Restoration and Waste Management, "Waste Acceptance Product Specifications for Vitrified High-Level Waste Forms," U.S. Department of Energy Report, EM-WAPS, Rev. 2, (December 1996).

[2]West Valley Nuclear Services Co., Inc., "Waste Form Compliance Plan for the West Valley Demonstration Project High-Level Waste Form (WCP)," WVDP-185, (April 1998).

[3]West Valley Nuclear Services Co., Inc., "Waste Form Qualification Report (WQR)," WVDP-186, (October 1998).

WASTE GLASS PROCESSING REQUIREMENTS OF THE HANFORD TANK WASTE TREATMENT AND IMMOBILIZATION PLANT

George Mellinger and Langdon Holton
Pacific Northwest National Laboratory
P.O. Box 999
Battelle Boulevard
Richland, WA 99352

Dr. Neil Brown
Office of River Protection
U.S. Department of Energy
P.O. Box 450
Richland, WA 99352

ABSTRACT

In December 2000, the Department of Energy awarded a contract to Bechtel National, Inc. for the design, construction and commissioning of a facility that will be used to pretreat and vitrify high-level tank waste currently stored in underground tanks at the Hanford Site in south-eastern Washington State. The facility will receive the waste, separate it into high-level waste (HLW) and low-activity waste (LAW) fractions. Both fractions will be vitrified; the HLW as borosilicate glass for off-site disposal in a licensed geologic repository, and the LAW as a silicate glass for disposal at Hanford. This paper discusses the glass processing requirements specified in the contract, and the technical baseline for the project with emphasis on the waste glass processing aspects of the baseline.

INTRODUCTION

The Hanford Site in southeastern Washington State has one of the largest concentrations of radioactive waste in the world. That waste is the legacy of 45 years of plutonium production for nuclear weapons, which began with the Manhattan Project in the 1940s. Approximately fifty-three million gallons of high-level radioactive waste are stored in 177 underground tanks near the Columbia River. Sixty-seven of the 149 older single-shell tanks have leaked an estimated one million gallons of waste. It is critical to treat, immobilize, and dispose of this waste before more waste leaks to the soil and groundwater. As directed by Congress in 1999, the U.S Department of Energy (DOE) established the Office of River Protection (ORP) at the Hanford Site. ORP is responsible for managing, treating and disposing of this waste. ORP has established the River Protection Project (formerly known as the Tank Waste Remediation System) for accomplishing this mission.

To the extent authorized under the laws of the United States of America, all copyright interests in this publication are the property of The American Ceramic Society. Any duplication, reproduction, or republication of this publication or any part thereof, without the express written consent of The American Ceramic Society or fee paid to the Copyright Clearance Center, is prohibited.

In December 2000, ORP awarded a contract to Bechtel National Inc. for design, construction, and commissioning of the Hanford Tank Waste Treatment and Immobilization Plant (WTP). The requirements specified in the contract were developed to ensure that the completed WTP will fulfill its role in completing the RPP mission while also allowing the contractor the flexibility to optimize the design and construction of this facility. When completed, the WTP will receive waste retrieved from the underground storage tanks, separate the waste into high-level and low-activity fractions, and vitrify these two fractions. The vitrified low-activity waste will be disposed on-site at Hanford while the high-level fraction will be stored on-site pending off-site disposal at a geologic repository. A copy of the WTP contract may be found on the Internet on the ORP website at: http://www.hanford.gov/orp/procure/contracts.html.

Under the current facility concept, the WTP consists of four major facilities: two pretreatment facilities, one HLW vitrification facility, and one LAW facility. The contract requires the WTP to have a design operational life of 40 years; for equipment or facilities with a design operational life of less than 40 years, the WTP design must allow for replacement of these items.

The schedule for completing the WTP contract is shown in Figure 1. Key dates for this contract include the Start of Hot Commissioning in December 2007 (when high-level waste will be transferred to the facility and pretreatment initiated); Completion of Acceptance Testing in November 2007 (when construction will be complete); and Completion of Hot Commissioning in January 2011 (when all commissioning requirements will have been met, including production of a substantial number of final high-level and low-activity waste products). Certain dates are shown as TBD. These are dates that were not specified in the initial contract, but will be established when ORP and the contractor have agreed upon a project baseline.

WTP PROCESSING REQUIREMENTS

The processing capacity requirements for the WTP are specified in Section C.7 *Facility Specification* of the contract. The contract specifies the processing capacity requirements for the plant during the initial phase of waste treatment to be completed by 2018. The contract also specifies that the contractor must design and build the WTP with features to provide for expanded operational capabilities. This will provide ORP with flexibility for completing the balance of the RPP mission once the first phase of waste treatment is completed. These requirements are summarized in Table I. As the table shows, the initial HLW vitrification capacity requirement is comparable to the design capacity for the HLW melter at the Defense Waste Processing Facility, and approximately twice that for the HLW melter at the West Valley Demonstration Project. The LAW vitrification capacity requirement is substantially larger.

Figure 1. Summary schedule for WTP contract

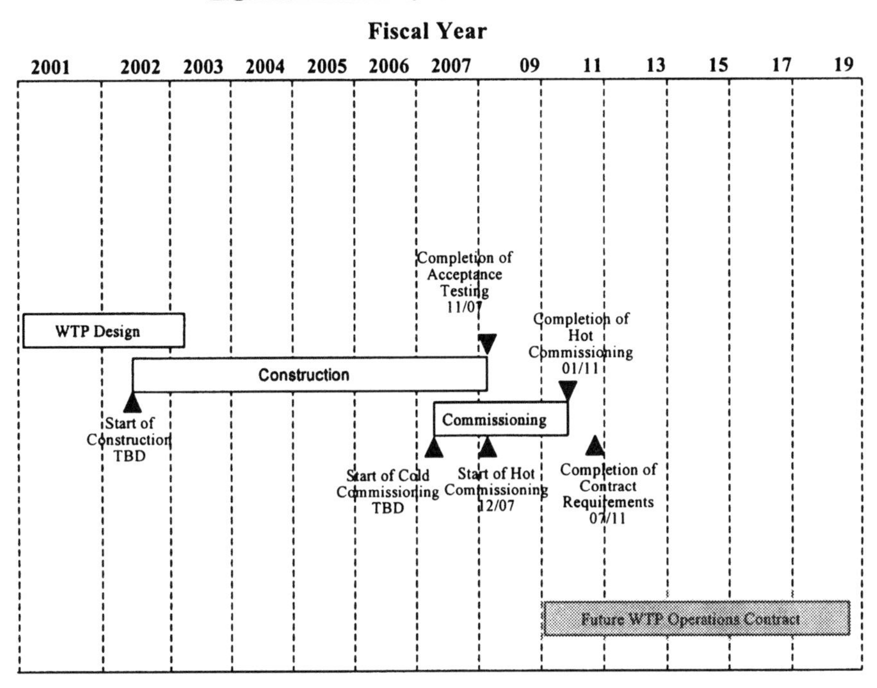

Table I. WTP processing requirements

Facility	WTP Initial Capacity	WTP Expanded Capacity	Other Sites in DOE Complex	
			DWPF	WVDP
Pretreatment (glass equivalent)	LAW – 60 MTG/d HLW – 6.0 MTG/d	N/A (initial capacity sufficient to cover expanded glass capacity)	Sufficient to Provide feed for HLW vitrification, LAW immobilization	Sufficient to provide feed for HLW vitrification, LAW immobilization
LAW Vitrification	30MTG/d	60MTG/d	N/A	N/A
HLW Vitrification	1.5MTG/d	6.0MTG/d	Single melter 2.0 MTG/d design capacity	Single melter 0.70 MTG/d design capacity

As shown in Table I the initial pretreatment capacity must be large enough to accommodate the expanded glass production requirement. This is due to the cost effectiveness of incorporating this capability in the original design. The expanded LAW capacity requirement would be accomplished by the addition of a second LAW vitrification facility. The expanded HLW capacity would be accomplished in two ways. First, although the HLW vitrification facility will have two process cells for HLW vitrification, a melter system will be installed in only one of these cells during the initial phase of waste treatment; the HLW capacity would be expanded by installing and operating a second melter in the unused cell. Second, at that time the capacity of each HLW melter would be increased from the original 1.5 metric tons of glass per day (MTG/d) to 3.0 MTG/d for a total production rate of 6.0 MTG/d. HLW melter capacity enhancements are currently being examined by DOE-EM's Office of Science and Technology

FACILITY COMMISSIONING

Requirements for commissioning of the WTP are specified in Section C, Standard 5, *Commissioning* of the contract. One of the key objectives of the commissioning is to demonstrate that process and facility performance meets or exceeds contract requirements. The contract specifies requirements for both "cold" and "hot" commissioning.

During cold commissioning the contractor will use non-radioactive simulated feeds to conduct tests to verify that the WTP will perform in accordance with design specifications while treating a broad range of tank wastes. During hot commissioning the contractor will treat actual Hanford tank waste. Two performance requirements must be met during hot commissioning while also producing fully compliant waste products. The first requirement is to process sufficient waste to produce 300 units of immobilized LAW (equivalent to ~450 containers containing ~2700 metric tons of LAW glass) and 60 high-level waste canisters containing ~180 metric tons of high-level waste glass. The second performance requirement is to complete tests of specified duration during which nominal and peak waste processing rates are demonstrated. These hot commissioning requirements are summarized in Table II.

WTP UNIT OPERATIONS

Pretreatment Unit Operations

Section C.7 *Facility Specification* specifies the unit process operations for the WTP facility. The pretreatment unit operations are intended to partition the waste in such a way as to minimize the mass of waste that is fed to HLW vitrification, while also minimizing the concentration of radionuclides fed to LAW vitrification. The pretreatment unit operations required by the contract and the technologies specified for these unit operations are shown in Figure 2.

Table II. Processing rate tests during WTP hot commissioning

System	Nominal		Peak	
	Duration	Requirement	Duration	Requirement
LAW Pretreatment HLW Pretreatment	40 days 30 days	30 MTG/d equiv. 1.5 MTG/d equiv.	5 days 10 days	60 MTG/d equiv. 6.0 MTG/d equiv.
LAW Vitrification	≤ 30 days	≥150 containers	10 days	60 containers
HLW Vitrification	≤ 30 days	≥12 canisters	10 days	5 canisters

Figure 2. Pretreatment Unit Operations

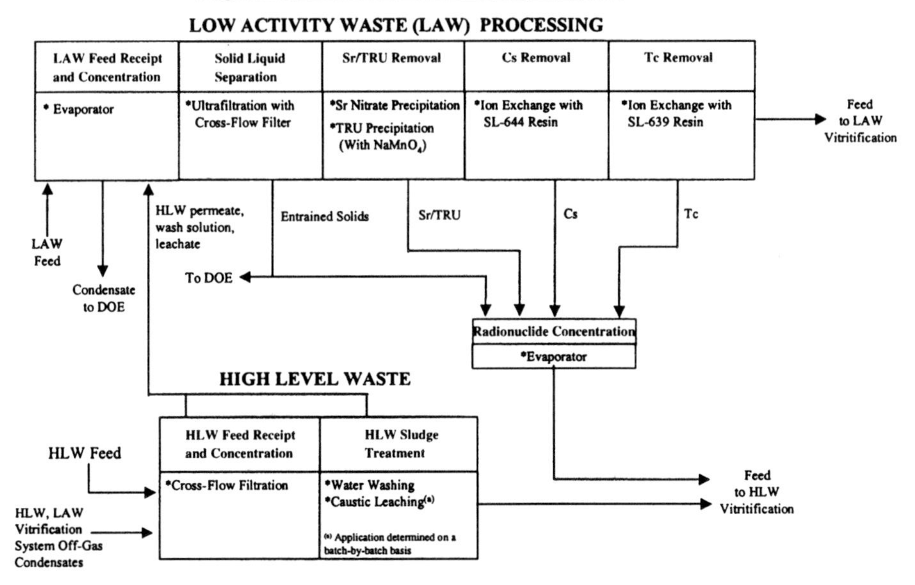

As shown in the figure the primary inputs to pretreatment are LAW and HLW feed and the primary outputs are pretreated feed that goes to LAW and HLW vitrification. Secondary outputs from LAW pretreatment include evaporator condensate that is returned to DOE for on-site treatment and disposal; and entrained solids removed from the LAW feed, if DOE elects to receive these rather than include them with the HLW feed. A secondary input to HLW pretreatment is off-gas condensates from HLW and LAW vitrification.

Waste entering LAW pretreatment is first concentrated by evaporation and then solids and liquids are separated. The solids may either be returned to DOE, or transferred to a concentration step prior to being added to the HLW feed. Following the removal of the solids, strontium and tranuranics, cesium, and technetium are removed from the liquid fraction and transferred to a concentrator prior to being added to the HLW feed.

Waste entering HLW pretreatment is first filtered and the permeate is transferred to LAW pretreatment. The filtrate is then water washed. It may also be caustic leached, depending on the chemistry of a particular batch of waste. The wash solution and leachate are transferred to LAW pretreatment.

Vitrification Unit Operations

The unit operations specified in the contract for HLW and LAW vitrification and off-gas treatment are similar.

The contract specifies the use of ceramic lined, joule-heated melters for both HLW and LAW vitrification. However, the contract does not require a specific melter design. The current technical baseline for these melters (for the initial treatment capacity) is summarized in Table III.

Table III. HLW and LAW melters – current technical baseline

Factor	HLW	LAW
Surface Area	3.75 m^2	10.0 m^2
Glass Temperature	1100-1200 °C	1100-1200 °C
Number	1	3
Nameplate Capacity	1.5 MTG/d	10 MTG/d
Anticipated Total Online Efficiency	0.68	0.82
Electrodes	Inconel® 690, parallel plate	Inconel® 690, parallel plate

The HLW and LAW melter off-gas are treated in their respective off-gas treatment systems. The melter off-gas treatment system unit operations are based upon technologies proven in other DOE projects with consideration for WTP treatment requirements. These requirements are comparable for the HLW and LAW vitrification facilities.

Melter off-gas first enters an air film cooler that reduces the off-gas temperature before it enters the submerged bed scrubber (SBS). In the SBS the steam is condensed, and larger particles, aerosols, and soluble constituents are removed. A wet electrostatic precipitator (WSEP) then removes additional aerosols. Following the WSEP the HLW off-gas system includes a high-efficiency mist eliminator (HEME) that further removes aerosols from the melter off-gas stream. Condensates from the SBS and HEME, and wash water from the WESP are collected in vessels and transferred back to the pretreatment facility. High efficiency particulate air (HEPA) filters further remove entrained particulates. Volatile organics are destroyed in a thermal catalytic oxidation process. The LAW off-gas treatment system also includes a selective catalytic reduction unit that uses a catalyst and ammonia to reduce the NO_x generated from the vitrification of the LAW feeds. Before exiting the facility, the off-gas passes through a caustic scrubber.

PRODUCT REQUIREMENTS

Vitrified High-Level Waste Product

The requirements for the vitrified HLW product are specified in Section C.8, Specification 1, *Immobilized High-Level Waste.* The contract specifies that the vitrified HLW product be a borosilicate glass contained in a 4.5-meter long by 0.61-meter diameter canister with a neck and flange design similar to that used at the West Valley Demonstration Project. The mass of glass in each canister is approximately 3,100 kg. The fundamental specification for the vitrified HLW product is that it must meet the requirements in the Waste Acceptance System Requirements Document (WASRD) [1], the Waste Acceptance Product Specifications for Vitrified High Level Waste Forms (WAPS) [2] and the Quality Assurance Requirements and Description for the Civilian Radioactive Waste Management Program (QARD) [3].

Vitrified Low-Activity Waste Product

The specifications for the vitrified LAW product are provided in Section C.8, Specification 2, *Immobilized Low-Activity Waste.* The specifications for the LAW were developed based on the requirements for shallow land disposal at the Hanford Site. The contract specifies that the vitrified LAW product be a glass contained in a 304 stainless-steel cylinder with a height of 2.3 meters and a diameter of 1.22 meters. The mass of glass in each package is approximately 6,000 kg. The contract also specifies additional requirements including:

Glass durability	Surface dose rate and contamination
Void space	External temperature
Chemical composition	Closure and sealing
Radionuclide concentration limits	Compressive strength

CONCLUSIONS

The requirements in the WTP contract address two needs. First, they are intended to be sufficient to ensure that the WTP, when completed, will meet the RPP mission requirements allocated to this facility. Second, they are intended to provide the contractor with sufficient flexibility to allow the contractor to optimize the design and construction of the WTP. This flexibility, combined with the incentive features of the WTP contract, is expected to drive the contractor to identify and implement actions that will minimize the cost and schedule for completion of the WTP.

REFERENCES

[1]DOE/RW-0351P. Rev. 3. DCN 02. April 1999. Waste Acceptance System Requirements Document (WASRD). U.S. Department of Energy, Office of Civilian Radioactive Waste Management, Washington, D.C.

[2]DOE/EM-0093. Rev. 3. December 1996. Waste Acceptance Product Specifications for Vitrified High Level Waste Forms (WAPS). U.S. Department of Energy, Office of Environmental Management, Washington, D.C.

[3]DOE/RW-0351P. Rev. 3. DCN 02. April 1999. Quality Assurance Requirements and Description for the Civilian Radioactive Waste Management Program (QARD). U.S. Department of Energy, Office of Civilian Radioactive Waste Management, Washington, D.C.

INFLUENCE OF GLASS PROPERTY RESTRICTIONS ON HANFORD HLW GLASS VOLUME

Dong-Sang Kim and John D. Vienna
Pacific Northwest National Laboratory,* Richland, WA 99352, (509) 373-0256

ABSTRACT

A systematic evaluation of Hanford High-Level Waste (HLW) loading in alkali-alumino-borosilicate glasses was performed. The waste feed compositions used were obtained from current tank waste composition estimates, Hanford's baseline retrieval sequence, and pretreatment processes. The waste feeds were sorted into groups of like composition by cluster analysis. Glass composition optimization was performed on each cluster to meet property and composition constraints while maximizing waste loading. Glass properties were estimated using property models developed for Hanford HLW glasses. The impacts of many constraints on the volume of HLW glass to be produced at Hanford were evaluated. The liquidus temperature, melting temperature, chromium concentration, formation of multiple phases on cooling, and product consistency test response requirements for the glass were varied one- or many-at-a-time and the resultant glass volume was calculated. This study shows clearly that the allowance of crystalline phases in the glass melter can significantly decrease the volume of HLW glass to be produced at Hanford.

INTRODUCTION

The Hanford Site retains 177 underground waste tanks containing 204,400 m^3 of HLW, which will be retrieved from the tanks, separated into HLW and low-activity waste (LAW) fractions, and separately vitrified. This study evaluates the effects of glass property constraints on the Hanford HLW loading in alkali-alumino-borosilicate glasses. The objective is to identify the factors that have strong effects on the waste loading of the Hanford waste glass and thus warrant further studies to minimize the waste glass volume.

The waste compositions used in this study were those estimated by the Hanford Tank Waste Optimization Simulator (HTWOS) [1] using current baseline waste-retrieval scenario described by Garfield et al. [2]. The actual Hanford retrieval schedules and separations methods will evolve with time. This particular "slice-in-time" is expected to adequately assess the impacts of major vitrification process/product changes on the scale of waste cleanup costs. The baseline scenario generated 89 batches of waste, which were later grouped into 17 waste clusters with like compositions by cluster analysis. Table I

---

* Pacific Northwest National Laboratory is operated for the U. S. Department of Energy by Battelle Memorial Institute under Contract DE-AC06-76RL01830.

To the extent authorized under the laws of the United States of America, all copyright interests in this publication are the property of The American Ceramic Society. Any duplication, reproduction, or republication of this publication or any part thereof, without the express written consent of The American Ceramic Society or fee paid to the Copyright Clearance Center, is prohibited.

shows the weighted average composition and total mass of waste for each cluster used in this study.

APPROACH

The method chosen to evaluate the effect of property constraints on the glass volume was first to define a reference case, which is summarized in Table II, then vary constraints from that reference. First-order models for normalized PCT releases (r_B, r_{Li}, r_{Ni}), viscosity (η), and liquidus temperature (T_L) from Hrma et al. [3] and for electrical conductivity (ε) and density (ρ) from Hrma et al. [4] were used in glass-property calculations.

The baseline melter technology assumed was a joule-heated melter operating at a nominal operating temperature of 1150°C. To avoid crystalline phases in the melter, a limit of $T_L \leq T_M$ - 150°C (T_M is the nominal operating temperature) was used. The 150°C margin was to allow temperature variations within the melter and uncertainty of model prediction and feed composition. The viscosity at T_M is limited to the range from 2 to 10 Pa·s. The electrical conductivity is restricted to between 10 and 100 S/m at T_M. We adopted limits of 0.8 mass% SO_3 to avoid salt segregation in the melter and 0.10 mass% RuO_2 and Rh_2O_3 combined to limit noble metal accumulation in the melter bottom.

When the concentration of chromium in glass exceeds its solubility, it precipitates as either spinel, $[Fe,Ni,Mn][Fe,Cr]_2O_4$, or eskolaite, Cr_2O_3. The liquidus temperature in both primary phase fields is a function of glass composition. The effect of glass composition on T_L is known within a reasonably large portion of the spinel primary phase field. However, since there is a scarcity of data on T_L in the eskolaite primary phase field, we assumed the solubility of Cr_2O_3 at 1 mass% as a reference based on preliminary tests on limited number of glasses.

The WAPS [5] requires that r_B, r_{Li}, and r_{Na} are less than those of the DWPF EA glass by the PCT [6]. The EA glass release values are 8.35, 4.78, and 6.67 g/m^2 for r_B, r_{Li}, and r_{Na}, respectively. We have used a limit of 2 g/m^2 for r_B, r_{Li}, and r_{Na} to account for model uncertainty. To ensure that the glass composition did not significantly deviate from the model validity ranges, limits were placed on the concentrations of (with mass% limits) B_2O_3 (5–15), Fe_2O_3 ($\leq$ 20), MnO ($\leq$ 4), Li_2O ($\leq$ 4), Na_2O ($\leq$ 20), SiO_2 ($\geq$ 35), and $Na_2O+Li_2O+K_2O$ ($\leq$ 22).

There are components that can limit the loading of waste in glass by their strong influence on amorphous phase separation or crystalline phase formation that in turn may negatively impact glass properties (PCT). The single component concentrations or composition rules were used as multi-phase constraints. The constraints of $P_2O_5 \leq 2.5$ mass% and F $\leq$ 2 mass% were used to avoid the formation of an immiscible phosphate phase [7] or a fluoride. The composition rules expressed by $[SiO_2]/([Na_2O] + [Al_2O_3] + [SiO_2]) \geq 0.62$ and $([Na_2O] + [Li_2O])/([Na_2O] + [Li_2O] + [B_2O_3] + [SiO_2]) \geq 0.12$ were used to make the glass less susceptible to nepheline formation [8] and liquid-liquid phase separation [9], respectively.

This set of constraints (Table II) is expected to yield reasonable, if a bit conservative, estimates of Hanford HLW loadings in glass. These constraints are not to be regarded as absolute requirements, but are to be used as the reference set of constraints from which sensitivity studies are performed.

Table I. Concentration of major[1)] waste component (mass%) and total mass ($\times 10^3$ kg) by cluster

Cluster#	1	2	3	4	5	6	7	8	9	10	11	12	13	14	15	16	17
Al_2O_3	19.6	12.7	19.6	18.5	17.7	27.2	18.1	23.0	6.6	6.7	7.5	7.3	2.2	15.0	8.3	3.3	2.6
B_2O_3	0.1	0.1	0.5	0.2	0.1	0.2	0.3	2.0	0.6	0.2	0.3	1.8	1.4	0.8	1.3	1.9	0.1
Bi_2O_3	6.5	12.1	1.2	3.0	9.6	6.2	0.4	2.1	3.8	0.2	0.0	0.3	0.0	1.6	0.0	0.0	0.1
CaO	2.8	3.2	2.3	5.5	3.4	2.2	2.0	1.5	0.7	1.0	2.9	0.9	1.4	0.8	0.9	1.9	0.4
CdO	0.0	0.0	0.0	0.0	0.0	0.0	0.0	0.0	0.0	0.0	0.1	0.1	0.3	0.1	4.5	0.4	0.0
Cr_2O_3	0.6	0.8	1.7	1.3	0.9	1.4	2.0	2.5	0.2	1.3	0.2	0.5	0.2	4.3	0.1	0.2	0.5
F	0.4	1.0	2.8	2.2	3.7	1.2	1.0	1.2	2.3	0.1	0.0	3.8	0.1	1.4	0.3	0.1	0.1
Fe_2O_3	15.7	21.5	7.3	13.5	17.3	12.1	8.4	7.3	14.1	5.2	45.7	1.4	11.8	11.0	47.8	15.6	2.2
K_2O	0.2	0.3	0.7	0.3	0.2	1.3	0.8	0.2	0.4	1.8	0.1	1.1	0.3	2.0	1.0	0.4	0.6
MgO	0.1	0.5	0.6	0.5	0.2	0.3	0.1	2.3	0.5	0.3	2.0	1.7	0.1	0.7	0.3	0.1	0.1
MnO	1.7	2.8	1.5	1.5	1.4	2.3	1.0	1.8	6.8	10.8	8.8	0.5	10.2	1.4	0.9	3.4	23.9
Na_2O	22.9	20.7	20.4	22.7	21.2	22.8	21.1	27.2	12.4	45.1	14.6	29.3	9.1	35.5	12.9	12.1	14.5
NiO	1.1	0.7	1.5	2.6	1.5	2.2	0.4	0.6	0.4	0.2	1.4	0.6	0.8	0.5	2.1	1.1	0.1
P_2O_5	3.1	4.5	2.8	3.8	4.3	2.9	2.1	6.4	0.7	1.3	0.6	0.2	0.2	2.4	0.2	0.3	0.5
SiO_2	17.3	10.7	7.4	9.5	8.5	8.0	27.5	4.4	4.6	1.3	5.2	1.4	3.1	2.0	0.9	4.2	0.5
SrO	0.3	0.3	0.4	1.0	0.6	0.4	0.1	0.3	13.7	21.1	0.2	0.1	17.1	0.1	0.1	0.0	52.3
ThO_2	0.1	0.1	0.3	0.2	0.1	0.3	0.4	0.2	0.1	0.2	0.0	0.2	2.8	0.2	0.2	3.7	0.1
Tl_2O	0.0	0.0	0.3	0.1	0.0	0.2	0.0	1.7	0.6	0.2	1.1	2.2	0.0	0.7	0.0	0.0	0.1
U_3O_8	6.0	4.8	10.8	10.1	7.8	6.4	10.7	6.9	6.1	1.8	1.4	9.9	15.8	5.1	3.6	20.9	0.7
ZrO_2	0.2	0.6	15.2	1.8	0.3	0.4	2.2	0.7	21.6	0.1	0.6	28.0	21.9	10.2	10.6	29.0	0.0
Others	1.3	2.6	3.0	1.7	1.4	2.1	1.4	7.7	3.7	1.2	7.3	9.1	0.9	4.3	3.8	1.2	0.6
Mass	2349	1749	1647	1395	1384	947	678	506	426	232	216	175	171	150	142	128	19

[1)] Only the components with at least 2 mass% in at lease one cluster are listed.

Table II. Reference set of constraints used in this study

Constraint	Value	Unit	Reason
T_M	1150	°C	nominal operating temperature
T_L	≤1000	°C	processability
η at T_M	2–10	Pa·s	processability
ε at T_M	10–100	S/m	processability
PCT releases (r_B, r_{Li}, r_{Na})	≤2	g/m^2	WAPS
[B_2O_3]	5–15	mass%	model validity
[Fe_2O_3]	≤20	mass%	model validity
[MnO]	≤4	mass%	model validity
[Li_2O]	≤4	mass%	model validity
[Na_2O]	≤20	mass%	model validity
[SiO_2]	≥35	mass%	model validity
[Na_2O]+[Li_2O]+[K_2O]=[Alk]	≤22	mass%	model validity
[Cr_2O_3]	≤1	mass%	Cr_2O_3 solubility (eskolaite T_L)
[P_2O_5]	≤2.5	mass%	phosphate phase immiscibility
[F]	≤2	mass%	F solubility
[SO_3]	≤0.8	mass%	salt segregation
[RuO_2]+[Rh_2O_3]	≤0.10	mass%	noble metals settling
[SiO_2]/([SiO_2]+[Na_2O]+[Al_2O_3])	≥0.62		nepheline formation on cooling
[Alk]/([Alk]+[SiO_2]+[B_2O_3])	≥0.12		Liquid-liquid immiscibility

The glass compositions for each cluster were numerically optimized. If a property constrained the loading of a particular waste in glass, the additive compositions (Al_2O_3, B_2O_3, Na_2O, Li_2O, and SiO_2) would be adjusted until at least one additional constraint was met. Therefore, waste loading was limited by more than one property constraint, unless a single component concentration constraint (e.g., [Cr_2O_3] or [P_2O_5]) was encountered. The [Fe_2O_3] ≤ 20 mass% constraint or [MnO] ≤ 4 mass% constraint is included as part of model validity constraints, but actually operates as a single component constraint because Fe_2O_3 and MnO are not additive components. The SO_3 and liquid-liquid immiscibility (expressed by normalized alkali content) constraints were never met in any clusters from any cases.

RESULTS AND DISSCUSSION

Effect of Constraint Changes on Total Glass Volume

Calculations using the reference set of constraints (See Table II) resulted in an estimated total glass mass of 26,539 Mg (1 Mg = 10^6 g) or glass volume of 9,720 m^3 from vitrification of the 17 Hanford waste clusters. With 12,314 Mg of waste, this represents a weighted average waste loading of 45.9 mass%, and assuming 1.15 m^3 glass/canister

Table III. Effect of constraint changes on estimated total waste loading, glass mass, and glass volume

Case	Constraint Changes from Reference Case	Loading (mass%)	Mass (Mg)	Volume (m^3)
1	Reference	46.4	26,539	9,720
2	$T_L \leq 1150°C$	51.0	24,166	8,808
3	$T_L \leq 1350°C$	52.8	23,317	8,462
4	PCT releases ≤ 6 g/m^2	46.5	26,476	9,700
5	Allow multi-phase formation	49.3	24,961	9,029
6	Remove model validity constraints	47.8	25,779	9,295
7	$T_M = 1350°C$ ($T_L \leq 1200°C$)	49.9	24,682	9,089
8	$T_M = 1350°C$, $T_L \leq 1350°C$	52.7	23,350	8,532
9	$T_M = 1350°C$, $T_L \leq 1550°C$	53.6	22,963	8,370
10	$[Cr_2O_3] \leq 0.5$ mass%	38.1	32,335	12,110
11	$[Cr_2O_3] \leq 1.5$ mass%	46.9	26,265	9,600
12	$[Cr_2O_3] \leq 1.5$ mass%, $T_L \leq 1150C$	52.0	23,679	8,592
13	$T_M = 1350°C$, $[Cr_2O_3] \leq 1.5$ mass%, $T_L \leq 1350°C$	54.1	22,757	8,331
14	$T_M = 1350°C$, $[Cr_2O_3] \leq 0.5$ mass%	39.0	31,546	11,854
15	Allow multi-phase, PCT releases ≤ 6 g/m2	49.6	24,834	8,975
16	Remove model validity constraints, $T_L \leq 1150°C$	52.5	23,465	8,383
17	Remove model validity constraints, $T_L \leq 1350°C$	55.9	22,039	7,833
18	$T_M = 1350°C$, $[Cr_2O_3] \leq 1.5$ mass%, $T_L \leq 1550°C$	55.4	22,243	8,130
19	$T_M = 1350°C$, $[Cr_2O_3] \leq 1.5$ mass%	50.9	24,200	8,875
20	$T_M = 1350°C$, $\eta = 0.1$-10 Pa·s, $\varepsilon = 10$-150 S/m	51.8	23,776	8,647
21	$T_M = 1350°C$, Remove model validity constraints	51.6	23,870	8,697

would yield 8,542 canisters. Of the 26,539 Mg of glass, 75% was limited by the T_L(spinel) $\leq$ 1000°C constraint, 8% was limited by $[Cr_2O_3] \leq 1.0\%$ constraint, 6% was limited by the T_L(zircon) $\leq$ 1000°C constraint, with the remaining 11% limited by a range of other constraints.

Table III shows how the constraints were varied from the reference set in a total of 21 cases with corresponding results on the estimated waste loadings and glass volumes. The total glass volume for each case is the sum of glass volumes calculated for each cluster, by dividing glass mass by estimated density value.

Compared to the reference case (Case 1), increasing the melting temperature from 1150 to 1350°C (Case 7) decreases the volume of glass by 631 m^3 or 6.5%. Here it should be noted that a possible increase of Cr_2O_3 solubility by increasing melting temperature is not included. Relieving the constraints on the viscosity and electrical conductivity at $T_M = 1350°C$ resulted in a moderate decrease of glass volume, by 442 m^3 or 4.9% (Case 20 compared to Case 7).

If the T_L constraints are relieved to allow the crystals (spinel and/or zircon) to be present in the melter, the volume of glass decreases considerably: 1258 m^3 or 12.9% decrease for T_M = 1150°C and T_L = 1350°C (Case 3 compared to Case 1) and 719 m^3 or 7.9% decrease for T_M = 1350°C and T_L = 1550°C (Case 9 compared to Case 7). Case 3 and Case 9 both are represented by the constraint, $T_L < (T_M + 200°C)$ or 350°C increase from the reference constraint, which is $T_L < (T_M - 150°C)$. Relieving the T_L constraint further had only minor effect on total glass volume (not included in Table III). The allowance of the T_L constraint to be increased is in effect allowing crystals to form in the melter. These crystals will primarily be spinel, which has been shown to have little effect on PCT release of the final waste form [10-12].

The volume of glass decreases significantly when the limit of the Cr_2O_3 concentration is increased from 0.5 to 1.0 mass%: 2390 m^3 or 19.7% decrease for T_M = 1150°C (Case 10 vs. Case 1) and 2765 m^3 or 23.3% decrease for T_M = 1350°C (Case 14 vs. Case 7). However, further change of the Cr_2O_3 concentration limit from 1.0 to 1.5 mass% has a smaller effect on estimated glass volume: 120 m^3 or 1.2% decrease of glass volume for T_M = 1150°C (Case 11) and 214 m^3 or 2.4% decrease for T_M = 1350°C (Case 19). Case 18 shows that increasing the T_M to 1350°C along with relieving the T_L constraint by 350°C and increasing the Cr_2O_3 constraint to 1.5 mass% results in a decrease of glass volume by 1383 m^3 or 16.4%. The solubility of Cr_2O_3 depends on glass composition and temperature, which is not accounted for in this study because of insufficient data. Further study is required to more accurately assess the temperature and composition impacts on Cr_2O_3 solubility, especially considering the strong effect of the Cr_2O_3 limit on the glass volume.

The constraints on the PCT releases have little impact on the glass volume: the decrease of volume achieved by changing the constraint from 2 to 6 g/m^2 is negligibly small (20 m^3 or 0.2%, Case 4). A moderate decrease of glass volume is achieved by relieving the constraints on multi-phase formations (691 m^3 or 7.1 %, Case 5). Case 15 shows the effects of relieving the constraints on the multiphase formation and PCT releases at the same time, which results in the decrease of glass volume by 745 m^3 or 7.7 %. This decrease of glass volume is only slightly larger than the effect of relieving the constraint on the multiphase formation alone (7.1%), which also shows that the PCT constraints have very small effect. Current models are not able to predict the impact of multi-phase formation on the glass durability. PCT releases of glasses forming nepheline and immiscible glass phases may be significantly higher than those predicted by the current models. Further study is required to more accurately estimate the impacts of multi-phase formation on glass volume while maintaining adequate waste form durability.

A moderate decrease of glass volume is achieved by relieving the constraints on model-validity composition ranges (461 m^3 or 4.7% for T_M = 1150°C - Case 6 compared to Case 1; 392 m^3 or 4.3% for T_M = 1350°C - Case 21 compared to Case 7). The model validity constraints are not the actual limitations imposed by any processing or product performance requirements. Therefore, expanding the model validity range through testing of more glasses has the potential to decrease glass volume. Relieving the T_L constraint by 150°C and 350°C together with removing the model validity constraint result in the decrease of glass volume by 1163 m^3 or 13.8% (Case 16) and 1640 m^3 or 19.4% (Case 17), which is the greatest decrease observed in the current study. However, the prediction of glass properties and therefore the estimation of glass volumes for compositions outside the ranges of model validity is subject to high uncertainties. Further testing would be required to better estimate glass volume if compositions are allowed to deviate significantly from current model validity ranges.

Figures 1 through 4 show the effect of certain constraint change (x-axis) for given constraint conditions (legend) on the number of canisters (y-axis) assuming 1.15 m^3/canister. Figure 1 shows the great effect of changing the Cr_2O_3 limit from 0.5 to 1.0 mass% for both baseline and high-temperature melter and small impact of changing the Cr_2O_3 limit from 1.0 to 1.5 mass%. The effect of changing the T_L constraint for both 1150°C and 1350°C-T_M melters is shown in Figure 2. It can be seen that the advantage of high-temperature melter observed at the reference T_L constraint ($T_M - T_L < 150$°C) fades as the T_L constraint is relieved allowing crystals in the melter. Figure 3 shows the effect of changing the T_L constraint for 1150°C melter at the conditions with and without model validity constraint. The effect of removing model validity constraint becomes greater as the T_L constraint is relieved. The moderate effect of removing the constraint on the multi-phase formation is shown in Figure 4, which also clearly shows the very small effect of PCT constraint.

The constraints that have largest impacts on the glass volume such as the Cr_2O_3 concentration limit and T_L constraints are all related to crystalline phase formation in the melter. For example, in the reference case (Case 1), the constraints related to crystallization in the melter limited waste loading in 12 clusters out of 17, which accounts for 89% in terms of glass volume. Considering that the constraints on the Fe_2O_3 and MnO within model validity constraints are also related to the crystallization of spinel, the numbers are even higher. Table IV summarizes the effect of changing the Cr_2O_3 limit and T_L constraints on the glass volume.

Effect of Constraint Changes on Glass Volume from Each Cluster

Figure 5 shows the range of glass volume in each cluster resulting from different constraints set in this study. For every cluster, Case 10 gives the maximum volume, which is the same as given by the reference case for those clusters that are not affected by the Cr_2O_3 constraint. The minimum volume is achieved from Case 17, which relieves the T_L constraint by 350°C together with removing the model validity constraint, for most of clusters. The exceptions are Clusters 2, 7, and 17, of which the primary constraint is either P_2O_5 or Cr_2O_3 concentration. In these clusters, minimum volume is obtained when the P_2O_5 or Cr_2O_3 constraints are relieved. It can be seen from Figure 5 that the glass volume from some waste clusters (like 9 to 17) are insensitive to the variation of the glass property constraints. Figure 5 in general gives the information on which are the major clusters that contribute to the total glass volume most and so should be the focus on studies aimed at decreasing total glass volume.

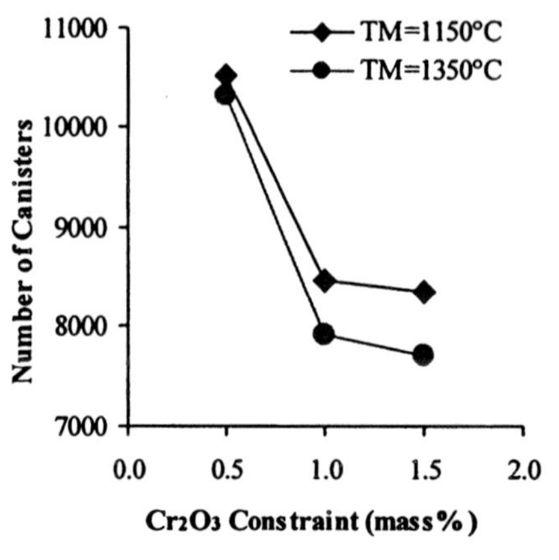

Figure 1. Effect of Cr_2O_3 constraint on the estimated number of canisters in different T_M Melters

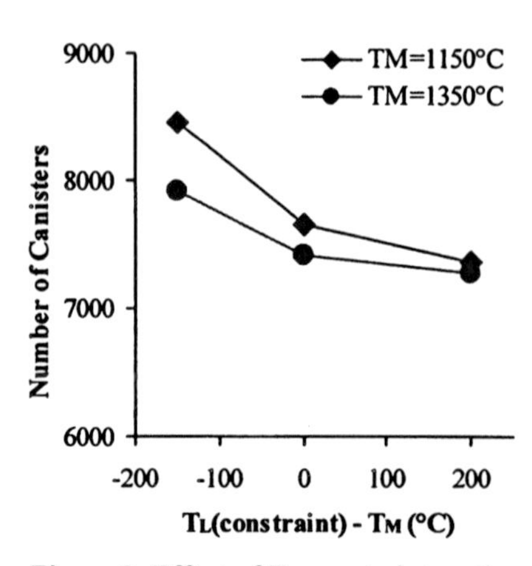

Figure 2. Effect of T_L constraint on the estimated number of canisters in different T_M melters

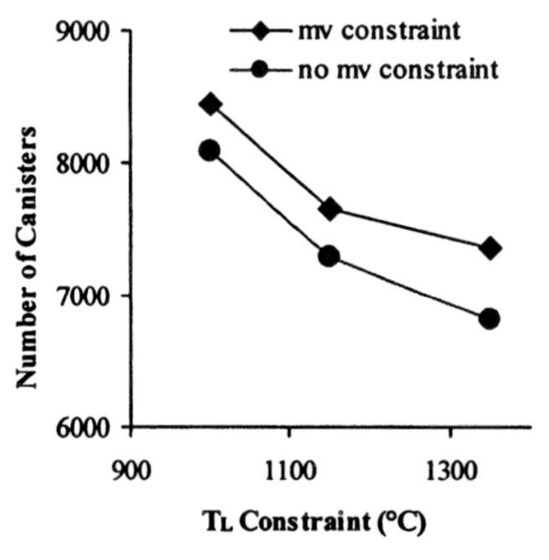

Figure 3. Effect of T_L constraint on the estimated number of canisters with and without model validity (mv) constraints

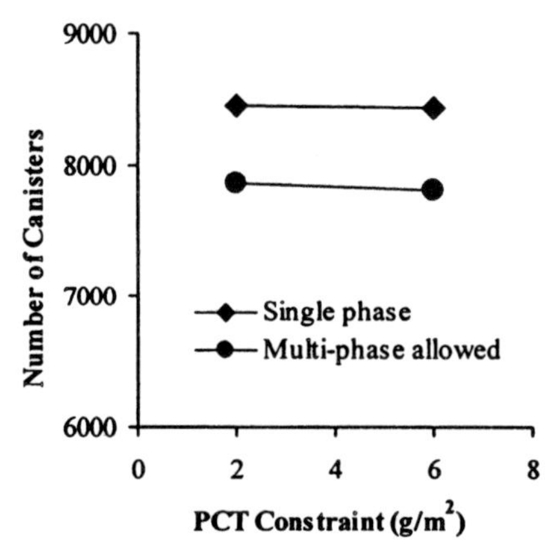

Figure 4. Effect of PCT release constraint on the estimated number of canisters with and without restrictions on multi-phase

Effect of Chromium leach factor

Estimated fraction of each chemical component that partition to the liquid or LAW by caustic leaching is known as the leach factor. The leach factor for chromium used in the baseline scenario was 0.770, which was taken from Colton [13]. Because chromium concentration in the waste was expected to have a substantial impact on glass volume, we estimated the influence of the chromium leach factor on glass volume. Two more

Table IV. Effect of T_L and Cr_2O_3 constraints on estimated glass volume (in m^3) (changes relative to a Reference Case are in parentheses)

Cr_2O_3 (mass%)	$T_L < (T_M - 150°C)$		$T_L < T_M$		$T_L < (T_M + 200°C)$	
	T_M=1150°C	T_M=1350°C	T_M=1150°C	T_M=1350°C	T_M=1150°C	T_M=1350°C
<0.5	12,110 (+24.6%)	11,854 (+22.0%)				
<1.0	9,720 (Reference)	9,089 (-6.5%)	8,808 (-9.4%)	8,532 (-12.2%)	8,462 (-12.9%)	8,370 (-13.9%)
<1.5	9,600 (-1.2%)		8,592 (-11.6%)	8,331 (-14.3%)		8,130 (-16.4%)

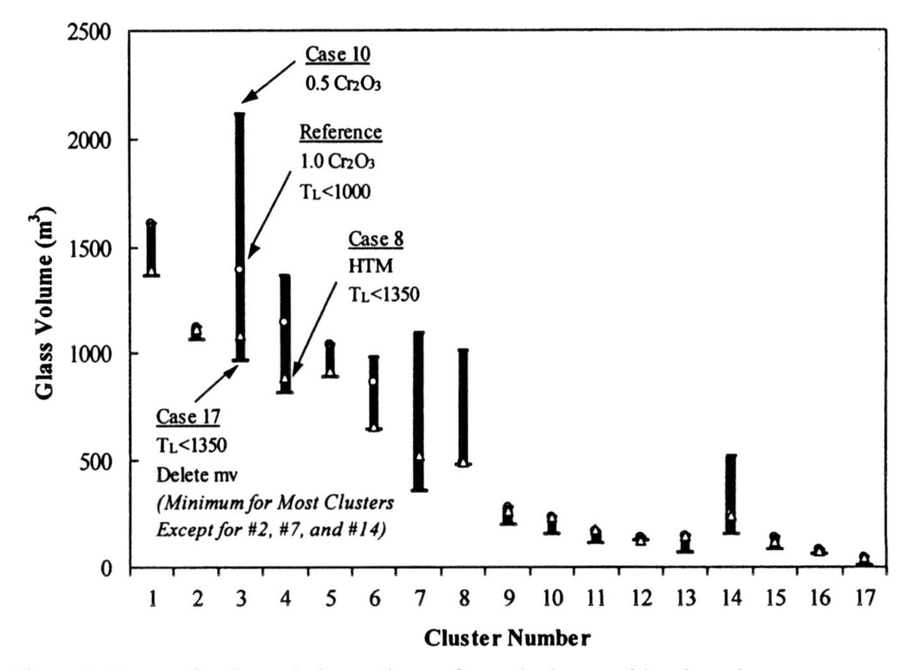

Figure 5. Range of estimated glass volumes for each cluster with selected cases highlighted

chromium leach factors were used: 0.385 and 0.924. Glass volume was estimated using HTWOS tools for a reference property constraint set (1.0 mass% Cr_2O_3 limit) and for a case with the Cr_2O_3 constraint of 0.5 mass%. The resulting estimates of glass canisters at 1.15 m^3/canister are presented in Figure 6, which shows that the chromium leach factor has an enormous impact on estimated glass volume provided that the low solubility limit for Cr_2O_3 is confirmed by experiments. The impact is much greater if the 0.5 mass% Cr_2O_3 limit is assumed.

This result shows that development of a high confidence in chromium concentrations in the waste, chromium leach factors, and chromium solubility in glass melts is important for Hanford to reduce uncertainty in glass volume estimates and more importantly to minimize glass volumes.

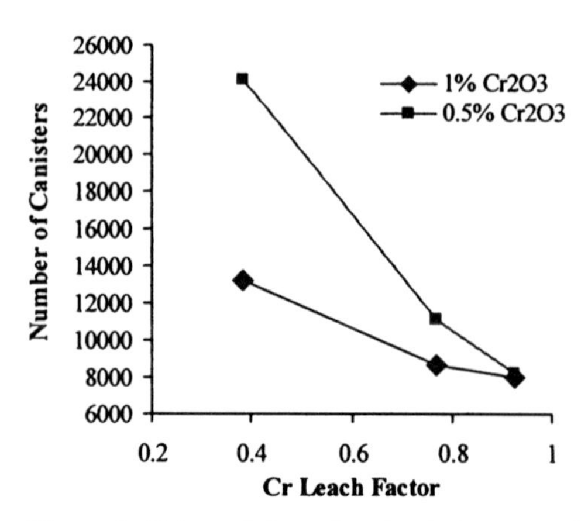

Figure 6. Effect of chromium leach factor on the estimated number of canisters at different Cr_2O_3 constraints

CONCLUSIONS

The key sources of uncertainty in the glass volume estimation are related to the amount of chromium in the waste and the allowable fraction of chromium in the glass. The combination of uncertainties in chromium inventories in tanks, partitioning of chromium to the HLW fraction (leach factors), and solubility of chromium in glass can change the glass volume estimates by several tens or even more than a hundred percents relative to the current reference. This impact on glass volume suggests that further testing is required to

- reduce the uncertainty in chromium concentration in waste
- improve the reliability of the chromium leach factor
- improve the understanding of chromium solubility in glass or composition effects on T_L in the eskolaite primary phase field.

In addition, the result of this study suggests that a melter technology that can tolerate crystalline phases (e.g., insoluble chromium in the form of eskolaite and spinel) would allow for not only increased waste loading in glass, but also would reduce the sensitivity of glass volume projections to these uncertainties. This study also suggests that it would be possible to decrease glass volume by expanding the model validity range through generation of more glass-property data and by studying the effect of multiphase on glass durability.

ACKNOWLEDGEMENTS

The authors are grateful to Steve Lambert and Rick Wittman for glass volume estimations using HTWO tools. We would like to thank Pavel Hrma for his critical review of the manuscript and Joe Perez, David Peeler, and Mark Triplett for their valuable comments. This work was supported by the U.S. Department of Energy (DOE) Office of Science and Technology through Tanks Focus Area (TFA).

REFERENCES

[1] R.A. Kirkbride, "Tank Farm Contractor Operation and Utilization Plan (TWRS-OUP)," HNF-SD-WM-SP-012, Rev. 2, CH2M Hill Hanford Group, Inc., Richland, Washington, 2000.

[2] J.S. Garfield, R.A. Kirkbride, T.M. Hohl, and W.J. Stokes, "Single-Shell Tank Retrieval Sequence: Fiscal Year 2000 Update," RPP-7087, Rev. 0, Numatec Hanford Company, Richland, Washington, 2000.

[3] P.R. Hrma, G.F. Piepel, J.D. Vienna, S.K. Cooley, D.S. Kim, and R.L. Russell, "Interim Glass Property Models for Hanford HLW Glasses," PNNL-13573, Pacific Northwest National Laboratory, Richland, Washington, 2001.

[4] P.R. Hrma, G.F. Piepel, M.J. Schweiger, D.E. Smith, D. Kim, P.E. Redgate, J.D. Vienna, C.A. LoPresti, D.B. Simpson, D.K. Peeler, and M.H. Langowski, "Property/Composition Relationships for Hanford High-Level Waste Glasses Melting at 1150°C," PNL-10359, Vol. 1 and 2, Pacific Northwest Laboratory, Richland, Washington, 1994.

[5] WAPS, "Waste acceptance Product Specifications for Vitrified High-Level Waste Forms", Office of Environmental Restoration and Waste Management, U.S. DOE, Report EM-WAPS Rev. 02, 1996.

[6] American Society for Testing and Materials (ASTM), "Standard Test Method for Determining Chemical Durability of Nuclear Waste Glasses, The Product Consistency Test (PCT)," ASTM-C-1285-97, in Annual Book of ASTM Standards, Vol. 12.01, Philadelphia, Pennsylvania, 1998.

[7] C.M. Jantzen, K.G. Brown, J.B. Pickett, and G.L. Ritzhaupt, "Crystalline Phase Separation in Phosphate Containing Waste Glasses: Relevancy to Vitrification of Idaho National Engineering and Environmental Laboratory (INEEL) High Activity Waste (U)," WSRC-TR-2000-00339, Westinghouse Savannah River Company, Aiken, South Carolina, 2000.

[8] H. Li, J.D. Vienna, P. Hrma, D.E. Smith, and M.J. Schweiger, "Nepheline Precipitation in High-Level Waste Glasses: Compositional Effects and Impact on the Waste Form Acceptability," in *Scientific Basis for Nuclear Waste Management XX*, pp. 261-268, Materials Research Society, Pittsburgh, Pennsylvania, 1997.

[9] D.K. Peeler and P.R. Hrma, "Predicting Liquid Immiscibility in Multicomponent Nuclear Waste Glasses," *Ceramic Transactions*, **45** 219 – 229 (1994).

[10] C.M. Jantzen, D.F. Bickford, "Leaching of Devitrified Glass Containing Simulated SRP Nuclear Waste," *Mat. Res. Soc. Symp. Proc.*, **44** 135-146 (1985).

[11] D.S. Kim, D.K. Peeler, and P. Hrma, "Effects of Crystallization on the Chemical Durability of Nuclear Waste Glasses," *Ceramic Transactions.* **61** 177-185 (1995).

[12] B.J. Riley, J.A. Rosario, and P. Hrma, "Impact of HLW Glass Crystallinity on the PCT Response," PNNL-13491, Pacific Northwest National Laboratory, Richland, Washington, 2001.

[13] N.G. Colton, "Status Report: Pretreatment Chemical Evaluation FY 1997- Wash and Leach Factors for the Single-Shell Tank Waste Inventory," PNNL-11646, Pacific Northwest National Laboratory, Richland, Washington, 1997.

VITRIFICATION AND TESTING OF HANFORD PRETREATED HLW SLUDGE MIXED WITH FLOWSHEET QUANTITIES OF SECONDARY WASTES

Gary L. Smith, Harry D. Smith, Michael J. Schweiger, and Rick J. Bates
Pacific Northwest National Laboratory
P.O. Box 999, MSIN K6-24
Richland, WA 99352

ABSTRACT

Actual pretreated HLW samples along with flowsheet quantities of secondary wastes were vitrified to demonstrate the RPP-WTP projects ability to satisfy the HLW product ORP Phase B-1 contract requirements concerning chemical and radionuclide reporting, waste loading, identification and quantification of crystalline and non-crystalline phases, and waste form leachability. Chemical composition of two HLW glasses (i.e., elements {excluding oxygen} present in concentrations greater than 0.5 percent by weight) were measured using ICP-AES of solutions generated by a KOH fusion and a Na_2O_2 fusion. The total waste percent oxide in the C-104 glass (all waste oxides (exclusive of Si) not identified in Table TS-1.1 of the contract) comes to 10.02%, which is greater than the required 8.0%. The total of Al_2O_3 + Fe_2O_3 + ZrO_2 for AZ-102 Melt 1 is 21.96%, which is greater than the required 21.0%. The inventory of radionuclides (in Curies) was measured for each glass. The total and fissile uranium and plutonium content of each canister of waste glass were calculated and are: 1161.5 grams for C-104, 642.1 grams for AZ-102 Melt 1. After canister centerline cooling, no crystals were observed in the C-104 sample by XRD, optical examination, and SEM analysis. XRD analysis combined with optical microscopy and SEM EDS results of the CCC heat-treated AZ-102 Melt 1 glass sample indicated the presence of a crystalline phase at approximately 1 to 2 volume percent, most likely trevorite and chromite. The normalized lithium, sodium and boron 90°C PCT releases for the C-104 and AZ-102 Melt 1 glasses are: 1) (for lithium) 0.5 g/m^2 and 0.4 g/m^2; 2) (for sodium) 0.4 g/m^2 and 0.4 g/m^2; and 3) (for boron) 0.4 g/m^2 and 0.3 g/m^2; respectively. The HLW product testing results from the C-104 and AZ-102 glasses show that in all cases they meet ORP Phase B-1 contract specifications.

To the extent authorized under the laws of the United States of America, all copyright interests in this publication are the property of The American Ceramic Society. Any duplication, reproduction, or republication of this publication or any part thereof, without the express written consent of The American Ceramic Society or fee paid to the Copyright Clearance Center, is prohibited.

INTRODUCTION

The U.S. Department of Energy (DOE) Office of River Protection (ORP) has contracted Hanford tank waste treatment services. The River Protection Project-Waste Treatment Plant (RPP-WTP) team is responsible for producing an immobilized (vitrified) high-level waste (IHLW) waste form. PNNL produced and tested a vitrified IHLW waste form from two Envelope D high-level waste (HLW) samples previously supplied to the RPP-WTP project by DOE.

One sludge sample each was taken from two underground storage tanks at the Hanford site in southeastern Washington. The two tank sludge samples (241-C-104 and 241-AZ-102) were processed through pretreatment chemical washing and leaching processes, and the pretreated sludges were converted to high-level waste (HLW) glass after flowsheet quantities of secondary wastes, i.e. Sr/TRU precipitate and Cs and Tc ion exchange eluants, generated from LAW supernatant pretreatment unit operations were added. Both sludge samples were processed through the following unit operations to simulate the RPP-WTP project flowsheet: 1) initial characterization; 2) washing; 3) leaching; and 4) filtration in a crossflow filtration system [1,2]. The analyzed compositions of the pretreated C-104 and AZ-102 wastes and secondary wastes were used by Catholic University of America's (CUA) Vitreous State Laboratory (VSL) to calculate the target glass composition.

The primary objective for vitrifying the two Envelope D (Tank C-104 and Tank AZ-102) pretreated HLW sludge samples was to characterize the glass produced from the crucible melts. Testing results of the waste glasses produced from actual tank waste shows compliance with the RPP-WTP contractual requirements such as chemical and radionuclide reporting, product loading and durability. The work reported includes: 1) Glass Fabrication, 2) Chemical Composition, 3) Radiochemical Composition, 4) Crystalline and Non-crystalline Phase Determination, and 5) Release Rate (PCT). The results show that good glass was formed having compositions in line with the target quantities, acceptable devitrification properties, with good leach resistance.

EXPERIMENTAL

HLW Glass Fabrication

Independently, each pretreated tank waste was vigorously blended in a stainless steel beaker using a magnetic stirrer and stir bar. Secondary wastes (Sr/TRU precipitate, the composite Cs ion exchange eluant, and Tc ion exchange eluants) were combined with the pretreated tank sludge waste into the same stainless steel beaker. A combination of glass former additives borax ($Na_2B_4O_7 \cdot 10H_2O$); lithium hydroxide monohydrate ($LiOH \cdot H_2O$); silica sand (SiO_2); zinc oxide (ZnO); and sugar) were added to each pretreated waste to produce a melter feed. The C-104 and AZ-102 melter feeds were dried, calcined, and melted at 1150°C for two hours. The glass melt was then poured onto a stainless steel plate (air quenched) and cooled to room temperature. A portion of the melt was poured into a small box crucible that was later heat-treated following the predicted canister centerline cooling (CCC) heat treatment of a Hanford HLW canister. The pretreated high-level waste sludges were processed and vitrified in the Radiochemical Processing Laboratory (RPL) in the High Level Radioactive Facility (HLRF) hot cells. Due to scheduling constraints and small initial sample size of the pretreated tank 241-AZ-102 sludge, this sample was divided into two samples that were vitrified separately (i.e., AZ-102, Melt 1 and AZ-102, Melt 2) so that physical and rheological testing of the pretreated waste and melter feed could be completed in parallel with vitrification testing.

Chemical Composition

Chemical compositions of the two HLW glasses (i.e., elements {excluding oxygen} present in concentrations greater than 0.5 percent by weight) were measured using a KOH and Na_2O_2 fusion preparation procedure. The KOH fusion used a nickel crucible and the Na_2O_2 fusion used a zirconium crucible. Cation analysis was performed using Inductively Coupled Plasma-Atomic Emission Spectrometry (ICP-AES). Two fusions and two ICP-AES analyses provided duplicate analyses for all cations except K, Ni, Na, and Zr.

Radiochemical Composition

Radiochemical analyses were performed on the HLW glass products. Analyses included ^{60}Co, ^{106}Ru, ^{125}Sb, ^{134}Cs, ^{137}Cs, ^{144}Ce, ^{154}Eu, ^{155}Eu, and ^{241}Am by gamma energy analysis (GEA), ^{90}Sr, $^{239/240}$Pu, ^{241}Pu, and total uranium. Concentration values of additional gamma emitters (i.e., ^{51}Cr, ^{59}Fe, ^{79}Se, ^{95}Nb, ^{103}Ru, ^{113}Sn, and ^{152}Eu) were obtained by GEA depending on concentrations and detection limits. Strontium was isolated from the initial fusion solutions and then beta-counted. Plutonium was isolated from the initial fusion solutions, precipitation plated, and then counted by alpha spectroscopy. Total uranium was determined on dilutions of the initial fusion solution using kinetic phosphorescence analysis. In addition, the following radioisotopes: ^{99}Tc, ^{237}Np, ^{239}Pu, ^{240}Pu, ^{233}U, ^{234}U, ^{235}U, ^{236}U and ^{238}U were measured by inductively coupled plasma mass spectroscopy (ICP-MS).

Crystalline and Non-Crystalline Phase Determination

Crystalline and non-crystalline phases were identified and measured using x-ray diffraction (XRD), optical microscopy, and scanning electron microscopy (SEM) on glass samples that had been heat-treated as indicated in Table I to simulate a Hanford HLW canister centerline cooling (CCC) curve. The HLW stainless steel canisters are basically right circular cylinders 4.5 m in height and 0.61 m in diameter. Glass canister filling and temperature profile was modeled with a batch target fill rate of 500 kg/hr for 30 minutes at a temperature of 1150°C with 3 hours between pours (12 MT/day). The modeled temperature, cooling curve for the centerline of a canister of glass 3/5 of the way from the bottom of the canister was used for the heat-treatment.

Table I. Temperature profile line segments for the Del Tech furnace controller to generate the Hanford HLW Canister centerline cooling profile

Hours	Temperature (°C)	dT/dt (deg./hr)
0.00 - 0.17	1004 -1050	+277
0.17 –2.17	1050 – 1003	-24
2.2 – 7.0	1003 – 844	-33
7.0 - 10.3	844 – 749	-28.4
10.3 - 15.5	749 – 617	-25.5
15.5 - 21.2	617 – 491	-22.3
21.2 – 25.8	491 – 400	-19.5

For XRD analysis, a piece of CCC heat treated glass was broken from a segment of sectioned glass, crushed in an alumina mortar and pestle grinder, sieved through a 75 micron (200 mesh) sieve and stored in a glass vial until analysis. The powder prepared for XRD analysis was weighed to between 8 and 9 mg, mixed into a solution of callodin, mounted on a plastic XRD sample mount, leveled to X-ray beam height, encapsulated in Mylar film, and transported to the XRD laboratory for analysis. The two-theta scan range was from 5 to 75 degrees at a step size of 0.05 degrees with a 1 second dwell at each step.

For optical and scanning electron microscopy, a thin slice of one of the cross sections of the CCC heat-treated glass was used. Conventional grinding and polishing techniques were followed (final polish was a 6 micron polishing paste). The thin slices of C-104 and AZ-102 glass were examined using an optical microscope in both reflected and transmitted light (magnification from 50 to 200×). Viewing of the samples was accomplished with a video camera with the image viewed on a monitor near the hot cell. For SEM analysis each slide was coated with a gold film and examined at low magnification (100×) and higher magnifications of 500×, 1500×, 2000×, and 5000×.

Release Rate, Product Consistency Testing (PCT) of HLW Glasses

PCT testing on the C-104 and AZ-102 glass samples was completed per Method A of ASTM C1285-97 [3]. Crushed glass of a particle size between 75 and 150 μm (-100 to +200 mesh) was used for testing. The glass was crushed and then sieved through 100- and 200-mesh stainless steel sieves. The crushed glass was cleaned by washing in deionized water (DIW) and ethanol using an ultrasonic cleaner. It was then dried and weighed, and approximately 1.5 g of glass added to a 22-mL desensitized Type 304L stainless steel container filled with DIW. The glass was precisely weighed and the leachant volume precisely controlled to achieve a solution volume to glass mass ratio of 10 mL/g glass. The ratio of the surface area of the sample to leachant volume is estimated to be 2000 m^{-1}. The container and its contents were held (without agitation) at a temperature of 90°C for 7 days. The initial and final pH values of the solution were measured. Aliquots of the solution were filtered through a 0.45-μm filter and submitted for ICP-AES analysis. Product Consistency Testing was performed on each HLW glass in triplicate to determine normalized release of sodium, silicon, lithium and boron.

RESULTS AND DISCUSSION

Glass samples of C-104, AZ-102 Melt 1, and AZ-102 Melt 2, were successfully processed and melted into a HLW glass form. Approximately 239.8 g of useable, i.e. poured from crucible, C-104 glass was produced. The amount of useable AZ-102 Melt 1 produced was about 115 g and the amount of AZ-102 Melt 2 produced was 35.86 g. The final C-104 melt pour was non-problematic with an estimated viscosity of about 15 Pa•s based on visual observation coupled with past experience. The final AZ-102 Melt 1 pour temperature was elevated to approximately 1200°C due to the smaller sample size; this pour was non-problematic with an estimated viscosity of about 5 Pa•s, based on visual observation. The final AZ-102 Melt 2 pour (see Figure 1) temperature was also elevated to approximately 1200°C due to the smaller sample size; the pour was non-problematic as well, with an estimated viscosity of about 5 Pa•s, based on visual observation coupled with past experience.

Figure 1. Molten AZ-102 Melt 2 glass being quenched on a stainless steel plate.

KOH and Na_2O_2 fusion preparations and ICP-AES analyses were performed on each of the radioactive glasses, C-104 and AZ-102 Melt 1 and 2, as well as a HLW glass reference standard, Analytical Reference Glass-1 (ARG-1) [4]. This process established elemental composition for contract compliance and allowed calculation of modified PCT normalized releases. Tables II and III provide analyzed chemical compositions in wt% oxide for C-104 and AZ-102 Melt 1 glasses, respectively (AZ-102 Melt 2 glass composition is not reported herein as it was not used for any of the testing reported in this paper). The reported wt% oxide values are analytical 'process blank' corrected. The analytical wt% values agreed with the target values for ARG-1 quite well, indicating good analytical results.

Adjustments have been made to the measured data to generate as realistic as possible estimates of the composition of each of the three glasses due to potential analytical problems caused by: analytical detection limits greater than target values and possible analytical bias. The approach taken for undetected elements/oxides was to use their target values as the measurement instead of a blank measurement. Use of target values and bias correction for reporting of glass compositions are quite common as it is extremely expensive to analyze for the large number of cations, anions, etc. along with the large range of masses and isotopes contained in nuclear waste glass. The approach taken for bias correction was to analyze a well characterized glass at the same time as these three glasses to evaluate potential biases between measured wt% oxides in a glass sample and the true wt% oxides in the glass. Using nominal wt% oxides and associated standard deviations for ARG-1 from the Materials Characterization Center (MCC) Round Robin [4], a 80% prediction interval for a single observation was formed for each oxide as discussed in Hahn and Meeker [5]. An 80% confidence level was used because the fact that ARG-1 was only analyzed once with AZ-102 Melt 1, AZ-102 Melt 2, and C-104 makes it statistically more difficult to declare significant biases when ARG-1 measured values differ from nominal values.

The approach taken for potential analytical bias was to decide which bias corrections to use for the three glasses, if any. Candidate oxides for bias correction met the following criteria:

- Detected in the ARG-1 glass and the other glass (necessary to allow bias calculation)
- The ratio of the nominal ARG-1 oxide concentration to the other glass target oxide concentration was within a reasonable range, selected as 1/5 to 5.
- For the given oxide, the unknown glass oxide concentration is outside the 80% prediction interval based on the ARG-1 historical data.

Bias assessments were performed separately for the solutions generated by the K/Ni and Na/Zr fusions. Based on the above criteria, bias corrections were made for the following for all three glasses:

- Na/Zr fusion: CaO, Na_2O
- K/Ni fusion: Al_2O_3, B_2O_3, CaO, Fe_2O_3, Li_2O, MnO_2, Na_2O, and SiO_2

The total wt% values for the adjusted C-104 and AZ-102 Melt 1 compositions are close enough to 100 wt% to renormalize the adjusted compositions so they total 100 wt%. Renormalization of unadjusted measured compositions to 100 wt% can be inappropriate in that: (1) biases for specific oxides may not be properly addressed by the renormalization of all oxides, and (2) renormalization to 100 wt% can induce biases in unbiased measured values. However, if after appropriate bias corrections or adjustments, the total wt% values are close enough to 100 wt% to suggest that all significant biases have been addressed, then renormalizing the adjusted compositions to 100 wt% is appropriate. In fact, it has been shown in the statistics literature that renormalization in such a case actually reduces the uncertainty in the estimated composition [6].

The "Target" and "Averaged Normalized" composition columns in Tables II and III agree quite well for oxides with higher target values with one apparent exception for AZ-102 Melt 1. For this glass the averaged normalized adjusted values of UO_2 is nearly twice the target value. Independent uranium isotopic analyses have been performed by ICP-MS and give values 1.72 wt% UO_2 for the AZ-102 Melt 1 glass consistent with its target value of 1.63 wt% UO_2. This result indicates that the measured values for AZ-102 Melt 1 are high by a factor of two. The apparent problem with the analytical results for uranium in the glass is believed to be a result of the very high detection limit (~2.0 wt%) for this element under the conditions of the analysis (a dilution factor of ~ 10000 times). The uranium target value was less than the detection limit and the detected values were less than 1.5 times the detection limit, so there was a very large uncertainty attached to this analysis. This is all consistent with using the target value, which was calculated on the basis of the AZ-102 waste analysis, which is also consistent with ICP-MS results. It is concluded that the target value is more accurate for uranium in AZ-102 Melt 1.

The total of all waste oxides (exclusive of Si) not identified in Table TS-1.1 of the ORP contract (DOE contract number DE-AC06-96RL13308) was calculated for the C-104 glass, taking into account the glass former minerals added, using the "Average Normalized" weight percent oxide values. The total waste percent oxide in the C-104 glass (B_2O_3 of 0.020; BeO of 0.005; CeO_2 of 0.079; Co_2O_3 of 0.002; CuO of 0.030; La_2O_3 of 0.019; MnO of 2.992; MoO_3 of 0.002; Nd_2O_3 of 0.044; SnO_2 of 0.068; SrO of 2.924; ThO_2 of 3.787; V_2O_3 of 0.004; Y_2O_3 of 0.003; and ZnO of 0.036) comes to 10.02%, which is greater than the required 8.0%. The total of Al_2O_3 + Fe_2O_3 + ZrO_2 for AZ-102 Melt 1 is 21.96% which is greater than the required 21.0%. Therefore, the C-104, and AZ-102 Melt 1 glasses meet the ORP contract specifications for waste product loading.

Table II. Target, Measured, Adjusted, and Normalized Adjusted compositions of C-104 radioactive glass

Glass Fusion		Na/Zr [1][2][3]			K/Ni [1][2][3]			
Oxide	Target wt%	Measured	Adjusted	Normalized	Measured	Adjusted	Normalized	Average Normalized
Ag_2O	0.0657	0.0333	0.0333	0.0327	0.0585	0.0585	0.0580	0.0453
Al_2O_3	2.3585	2.6257	2.6257	2.5791	2.4463	2.7247	2.6988	2.6389
B_2O_3	9.0081	9.1448	9.1448	8.9823	8.4203	9.5019	9.4116	9.1970
BaO	0.0191	0.0246	0.0246	0.0241	0.0234	0.0234	0.0232	0.0237
BeO	0.0051		0.0051	0.0050		0.0051	0.0051	0.0050
CaO	0.4565	1.0143	0.9057	0.8896	0.4267	0.4607	0.4563	0.6729
CdO	0.0611	0.0582	0.0582	0.0572	0.0554	0.0554	0.0549	0.0560
CeO_2	0.0810		0.0810	0.0796		0.0810	0.0802	0.0799
Co_2O_3	0.0023		0.0023	0.0023		0.0023	0.0023	0.0023
Cr_2O_3	0.1385	0.1534	0.1534	0.1507	0.1388	0.1388	0.1375	0.1441
CuO	0.0256		0.0256	0.0251	0.0351	0.0351	0.0347	0.0299
Dy_2O_3			0	0		0	0	0
Eu_2O_3			0	0		0	0	0
Fe_2O_3	4.7179	4.4330	4.4330	4.3543	4.1327	4.7629	4.7176	4.5359
K_2O	0.0606		0.0606	0.0595		0.0606	0.0600	0.0598
La_2O_3	0.0190		0.0190	0.0187		0.0190	0.0188	0.0187
Li_2O	5.0045	5.4256	5.4256	5.3292	4.9734	6.0157	5.9585	5.6438
MgO	0.0592		0.0592	0.0581		0.0592	0.0586	0.0584
MnO_2	2.5486	3.0077	3.0077	2.9543	2.7228	3.0588	3.0297	2.9920
MoO_3	0.0020		0.0020	0.0020		0.0020	0.0020	0.0020
Na_2O	8.5711	7.9613	9.1795	9.0164	7.9613	9.1795	9.0922	9.0543
Nd_2O_3	0.0446		0.0446	0.0438		0.0446	0.0442	0.0440
NiO	0.2365	0.2423	0.2423	0.2379	0.2104	0.2400	0.2377	0.2378
P_2O_5	0.3301	0.3438	0.3438	0.3377	0.4011	0.4011	0.3973	0.3675
PbO	0.1536	0.2046	0.2046	0.2010	0.1992	0.1992	0.1974	0.1992
PdO	0.0109		0.0109	0.0107		0.0109	0.0108	0.0108
Rh_2O_3	0.0323		0.0323	0.0317		0.0323	0.0320	0.0319
RuO_2	0.0164		0.0164	0.0161		0.0164	0.0162	0.0162
Sb_2O_3			0	0		0	0	0
SiO_2	47.8667	48.1500	48.1500	47.2947	43.8700	47.0325	46.5854	46.9400
SnO_2	0.0687		0.0687	0.0675		0.0687	0.0680	0.0678
SrO	3.3905	2.9618	2.9618	2.9092	2.9677	2.9677	2.9395	2.9243
ThO_2	4.1008	4.1537	4.1537	4.0799	3.5278	3.5278	3.4943	3.7871
TiO_2	0.0166	0.0450	0.0450	0.0442		0.0166	0.0164	0.0303
UO_2	3.6353	3.9123	3.9123	3.8428	3.9123	3.9123	3.8751	3.8590
V_2O_3	0.0038		0.0038	0.0037		0.0038	0.0038	0.0037
Y_2O_3	0.0030		0.0030	0.0029		0.0030	0.0030	0.0030
ZnO	2.0017	2.0729	2.0729	2.0361	1.9422	1.9422	1.9237	1.9799
ZrO_2	4.8440	4.2962	4.2962	4.2199	4.2962	4.2962	4.2553	4.2376
Total	99.960[4]	100.2644	101.8085	100.0000[4]	92.7215	100.9598	100.0000[4]	100.0000[4]

1. See text for description of how adjusted values were determined.
2. Adjusted values normalized to total 100 wt%.
3. Average of normalized Ni fusion and Zr fusion compositions.
4. Total is prior to rounding entries to four decimal places.

Table IV provides radiochemical data from the C-104 and AZ-102 Melt 1 glass analyses. To demonstrate that the HLW glass product, radionuclide compositional contract criteria were met, it was assumed that each HLW glass, i.e. C-104[1] and AZ-102 Melt 1[2], are separate "waste types" and as such would fill

[1] Density of HLW98-51R glass (which is the equivalent C-104 simulant glass), provided by VSL, was measured at 20°C using ASTM D854-83 and is 2.888 g/cm^3.

Table III. Target, Measured, Adjusted, and Normalized Adjusted compositions of AZ-102 Melt 1 radioactive glass

Glass Fusion		Na/Zr [1][2][3]			K/Ni [1][2][3]			
Oxide	Target wt%	Measured	Adjusted	Normalized	Measured	Adjusted	Normalized	Average Normalized
Ag_2O	0.0190		0.0190	0.0192	0.0349	0.0349	0.0352	0.0272
Al_2O_3	7.6880	8.1605	8.1605	8.2295	7.3199	7.3199	7.3781	7.8038
B_2O_3	3.9793	3.7030	3.7030	3.7343	3.3649	3.7971	3.8273	3.7808
BaO	0.0383	0.0446	0.0446	0.0450	0.0430	0.0505	0.0509	0.0480
BeO	0.0026		0.0026	0		0	0	0
CaO	0.4779	0.2168	0.1936	0.1953		0.4779	0.4817	0.3385
CdO	1.3974	1.3076	1.3076	1.3187	1.2448	1.2448	1.2547	1.2867
CeO_2	0.0581		0.0581	0.0586		0.0581	0.0586	0.0586
Co_2O_3	0.0409		0.0409	0.0412		0.0409	0.0412	0.0412
Cr_2O_3	0.0071	0.1147	0.1147	0.1157	0.1059	0.1059	0.1068	0.1112
CuO	0.0922	0.0313	0.0313	0.0316	0.0651	0.0651	0.0656	0.0486
Dy_2O_3	0.0296		0.0296	0.0299		0.0296	0.0298	0.0298
Eu_2O_3	0.0092		0.0092	0.0093		0.0092	0.0093	0.0093
Fe_2O_3	12.1492	12.1479	12.1479	12.2506	10.8037	12.4511	12.5501	12.4004
K_2O	0.0258		0.0258	0.0260		0.0258	0.0260	0.0260
La_2O_3	0.2983	0.2874	0.2874	0.2898	0.3108	0.3108	0.3133	0.3016
Li_2O	5.0000	4.9734	4.9734	5.0155	4.3921	5.3125	5.3548	5.1851
MgO	0.1198		0.1198	0.1208	0.2321	0.2321	0.2340	0.1774
MnO_2	0.9225	1.1223	1.1223	1.1318	0.9672	1.0866	1.0952	1.1135
MoO_3	0		0	0		0	0	0
Na_2O	13.2719	11.7829	13.5858	13.7007	11.7829	13.5858	13.6938	13.6973
Nd_2O_3	0.2100	0.2624	0.2624	0.2646	0.3381	0.3381	0.3408	0.3027
NiO	0.7677	0.7803	0.7803	0.7869	0.7800	0.7800	0.7862	0.7866
P_2O_5	0.4624	0.5615	0.5615	0.5663	0.5730	0.5730	0.5776	0.5719
PbO	0.0932	0.0991	0.0991	0.0999	0.1992	0.1992	0.2008	0.1504
PdO	0		0	0		0	0	0
Rh_2O_3	0		0	0		0	0	0
RuO_2	0		0	0		0	0	0
Sb_2O_3	0.0212		0.0212	0.0214		0.0212	0.0214	0.0214
SiO_2	47.9950	46.6520	46.6520	47.0466	43.1210	46.2295	46.5971	46.8219
SnO_2	0.1638		0.1638	0.1652		0.1638	0.1651	0.1651
SrO	1.4921	1.2154	1.2154	1.2257	1.2390	1.2390	1.2489	1.2373
ThO_2	0		0	0		0	0	0
TiO_2	0.0107		0.0107	0.0108		0.0107	0.0108	0.0108
UO_2	1.6230	2.4381	1.6230[4]	1.6367	3.7422	1.6230[4]	1.6359	1.6363
V_2O_3	0.0036		0.0036	0.0036		0.0036	0.0036	0.0036
Y_2O_3	0.0146		0.0146	0.0147		0.0146	0.0147	0.0147
ZnO	0.0405		0.0405	0.0408		0.0405	0.0408	0.0408
ZrO_2	1.4700	1.7360	1.7360	1.7507	1.7360	1.7360	1.7498	1.7503
Total	99.995[5]	97.6372	99.1612	100.0000[5]	92.3959	99.2110	100.0000[5]	100.0000[5]

1. See text for description of how adjusted values were determined.
2. Adjusted values normalized to total 100 wt%.
3. Average of normalized Ni fusion and Zr fusion compositions.
4. Set to target value as best available estimate of actual value.
5. Total is prior to rounding entries to four decimal places.

multiple Hanford HLW canisters and that the HLW canister can be modeled as a right circular cylinder of 4.5 m height and 0.61 m diameter with a 100% glass fill of approximately 1.27 m^3. The primary success objectives accomplished with this

[2] Density of HLW98-61 (which is the equivalent AZ-102 simulant glass), provided by VSL, was measured at 20°C using ASTM D854-83 and is 2.703 g/cm^3.

work are: 1) "the inventory of radionuclides (in Curies) that have half-lives longer than 10 years and that are, or will be, present in concentrations greater than 0.05 percent of the total radioactive inventory for each waste type, indexed to the years 2015 and 3115" are reported; 2) the total and fissile uranium and plutonium (U-233, U-235, Pu-239, and Pu-241) contents of each canister of waste glass were calculated (the radiochemistry values for Pu, ratioed per ICP-MS values, were used to be conservative as they were twice as large at the ICP-MS valuses) and are: 1161.5 grams for C-104 and 642.1 grams for AZ-102 Melt 1; and 3) the

Table IV. Radiochemical Composition of C-104 and AZ-102 Melt 1 Glasses

	Analysis of C-104 Glass (µCi/g glass)		Analysis of AZ-102 Melt 1 Glass (µCi/g glass)	
Radioisotopes	Radiochemistry	ICP-MS	Radiochemistry	ICP-MS
Cr-51	<5.0		<5.0	
Fe-59	<0.1		<0.3	
Co-60	0.127		2.67	
Sr-90	519		8900	
Nb-95	<0.06		<0.2	
Tc-99	NM	0.0125	NM	0.00939
Ru-103	<0.06		<0.6	
Ru-106	<4.0		<0.4	
Sn-113	<0.9		<0.8	
Sb-125	<3.0		20.9	
Cs-134	0.12		<0.3	
Cs-137	1160		705	
Ce-144	<3.0		<4.0	
Eu-152	<0.3		<0.5	
Eu-154	2.68		26.2	
Eu-155	1.30		46.6	
U-233	NM	0.328	NM	0.0062
U-234	NM	0.0137	NM	0.00688
U-235	NM	0.000468	NM	0.000278
U-236	NM	0.000627	NM	0.000498
U-238	NM	0.00978	NM	0.00506
Pu-236	<0.005		<0.03	
Np-237	NM	0.00298	NM	0.0356
Pu-238	0.443		0.44	
Pu-239	NM	1.203	NM	1.125
Pu-240	NM	0.449	NM	0.343
Pu-239 + Pu-240	3.33	1.652	3.37	1.468
Pu-241	11.6		16.1	
Am-241	4.8		71.5	

<x.xx = indicates that the radioisotope is below the detection limit, detection limit value is provided for those radioisotopes.
NM = not measured

concentration of plutonium in grams per cubic meter of each waste glass are: 124.7 g/m^3 for C-104 and 122.4 g/m^3 for AZ-102 Melt 1, none of which exceed the maximum contract plutonium loading of 2500 grams per cubic meter.

Identification and quantification of crystalline and non-crystalline phases were completed by using x-ray diffraction (XRD), optical microscopy, and scanning electron microscopy (SEM) on samples given a slow cool down heat treatment which simulated the calculated cooling profile for glass at the centerline of a Hanford HLW canister during filling. No crystals were observed in the C-104 sample by XRD, optical examination, and SEM analysis. XRD analysis combined with optical microscopy and SEM EDS results of the CCC heat-treated AZ-102 Melt 1 glass sample indicated the presence of a crystalline phase at approximately 1 to 2 volume percent. An XRD search match analysis of the major peaks found in the XRD pattern of the CCC heat-treated AZ-102 Melt 1 glass sample indicated trevorite ($NiFe_2O_4$) and chromite ($FeCr_2O_4$) as the most likely spinel phases. The small amount of crystalline material in the CCC heat-treated AZ-102 Melt 1 glass sample does not significantly alter the leaching resistance of the glass as indicated by the PCT test results. SEM examination of both the C-104 and AZ-102 Melt 1 CCC heat-treated glasses at magnifications up to 5000× showed homogenous glass with no evidence of any phase separation.

The Product Consistency Test (PCT) was employed to gauge the HLW glass chemical durability. The PCT was run at 90°C, using HLW glass samples given a slow cool down heat treatment that simulates the cooling profile for glass at the centerline of a canister being filled with waste glass, to determine the normalized release of lithium, sodium, and boron. Parallel tests were run with the environmental assessment glass (EA glass) [7] benchmark standard glass to provide a reliable baseline of results by which to judge the quality of the PCT results for the C-104 and AZ-102 Melt 1 glasses. The normalized lithium, sodium and boron 90°C PCT releases for the C-104, AZ-102 Melt 1, and EA glasses are: 1) (for lithium) 0.5 g/m^2, 0.4 g/m^2, and 3.75 g/m^2; 2) (for sodium) 0.4 g/m^2, 0.4 g/m^2, and 5.1 g/m^2; and 3) (for boron) 0.4 g/m^2, 0.3 g/m^2, and 6.9 g/m^2; respectively. More importantly, as the C-104 and AZ-102 Melt 1 glasses average, normalized elemental release values are an order of magnitude lower, i.e., more durable, for Na, and B, and just slightly less than an order of magnitude lower for Li when compared to the reported results of the benchmark EA glass, the ORP Phase B-1 contract criteria were easily met.

CONCLUSIONS

The HLW product testing results from the C-104 and AZ-102 Melt 1 glasses show that in all cases they meet ORP Phase B-1 contract specifications for waste loading, chemical composition documentation, radionuclide concentration limitations, and waste form product consistency testing (i.e. chemical durability).

ACKNOWLEDGMENTS

This work was sponsored by the U.S. Department of Energy through a contract with the Office of River Protection and the River Protection Project-Waste Treatment Plant (DOE contract number DE-AC06-96RL13308). Pacific Northwest National Laboratory is operated for DOE by Battelle Memorial Institute under contract DE-AC06-76RLO 1830.

REFERENCES

[1]Brooks, K. P., P. R. Bredt, G. R. Golcar, S. A. Hartley, L. K. Jagoda, K. G. Rappe, and M. W. Urie. 2000a. *Characterization, Washing, Leaching, and Filtration of C-104 Sludge*. BNFL-RPT-030, Rev. 0. PNWD-3024. Pacific Northwest National Laboratory, Richland, Washington.

[2]Brooks, K. P., P. R. Bredt, S. K. Cooley, G. R. Golcar, L. K. Jagoda, K. G. Rappe, and M. W. Urie. 2000b. *Characterization, Washing, Leaching, and Filtration of AZ-102 Sludge*. BNFL-RPT-038, Rev. 0. PNWD-3045. Pacific Northwest National Laboratory, Richland, Washington.

[3]American Society for Testing and Materials, ASTM C1285-97 "*Standard Test Methods for Determining Chemical Durability of Nuclear, Hazardous, and Mixed Waste Glasses: The Product Consistency Test (PCT)*", West Conshohoken, Pennsylvania.

[4]Smith, G. L. 1993. *Characterization of Analytical Reference Glass-1 (ARG-1)*. PNL-8992. Pacific Northwest National Laboratory, Richland, Washington.

[5]Hahn, Gerald J. and William Q. Meeker. Statistical Intervals: A Guide for Practitioners. John Wiley & Sons, 1991.

[6]Deming, W. Edwards. "Statistical Adjustment of Data". Dover Publications, Inc. 1964.

[7]Jantzen, C. M., N. E. Bibler, D. C. Beam, C. L. Crawford, and M. A. Pickett. 1993. *Characterization of the Defense Waste Processing Facility (DWPF) Environmental Assessment (EA) Glass Standard Reference Material*. WSRC-TR-92-346, Rev. 1. Westinghouse Savannah River Company, Aiken, South Carolina.

VITRIFICATION AND TESTING OF HANFORD PRETREATED LOW ACTIVITY WASTE

Gary L. Smith, Harry D. Smith, Michael J. Schweiger, and Gregory F. Piepel
Pacific Northwest National Laboratory
P.O. Box 999, MSIN K6-24
Richland, WA 99352

ABSTRACT

Actual pretreated LAW samples were vitrified to demonstrate the RPP-WTP projects ability to satisfy the LAW product ORP Phase B-1 contract requirements concerning, chemical and radionuclide reporting, waste loading, identification and quantification of crystalline and non-crystalline phases, and waste form leachability. Chemical compositions of two LAW glasses (i.e., elements {excluding oxygen} present in concentrations greater than 0.5 percent by weight) were measured using KOH and Na_2O_2 fusion preparation procedures. The measured wt% sodium oxide content for the AW-101 and AN-107 glasses are 17.7 and 18.3, respectively; however, it is argued herein that process knowledge, i.e., the target sodium oxide content, is better than the analytical measurement. Therefore, for both LAW glasses the target oxide loading for sodium of 20 wt% is accepted. At these levels the glass meets or exceeds both the RPP-WTP glass specification and the DOE ORG contract requirement for waste sodium loading. The concentrations of ^{137}Cs, ^{90}Sr, ^{99}Tc and transuranic (TRU) radionuclides for AW-101 and AN-107 are: 1) 0.231 and 0.292 Ci/m^3, 0.435 and 0.005 Ci/m^3, 0.019 and 0.129 Ci/m^3, and < 0.16 and < 2.6 nCi/g, respectively. The ORP contract criteria for ^{137}Cs, ^{90}Sr and TRU (shall be < 3 Ci/m^3, < 20 Ci/m^3, and < 100 nCi/g, respectively) are met in both glasses. The ORP contract criteria for ^{99}Tc (shall be less than 0.1 Ci/m^3) is met explicitly by AW-101 and will be met for the AN-107 glass by averaging its ^{99}Tc content over the previous LAW glasses produced to meet the contract. After canister centerline cooling, no crystals were observed in the AW-101 and AN-107 glasses by XRD, optical examination, and SEM analysis. The normalized PCT releases of sodium, silicon, and boron, at both 40 and 90°C, from the AW-101 and AN-107 glasses are less than 2.0 g/m^2, the ORP contract criterion. The LAW product testing results from the AW-101 and AN-107 glasses show that in all cases they meet ORP contract specifications.

INTRODUCTION

The U.S. Department of Energy (DOE) Office of River Protection (ORP) has contracted Hanford tank waste treatment services. The River Protection Project

To the extent authorized under the laws of the United States of America, all copyright interests in this publication are the property of The American Ceramic Society. Any duplication, reproduction, or republication of this publication or any part thereof, without the express written consent of The American Ceramic Society or fee paid to the Copyright Clearance Center, is prohibited.

Waste Treatment Plant (RPP-WTP) team is responsible for producing an immobilized (vitrified) low activity waste (ILAW) waste form. PNNL produced and tested a vitrified ILAW waste form from Envelope A and C LAW samples previously supplied to the RPP-WTP by DOE.

One supernatant sample each was taken from two underground storage tanks at the Hanford site in southeastern Washington. The two tank supernatant samples (241-AW-101 and 241-AN-107) were processed through representative unit operations such as entrained solids removal, Sr/TRU precipitation, and Cs and Tc ion exchange to remove most of the radioactivity. The decontaminated supernatant samples were converted into low-activity waste (LAW) glass.

The primary objective for vitrifying the Envelope A (Tank AW-101) and Envelope C (Tank AN-107) pretreated waste samples was to characterize the glass produced from the crucible melts. Testing results of the waste glasses produced from actual tank waste shows compliance with the RPP-WTP contractual requirements such as chemical and radionuclide reporting, product loading and durability. The work reported includes: 1) Glass Fabrication, 2) Chemical Composition, 3) Radiochemical Composition, 4) Crystalline and Non-crystalline Phase Determination, and 5) Product Consistency Test (PCT) data. The results show that good glass was formed having compositions in line with the target quantities, acceptable devitrification properties, with good leach resistance.

EXPERIMENTAL

LAW Glass Fabrication

Independently, each pretreated tank waste was vigorously blended in a glass beaker using a magnetic stirrer and stir bar. A combination of glass former additives (Kyanite (Al_2SiO_5); Orthoboric acid, (H_3BO_3); Wollastonite ($CaSiO_3$); Red Iron Oxide Pigment (Fe_2O_3); Olivine (Mg_2SiO_4); Silica sand (SiO_2); Rutile Ore (TiO_2); Zinc Oxide (ZnO); Zircon sand ($ZrSiO_4$); and sugar) were added to each pretreated waste to produce a melter feed. The AW-101 and AN-107 melter feeds were dried, calcined, and melted at 1150°C for one hour. Each melt was then poured onto a stainless steel plate, cooled, crushed to a fine powder, mixed, and added back into the crucible, and melted for an additional hour at 1150°C. The glass melt was then poured onto a stainless steel plate (air quenched) and cooled to room temperature. A portion of the melt was poured into a small platinum crucible that was later heat-treated following the predicted canister centerline cooling (CCC) heat treatment of a LAW canister. The pretreated low-activity waste were processed and vitrified in the Radiochemical Processing Laboratory (RPL) inside radiological fume hoods.

Chemical Composition

Chemical composition of the two LAW glasses, along with an ARG-1 powdered glass reference standard [1], (i.e., elements {excluding oxygen} present in concentrations greater than 0.5 percent by weight) were dissolved using KOH and Na_2O_2 fusion preparation procedures. The KOH fusion was done in a nickel crucible and the Na_2O_2 fusion used a zirconium crucible. Cation analysis was performed using inductively coupled plasma-atomic emission spectrometry (ICP-AES). All sample material after processing appeared to go into solution (no apparent residue remained in fusion crucibles or as precipitate in final solution). Analytical dilutions of 5, 10, and 50-fold were prepared from each fusion preparation and analyzed by ICP-AES. The fusion procedure was modified slightly by including additional hydrochloric acid to assist solubilization of silver, if present. Before ICP-AES analysis a small amount (0.1 ml) of hydrofluoric acid was added to the prepared samples. Two fusions and two ICP-AES analyses provided duplicate analyses for all cations except K, Ni, Na, and Zr.

Radiochemical Composition

Radiochemical analyses were performed on each ILAW product, i.e. AW-101 and AN-107 glasses. Analyses included ^{137}Cs by gamma emission spectroscopy (GEA), ^{90}Sr, ^{99}Tc, ^{238}Pu, ^{239}Pu, ^{240}Pu, ^{237}Np, ^{241}Am and ^{244}Cm. Concentration values of additional gamma emitters (i.e., Cr-51, Fe-59, Se-79, Nb-95, Ru-103, Sn-113, and Eu-152) that may be obtained by GEA, depending on concentrations and detection limits, were also looked for but not detected. Samples of powdered waste glass AW-101 and AN-107 were analyzed for gamma emitters, ^{90}Sr, Pu, and Am/Cm. Duplicate samples of the powdered waste glass were solubilized in the laboratory using a Na_2O_2-NaOH fusion in a Zr crucible. About 0.1 g of material was fused and then dissolved in acid and brought to a volume of 100 ml. This fused material preparation was sampled directly for gamma energy analysis (GEA). A 10-ml aliquot was evaporated to dryness to remove Cl^- (solution had been acidified with nitric acid, chloride is removed as HCl), then brought back to volume and filtered through a 0.45-micron filter. This matrix-adjusted material was used for Pu, Am, Cm, and Sr analyses. Strontium was isolated from the initial fusion solutions and then beta-counted. Plutonium and Am/Cm samples were isolated from the initial fusion solutions, precipitation plated, and then counted by alpha spectroscopy. In addition, the following radioisotopes: ^{99}Tc, ^{237}Np, ^{239}Pu, and ^{240}Pu were measured by inductively coupled plasma mass spectroscopy (ICP-MS).

Crystalline and Non-Crystalline Phase Determination

Crystalline and non-crystalline phases were identified and measured using x-ray diffraction (XRD), optical microscopy, and scanning electron microscopy

(SEM) on glass samples that had been heat-treated as indicated in Table I to simulate a LAW canister centerline cooling (CCC) curve. The LAW stainless steel canisters are basically right circular cylinders 2.29 m in height and 1.22 m in diameter. Glass canister filling and temperature profile was modeled with a batch target fill rate of 2,080 kg/hr at a temperature of 1150°C (50 MT/day). The modeled temperature, cooling curve for the centerline of a canister of glass 2/5 of the way from the bottom of the canister was used for the heat-treatment.

Table I. Temperature profile line segments for the Del Tech furnace controller to generate the LAW Canister centerline cooling profile

Hours	Temperature (°C)	dT/dt (deg./hr)
0.06 - 0.6	1021.26 -1000.95	-37.60
0.6 -1.80	1000.95 - 976.94	-20.01
1.80 - 2.80	976.94 - 969.68	-7.26
2.80 - 9.00	969.6 - 964.16	-0.89
9.00 - 16.00	964.16 - 909.73	-7.78
16.00 - 24.00	909.73 - 780.63	-16.14
24.00 - 38.00	780.63 - 536.13	-17.46
38.00 - 48.60	536.13 - 396.59	-13.16

Powder XRD was also used to characterize the heat-treated glass samples. The two-theta scan range was from 5 to 75 degrees at a step size of 0.04 degrees with a minimum of 2-second dwell at each step. Both the AW-101 and AN-107 glasses were powdered in a tungsten carbide grinding chamber using a disc mill. An approximately 100 mg sample of each glass was mounted on a plastic XRD sample mount, leveled to X-ray beam height, encapsulated in Mylar film, transported to the XRD facility, and analyzed.

For optical and scanning electron microscopy (SEM), a thin slice of one of the cross sections of the CCC heat-treated glass was used. Conventional grinding and polishing techniques were followed. Thin section samples were examined with transmitting light microscope at magnifications up to 250×. For SEM analysis, approximately 1 square cm by 4 mm thick samples of the LAW glasses were polished and then mounted on aluminum SEM specimen holders for microscopy. Both glass samples were polished to a minimum of a 600 grit finish. Each mount was then coated with a gold film and examined at low magnification (15× and 100×) and higher magnifications such as 500×, 1000×, 3000×, 10,000×, and 20,000×.

Release Rate, Product Consistency Testing (PCT) of LAW Glasses

PCT testing on the AW-101 and AN-107 glass samples were completed per ASTM C1285-97 [2]. Crushed glass of a particle size between 75 and 150 μm

(-100 to +200 mesh) was used for testing. The glass was crushed and then sieved through 100- and 200-mesh stainless steel sieves. The crushed glass was cleaned by washing in deionized water (DIW) and ethanol using an ultrasonic cleaner. It was then dried and weighed, and approximately 1.5 g of glass added to a 22-mL desensitized Type 304L stainless steel container filled with DIW. The glass was weighed and the leachant volume controlled to achieve a solution volume to glass mass ratio of 10 mL/g glass. The ratio of the surface area of the sample to leachant volume is estimated to be 2000 m^{-1}. The container and its contents were held (without agitation) at a temperature of 40 and 90°C for 7 days. The initial and final pH values of the solution were measured. Aliquots of the solution were filtered through a 0.45-μm filter and submitted for ICP-AES analysis. Product Consistency Testing was performed on each LAW glass in triplicate to determine normalized release of boron, sodium, and silicon. The low-activity test reference material (LRM) was included in these tests [3]. It has been extensively tested using the PCT and gives a reliable baseline of results by which to judge the quality of the PCT data for the AW-101 and AN-107 glasses.

RESULTS AND DISCUSSION

Glass samples of AW-101 and AN-107 were successfully processed and melted into a LAW glass form. Approximately 282 g of useable, i.e. poured from crucible, AW-101 glass was produced. The amount of useable AN-107 produced was about 230 g. The final AW-101 melt pour was excellent, estimated viscosity of about 5 Pa•s based on visual observation coupled with past experience, and bubbles present in the meniscus burst while being poured. The final AN-107 melt pour was excellent as well, with an estimated viscosity of about 8 to 10 Pa•s, based on visual observation. Some bubbles present in the meniscus were observed during pouring and a slight vapor of volatile components was observed when the lid was removed from the crucible during pouring. The first portion of each pour went into a platinum crucible for the canister centerline cooling test and the remainder of the melt was quenched on a stainless steel plate.

KOH and Na_2O_2 fusion preparations and ICP-AES analyses were performed on each of the radioactive glasses, AW-101 and AN-107, as well as a HLW glass reference standard, Analytical Reference Glass-1 (ARG-1) [1]. This process established elemental composition for contract compliance and allowed calculation of PCT normalized releases. Tables II and III provide analyzed chemical compositions in wt% oxide for AW-101 and AN-107, respectively. The reported wt% oxide values are analytical 'process blank' corrected. The analytical reference glass wt% values agreed with the target values for ARG-1 quite well, indicating good analytical results. Summation of measured wt% oxides in the ARG-1 laboratory controls standard was about 98%. The total accountability of mass in the glass by ICP-AES is 94.3% for Envelope A

Table II. Target versus measured composition of AW-101 radioactive LAW glass

Oxide or Element	Target (wt%)	Measured (wt%)	Adjusted[a] (wt%)	Normalized Adjusted[b] (wt%)
Al_2O_3	6.08	6.06	6.2198[e]	6.2289
B_2O_3	9.71	9.42	9.42	9.4337
BaO	0.0	<0.005	0.0[c]	0.0
CaO	1.99	2.02	1.8387[e]	1.8413
CdO	0.0	<0.008	0.0[c]	0.0
Co_2O_3	0.0	<0.033	0.0[c]	0.0
Cr_2O_3	0.009	0.04	0.04	0.0401
Cs_2O	0.0012	NM	0.0012[c]	0.0012
CuO	0.0	<0.015	0.0[c]	0.0
Fe_2O_3	5.54	5.05	5.05	5.0574
K_2O	2.58	3.07	3.07	3.0745
MgO	1.48	1.56	1.56	1.5623
MoO_3	0.0	0.0	0.0	0.0
Na_2O	20.0	17.66	20.0[d]	20.0291
NiO	0.0	0.0	0.0	0.0
P_2O_5	0.0699	0.15	0.15	0.1502
PbO	0.0	<0.050	0.0[c]	0.0
SO_3	0.2139	NM	0.2139[c]	0.2142
Sb_2O_3	0.0	0.0	0.0	0.0
SiO_2	44.05	42.16	44.05[d]	44.1141
SrO	0.0	0.0024	0.0029[e]	0.0029
TiO_2	1.9939	1.72	1.8698[e]	1.8725
WO_3	0.0	<1.181	0.0[c]	0.0
ZnO	2.95	2.83	2.83	2.8341
ZrO_2	2.99	2.56	3.1963[e]	3.2010
Br	0.0	NM	0.0[c]	0.0
Cl	0.0784	NM	0.0784[c]	0.0785
F	0.0	NM	0.0[c]	0.0
Unknown	0.2637	NM	0.2637[c]	0.2641
Total	100.0000	94.3	99.8546	100.0000[f]

(a) See text for description of how adjusted values were determined.
(b) Adjusted values normalized to total 100 wt%.
(c) Values less than detection limits or not measured (NM) values were adjusted to target values.
(d) Set to target value as best available estimate of actual value.
(e) Bias correction based on ARG-1 applied to measured value.
(f) Total is 100.0000 prior to rounding entries to four decimal places.
<x.xx = indicates that the analyte is below the detection limit, detection limit value is provided for those analytes.

Table III. Target versus measured composition of AN-107 radioactive LAW glass

Oxide or Element	Target (wt%)	Measured (wt%)	Adjusted[a] (wt%)	Normalized Adjusted[b] (wt%)
Al_2O_3	6.2311	6.11	6.2682[e]	6.3144
B_2O_3	8.9433	8.52	8.52	8.5827
BaO	0.0191	0.03	0.03	0.0302
CaO	2.0095	1.97	1.7941[e]	1.8073
CdO	0.004	<0.008	0.004[c]	0.0040
Co_2O_3	0.0003	<0.033	0.0003[c]	0.0003
Cr_2O_3	0.0029	0.03	0.03	0.0302
Cs_2O	0.0	0.0	0.0	0.0
CuO	0.0021	<0.015	0.0021[c]	0.0021
Fe_2O_3	7.018	6.40	6.40	6.4471
K_2O	0.1418	<1.138	0.1418[c]	0.1248
MgO	2.0117	2.11	2.11	2.1255
MoO_3	0.0031	<0.035	0.0031[c]	0.0031
Na_2O	20.0	18.33	20.0[d]	20.1473
NiO	0.0351	0.05	0.05	0.0504
P_2O_5	0.0153	<0.108	0.0153[c]	0.0154
PbO	0.0038	<0.051	0.0038[c]	0.0038
SO_3	0.1267	NM	0.1267[c]	0.1276
Sb_2O_3	0.0	0.0	0.0	0.0
SiO_2	44.7841	42.80	44.7841[d]	45.1139
SrO	0.0013	0.0041	0.0049[e]	0.0050
TiO_2	2.0039	1.87	2.0332[e]	2.0482
WO_3	0.0115	<1.190	0.0115[c]	0.0116
ZnO	2.9949	2.91	2.91	2.9314
ZrO_2	3.0097	2.72	3.3988[e]	3.4239
Br	0.0784	NM	0.0784[c]	0.0790
Cl	0.0784	NM	0.0784[c]	0.0790
F	0.4701	NM	0.4701[c]	0.4736
Unknown	0.0	NM	0.0[c]	0.0
Total	100.0001	93.85	99.2689	100.0000[f]
(a, b, c, d, e, f) See Table II for notes.				

(AW-101) and 93.9% for Envelope C (AN-107). One reason for the approximately 6% discrepancy in total wt% oxides is because certain elements (such as SO_3; the halides Br, Cl, and F; and trace metals) were not included in the analyses. Another reason is the lack of complete recovery of SiO_2 and Na_2O during the preparation of the sample for analysis. It will be discussed shortly that when omitted or discrepant components are adjusted, the total wt% values for AW-101 and AN-107 are quite close to 100 wt%.

Using nominal wt% oxides and associated standard deviations for ARG-1, an 80% prediction interval for a single observation was formed for each oxide as discussed in Hahn and Meeker [4]. If the weight percent for a particular oxide in the ARG-1 glass (measured along with AW-101 and AN-107) was found to be outside the prediction interval for that oxide, then the bias for that oxide was deemed to be statistically significant. CaO, SrO, and ZrO_2 were found to have statistically significant biases at the 80% confidence level. Although ARG-1 measured versus nominal differences in Al_2O_3 and TiO_2 were not statistically significant at the 80% confidence level, values of Al_2O_3 and TiO_2 in AW-101 and AN-107 were also bias corrected. Their measured values in AW-101 and AN-107 were consistently and non-negligibly below their target values, agreeing with the relative difference in measured and nominal ARG-1 values for those oxides. In summary, bias corrections (on a relative basis) were made to the measured wt% oxide values of CaO, SrO, ZrO_2, Al_2O_3, and TiO_2 for both AW-101 and AN-107.

For other oxides and elements, target values were used as the adjusted values when the oxide or element was not analyzed, or when the analyzed value was less than the detection limit (< DL). For the remaining components (except for Na_2O and SiO_2, discussed below), the analyzed value was used as the adjusted value (i.e., no adjustment). Based on past experience, the measured weight percentages for Na_2O and SiO_2 are typically lower than their true weight percentages. There are different reasons for this for each oxide. For silica, it is the difficulty in getting silica into solution and keeping it there; and the precipitate is not easily observed. Hence the solution analyzed by ICP-AES is actually low in silica. For sodium oxide the possible reasons include matrix effects (i.e., other elements present with sodium in the plasma flame) in the ICP plasma flame, temperature of the plasma flame, and stability of the flame. Matrix effects are important because they are known to affect the easily ionized elements such as the alkalies and typical standards do not duplicate the matrix effects well and as a result the sodium response in the sample and standard are different. In addition, the sodium emission intensity is sensitive to the temperature of the plasma, so any shifting of the plasma flame relative to the optical detectors can change the sodium signal. Though the effect of these factors can theoretically shift the sodium-analyzed value up or down, the observed shift is generally down. Therefore, the target values for Na_2O and SiO_2 were used as their adjusted values. Notice that this action is supported by the fact that the increase of Na_2O and SiO_2 to their target values increases the oxide total for the adjusted analyses closer to 100% without overshooting 100%. After all adjustments, the total wt% values for AW-101 and AN-107 are 99.85% and 99.27%, respectively.

Table II and III compare the measured and the adjusted compositions to target glass compositions and shows that both the AW-101 and AN-107 glasses are fairly close to their target compositions. Because the total wt% values for the adjusted AW-101 and AN-107 compositions are quite close to 100 wt%, it is also appropriate to renormalize the adjusted compositions to total 100 wt%. The renormalized, adjusted compositions of AW-101 and AN-107 are shown in Tables II and III. Renormalization of unadjusted measured compositions to 100 wt% can be inappropriate, in that: (1) biases are not properly addressed by the renormalization, and (2) renormalization to 100 wt% can induce biases in unbiased measured values. However, if after appropriate bias corrections or adjustments, the total wt% values are close enough to 100 wt% to suggest that all significant biases have been addressed, then renormalizing the adjusted compositions to 100 wt% is appropriate. In fact, it has been shown in the statistics literature that renormalization in such a case actually reduces the uncertainty in the estimated composition [5].

The measured wt% sodium oxide content for the AW-101 and AN-107 glasses are 17.7 and 18.3, respectively. As all of the sodium oxide content for the AW-101 glass originated from the initial tank waste, the AW-101 glass exceeds the RPP-WTP Task Specification, sodium oxide concentration level of 16 weight percent considering only the measured wt% sodium oxide content. Again, as just discussed, the measured weight percentage for Na_2O is almost always lower than its true weight percentage, which provides an even larger margin of passing the Task Specification requirement. However, not all of the sodium oxide content for the AN-107 glass originated from the initial tank waste. As 79.2% of the sodium oxide content for the AN-107 glass originated from the initial tank waste, for the AN-107 glass to exceed the RPP-WTP Task Specification, the wt% sodium oxide content of the glass would need to be 20 wt%, which is the target concentration. As just discussed, the measured weight percentage for Na_2O is almost always lower than its true weight percentage. For this reason, the target Na_2O value is used as the true weight percent oxide value for the ILAW glasses. Therefore, the original, as-received AN-107 waste is considered to meet the Task Specification for waste loading of the AN-107 glass. The ORP Contract (DOE contract number DE-AC06-96RL13308) states: "The loading of waste sodium from Envelope A in the ILAW glass shall be greater than 14 weight percent based on Na_2O. The loading of waste sodium from Envelope B in the ILAW glass shall be greater than 5.0 weight percent based on Na_2O. The loading of waste sodium from Envelope C in the ILAW glass shall be greater than 10 weight percent based on Na_2O." Therefore, both the AW-101 (Envelope A) and AN-107 (Envelope C) glasses also easily meet the ORP contract specifications for waste sodium loading.

Table IV provides radiochemical data from the AW-101 and AN-107 glass analyses. One of the primary objectives for this work was that for the ILAW glasses, "The concentrations of ^{137}Cs, ^{90}Sr, ^{99}Tc and transuranic (TRU) radionuclides shall be less than 3 Ci/m^3, 20 Ci/m^3, 0.1 Ci/m^3 and 100 nCi/g, respectively." The amount of each of these radionuclides was determined by multiplying the weight of a cubic meter of AW-101[1] and AN-107[2] glass by the

[1] Density of LAWA88 glass (which is equivalent to AW-101simulant glass), provided by VSL, was measured at a temperature of 20°C per ASTM D854-83 (density corrected to 25°C would be less than 0.005 g/cm^3 lower than this number) and is 2.668 g/cm^3.

measured concentration of the radionuclide per gram. The concentrations of ^{137}Cs, ^{90}Sr, ^{99}Tc and transuranic (TRU) radionuclides for AW-101 are 0.231 Ci/m^3, 0.435 Ci/m^3, 0.019 Ci/m^3, and < 0.16 nCi/g, respectively. The concentrations of ^{137}Cs, ^{90}Sr, ^{99}Tc and transuranic (TRU) radionuclides for AN-107 are 0.292 Ci/m^3, 0.005 Ci/m^3, 0.129 Ci/m^3, and < 2.6 nCi/g, respectively.

Table IV. Radiochemical composition of AW-101 and AN-107 LAW glasses

Radionuclide	Analysis of Glass (μCi/g glass)	
	AW-101 Glass	AN-107 Glass
Co-60	<2.E-3	5.64E-02
Sr-90	1.63E-01	<2.E-3
Nb-95	<1.E-03	<2.E-3
Tc-99	7.14E-03	4.82E-02
Sn-113	<2.E-3	<3.E-3
Sb-125	<4.E-3	<5.E-3
SnSb-126	<2.E-3	<2.E-3
Cs-137	8.65E-02	1.09E-01
Eu-154	<4.E-3	5.40E-03
Eu-155	<3.E-3	<2.E-3
Pu-236	<3.E-6	<8.E-6
Np-237	<5.6E-5	<6.9E-05
Pu-238	<5.E-6	1.00E-04
Pu-239	<7.8E-3	<9.6E-3
Pu-239 + Pu-240	3.55E-05	4.1E-04
Pu-240	<1.4E-1	<1.7E-2
Am-241	3.80E-05	1.95E-03
Cm-242	<7.E-6	<7.E-6
Cm-243 + Cm-244	<2.E-5	3.08E-05

<xxx = indicates that the radioisotope is below the detection limit, detection limit value is provided for those radioisotopes.

The ORP contract criteria for ^{90}Sr and ^{137}Cs are met in both glasses. AW-101 also passes for ^{99}Tc and transuranic (TRU) radioisotopes. AN-107 passes for (TRU) radioisotopes and fails for ^{99}Tc. The failure of AN-107 for ^{99}Tc was not unexpected because the pretreatment process for removing ^{99}Tc selectively removes the pertechnetate anion from LAW solutions and the AN-107 waste only contains approximately 15 to 20% of its ^{99}Tc as the pertechnetate anion [6]; the majority is Tc(IV) most likely complexed with organics. However, the ORP Contract Specification 2.2.2.8, states: "The average concentrations shall be calculated by summing the actual inventories of each of the above radionuclides in the packages that have been presented to date for acceptance and dividing by

---

[2] Density of LAWC15 glass (which is equivalent to AN-107simulant glass), provided by VSL, was measured at a temperature of 20°C per ASTM D854-83 (density corrected to 25°C would be less than 0.005 g/cm^3 lower than this number) and is 2.677 g/cm^3.

the total volume of waste in these packages. The Contractor shall remove on average a minimum of 80% of the ^{99}Tc present in the feed." therefore, it is believed that the RPP-WTP will be able to meet or exceed the contract ^{99}Tc ILAW glass content requirements by decontaminating the Envelope A and B wastes to a greater extent than needed to meet the contract ILAW requirements so that on average the ^{99}Tc concentration in the combined ILAW glasses produced, i.e. Envelope A, B, and C waste glasses, will be less than or equal to the contract ILAW glass limit.

Identification and quantification of crystalline and non-crystalline phases were completed by using x-ray diffraction (XRD), optical microscopy, and scanning electron microscopy (SEM) on samples given a slow cool down heat treatment which simulated the calculated cooling profile for glass at the centerline of a LAW canister during filling. No crystals were observed in the samples during both SEM and optical examination though bubbles were present in all samples and a few irregularly shaped, opaque particles (10-30 μm) were observed but could not be identified. The SEM-EDS survey did not observe any phase separation or glass heterogeneity. XRD analysis showed only broad amorphous peaks indicating that the glasses are completely amorphous, consistent with the optical and SEM results.

The Product Consistency Test (PCT) was employed as a measure of ILAW glass chemical durability. The PCT was run at 40 and 90°C, using glass samples given a slow cool down heat treatment simulating the cooling profile for glass at the center line of a canister being filled with waste glass, to determine the normalized release of sodium, silicon, and boron. Parallel tests were run with the low-activity test reference material (LRM) standard glass to provide a reliable baseline of results by which to judge the quality of the PCT results for the AW-101 and AN-107 glasses. Both the AW-101 and AN-107 glasses and the LRM glass gave a normalized sodium, silicon, and boron release rates of less than 1 g/m^2 for the 90°C PCT test, which is generally considered to indicate a durable glass. The normalized sodium, silicon, and boron 90°C PCT releases for the AW-101, AN-107, and LRM glasses are: 1) (for sodium) 0.6 g/m^2, 0.2 g/m^2, and 0.6 g/m^2; 2) (for silicon) 0.4 g/m^2, 0.2 g/m^2, and 0.4 g/m^2; and 3) (for boron) 0.5 g/m^2, 0.2 g/m^2, and 0.5 g/m^2, respectively. The normalized sodium, silicon, and boron 40°C PCT release rates for the AW-101, AN-107, and LRM glasses are: 1) (for sodium) 0.09 g/m^2, 0.03 g/m^2, and 0.05 g/m^2; 2) (for silicon) 0.07 g/m^2, 0.03 g/m^2, and 0.04 g/m^2; and 3) (for boron) 0.07 g/m^2, 0.02 g/m^2, and 0.02 g/m^2, respectively. More importantly, the normalized release rates of sodium, silicon, and boron from the AW-101 and AN-107 glasses are less than 2.0 g/m^2, the ORP contract criterion.

CONCLUSIONS

The ILAW product testing results from the AW-101 and AN-107 glasses show that in all cases they meet or exceed ORP contract specifications for waste loading, chemical composition documentation, radionuclide concentration limitations, and waste form product consistency testing (i.e. chemical durability).

ACKNOWLEDGMENTS

This work was sponsored by the U.S. Department of Energy through a contract with the Office of River Protection and the River Protection Project-Waste Treatment Plant (DOE contract number DE-AC06-96RL13308). Pacific Northwest National Laboratory is operated for DOE by Battelle Memorial Institute under contract DE-AC06-76RLO 1830.

REFERENCES

[1]Smith, G. L. 1993. *Characterization of Analytical Reference Glass-1 (ARG-1)*. PNL-8992. Pacific Northwest National Laboratory, Richland, Washington.

[2]American Society for Testing and Materials, ASTM C1285-97 "*Standard Test Methods for Determining Chemical Durability of Nuclear, Hazardous, and Mixed Waste Glasses: The Product Consistency Test (PCT)*", West Conshohoken, Pennsylvania.

[3] Ebert, W. L. and S. F. Wolf. 1999. *Round-Robin Testing of a Reference Glass for Low-Activity Waste Forms*. ANL-99/22, Argonne National Laboratory, Argonne, Illinois.

[4]Hahn, Gerald J. and William Q. Meeker. Statistical Intervals: A Guide for Practitioners. John Wiley & Sons, 1991.

[5]Deming, W. Edwards. "Statistical Adjustment of Data". Dover Publications, Inc. 1964.

[6] Blanchard Jr., D. L., D. E. Kurath, and B. M. Rapko. 2000. *Small Column Testing of Superlig 639 for Removing ^{99}Tc from Hanford Tank Waste Envelope C (Tank 241-AN-107)*. PNWD-3028, Battelle, Pacific Northwest Division, Richland, Washington.

CORROSION OF Ni-Cr ALLOYS IN MOLTEN SALTS AND HANFORD LAW WASTE GLASS

Igor Vidensky, Hao Gan, and Ian L. Pegg
Vitreous State Laboratory, The Catholic University of America
Washington, DC 20064

ABSTRACT

Corrosion of metallic components is an important issue in radioactive joule-heated ceramic melter vitrification, particularly for wastes that are rich in halides and sulfates. Crucible-scale corrosion experiments were performed in which eight Ni-Cr alloys were tested at 1130°C in a candidate Hanford LAW glass, molten sulfate/chloride salts, and salt vapors. The coupons were characterized by optical and electron microscopy in combination with electrochemical etching and the results compared to those for Inconel 690. The corrosion resistance of the alloys was found to vary considerably, depending on the alloy composition, alloying method, and the minor alloy additives. It is suggested that the grain boundary structure and its modification after short-term corrosion tests are critical parameters that determine the service life of a metal component in molten glass/molten salt/salt vapor environments. In contrast, simple dimensional loss measurements alone can be misleading. The "mechanically alloyed ODS superalloys," Inconel MA758 and 3001, as well as Inconel 693 (Al-bearing), are recommended as potential alternative materials in vitrification applications in which Inconel 690 has typically been used.

INTRODUCTION

In waste vitrification, the compatibility of melter materials with the glass to be processed is an important factor for system lifetime and maintenance. In the treatment of Hanford tank wastes, corrosive salts (halides and sulfates) partition preferentially toward the Low Activity Waste (LAW) fraction, with the result that melter material corrosion issues are particularly important for LAW vitrification[1, 2].

Inconel 690 (a Ni-Cr alloy) has been the material of choice for glass contact components in joule-heated radioactive waste glass melters operated both in the US and elsewhere (e.g., at PAMELA, West Valley, and DWPF, as well as in

To the extent authorized under the laws of the United States of America, all copyright interests in this publication are the property of The American Ceramic Society. Any duplication, reproduction, or republication of this publication or any part thereof, without the express written consent of The American Ceramic Society or fee paid to the Copyright Clearance Center, is prohibited.

Duratek, Inc. DuraMelter™ systems used at M-Area and the Hanford LAW Pilot Melter). While Inconel 690 performs very well when fully-immersed (e.g., electrodes in Joule-heated melter systems[3]), it fails more rapidly when passing through the surface of the glass melt (melt line) and, particularly when exposed to the cold-cap and melt vapors.

Selection of materials able to withstand the corrosive environments inside radioactive waste glass melters has long been recognized as an important aspect of system design. Rankin conducted laboratory-scale tests with several Inconel alloys in borosilicate waste glass at 1150°C and found that Inconel 617 was slightly more corrosion-resistant than Inconel 690[4]. Bickford et al. performed corrosion tests at 1150°C on a range of metals and alloys with molten borosilicate glass and molten halide salts. Those results indicated that high-Cr alloys were the most corrosion-resistant materials, probably due to the formation of chromia protective layers[5]. More recently, Marra et al. conducted corrosion coupon tests on Inconel 690 and Inconel MA-758 using glass compositions high in salt content and found that the presence of Cl in the melt caused the most severe attack[6]. In this laboratory, Inconel 690 electrodes were tested in several simulated Na-rich aluminosilicate waste glasses over a range of conditions[7,8]. The corrosion rate of Inconel-690 electrodes was found to depend on AC current density, glass compositions, temperature, and electrical waveform. Corrosion of Inconel electrodes in glass melts results in Cr depletion, grain boundary damage, and formation of Cr_2O_3 and spinel scales on the electrode surface. An important observation from the work of Gan et al. is that the grain boundary structure of Inconel 690 was severely damaged while in contact with a S- and Cl-containing glass melt. Excessive levels of Cr were depleted from Inconel 690 electrodes near the metal-glass melt interface[8].

Metal corrosion in molten salts has also been studied at temperatures below 1000°C for other applications. Various authors have reported that addition of Al or minor amounts of the oxide, increasing Ni or Cr content, or mechanical alloying may reduce corrosion damage and slow the corrosion rate[9].

The principal test objectives of the present work were to evaluate potential alternative Ni-Cr-based alloys in molten glass/molten salt/vapor environments and to determine the effects of alloy material characteristics (e.g., composition, additives, grain boundary properties, and fabrication methods) on their corrosion behavior.

TEST MATERIAL SELECTION AND EXPERIMENTAL METHOD

Eight Ni-Cr-based alloys were selected for screening tests (Table I). Alloy material selection was made based on evaluation of performance against high-temperature oxidation or corrosion in various gaseous or aqueous environments and by experience with crucible or melter tests in molten glasses or salts. Inconel

690 was the baseline alloy for these tests. Inconel 601, a widely used corrosion resistant alloy, was selected as the lower-Cr variation from Inconel 690. Inconel 693, and experimental alloys EA-3002 and EA-4018, contain higher contents of Al and might be expected to perform better in chloride-sulfate and molten glass environments. Inconel MA758 is an oxide dispersion strengthened (ODS) Ni-Cr superalloy produced by mechanical alloying. The oxide dispersion phase is Y_2O_3 and ODS remains effective at temperatures close to the alloy's melting point, 1400°C. Inconel 740 has a high cobalt content and thus may have a higher oxidation resistance at elevated temperatures. Experimental alloy EA-3001 is a mechanical alloy containing considerable cobalt. Alloy materials made by mechanical alloying usually have an enhanced grain-boundary attack resistance[10].

Table I. Composition of Alloys Tested in this Work (wt%).

Alloys	Cr	Ni	Co	Fe	Al	Ti	Si	Mn	C	Others
Inconel-690	30	58	--	10	0.8	0.3	0.5	0.5	--	--
Inconel-693	29	61	--	5.5	3.2	0.4	0.1	0.08	0.02	Nb0.94, S0.001
Inconel-740	25	49	20	1	0.5	1.6	0.4	--	--	Nb2, Mo0.5
Inconel-601	25	60	--	12	1	0.4	0.5	1	--	--
Inconel-758*	30	68	--	1	0.3	0.5	0.5	1.6	--	Y_2O_3 0.6
EA-3001*	36	44	12	4.4	0.5	0.5	--	--	0.05	Y_2O_3 0.5
EA-3002*	20	75	--	--	4.0	0.5	--	--	0.05	Y_2O_3 0.5
EA-4018	21	50	--	20	3.9	0.2	0.7	--		--

*mechanically alloyed

All alloy corrosion tests were conducted using a modified ASTM refractory corrosion procedure (ASTM-C-621-84) with an experimental setup consisting of a 55-ml Inconel 601 crucible placed in a refractory quartz holder. On top of the crucible was a cover lid with a 5 mm slot. The rectangular coupon was placed through the slot and suspended by a wire inserted through a hole drilled through one end of the coupon. For each corrosion test, 40 g of glass (6% Al_2O_3, 11.6% B_2O_3, 5.2% CaO, 5% Fe_2O_3, 2.5% Li_2O, 13.8% Na_2O, 42.8% SiO_2, 2.5% ZrO_2, (wt%) with the balance others) and 30 g of salt (48% Na_2SO_4, 45% of Li_2SO_4, and 7% NaCl (wt%)) were heated at 1130°C for 30 minutes, giving a coupon surface area to a melt volume ratio (S/V) for these tests of 0.14 cm^{-1}. Half of the immersed coupon was in contact with the molten glass and the other half with the molten salt. The alloy coupon with the lid was preheated to 800°C for about 10 minutes and then placed into the crucible with the molten glass. The entire setup was then placed in a furnace preheated to 1130°C. After a designated period (up to 7 days), the setup was removed from the furnace, the coupon extracted from the glass, and the glass melt poured into a graphite mold for further analysis. For 7-day tests, the Inconel 601 crucible and glass/salt mixture were replaced after 3 days to avoid excessive damage to Inconel 601 crucible and vaporization of salt phase. The coupon with the glass coating was then annealed at 450°C for 2 hours to keep the glass coating intact.

After the test, each coupon was cut in half along the vertical axis, mechanically polished, and electrochemically etched for metallographic analysis. The corrosion loss of the coupon was measured with an optical microscope. The microstructural alteration was studied by optical microscopy. Scanning Electron Microscope with Energy Dispersive X-ray Spectroscopy (SEM/EDS) was used for characterization of phase composition changes and formation of the oxide scales in the corroded part of the metal.

RESULTS AND DISCUSSION

High-temperature metal corrosion can be generally characterized by the levels of the metal loss and depth of internal penetration. Metal loss is estimated from dimension measurement, and the internal corrosion penetration by the structural alteration inside the alloy and the scale formed on or near the surface. Analysis of failed melter components indicates that resistance to intergranular attack is more important than dimension loss. As a result, the grain boundary properties of the test alloys were examined in this study, in addition to the routine dimension loss measurements. All eight alloys were first tested for 72 hours for screening purposes. Four alloys, Inconel 601, Inconel 740, experimental alloy 3002 and experimental alloy 4018, were outperformed by Inconel 690 and were, therefore, dropped from further testing. The remaining four alloys, Inconels MA758, 693, 690 and experimental alloy 3001 were subjected to further testing up to 165 hours. The characterization results of the four reacted alloys are summarized below

1) Inconel 690

The Inconel 690 coupon lost about 20 μm at the half-down position after the 7-day test (Figure 1). Optical micrographs show that the intergranular attack is visible throughout the cross section and is about 400 μm deep for the metal in contact with the molten glass, up to 1 mm deep for the metal in contact with the molten salt, and more

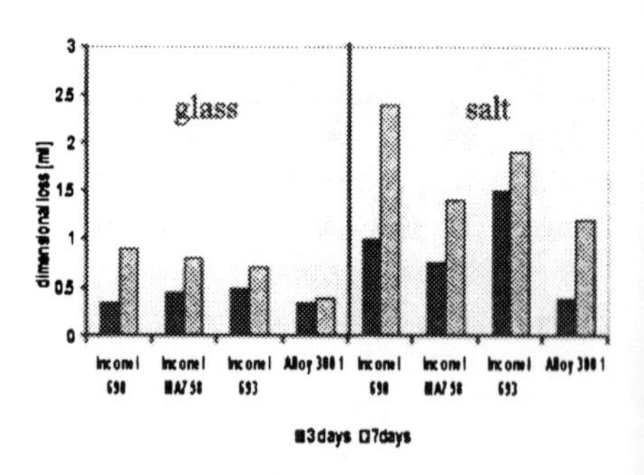

Figure 1. Dimensional loss of Inconels (690, 693, and MA758) and Alloy 3001 after 3- and 7-day corrosion tests; the estimated uncertainty is 0.2 mil.

than 1.9 mm for the part in contact with the vapor (Figures 2a-c). The microstructure of the alloy was modified and degraded greatly, especially as a result of attack by the molten salt and salt vapor. The severe and extensive intergranular corrosion damage would be a likely source of component failure.

2) Inconel 693

The 3.2% of Al in Inconel 693 appears to have limited the progress of grain boundary attack by the molten glass/molten salt phases. The dimensional loss is less then 20 μm at half-down after the 7-day test (Figure 2). While the alteration depth of the alloy is 100 μm in the glass contact area, which is about 25% of the depth observed in Inconel 690, the metal grain structure does not show significant alteration beyond 100 μm (Figure 3). Within the altered alloy, SEM/EDS revealed an oxide-phase layer composed predominantly of Al_2O_3 (Figure 4a). This alumina layer sandwiched inside the Ni-Cr based alloy probably acts as a barrier against continued attack, with the result that the internal penetration in salt and salt vapor decelerated considerably as compared to Inconel 690. The metal exposed to the molten salt (Figure 4b) exhibits severe (but shallower than in Inconel 690) internal penetration (200 μm deep). The grain boundary attack is only about 100 μm deep in the region exposed to salt vapor, which is less than 5% of the depth in the vapor-contact region for Inconel 690.

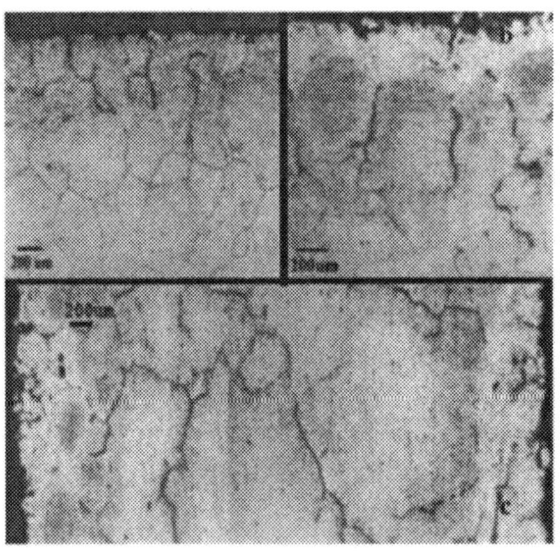

Figure 2. Microstructure of Inconel 690 after corrosion test for 165 hours in a) molten glass, b) molten salt and c) salt vapor (the dark area represent the glass).

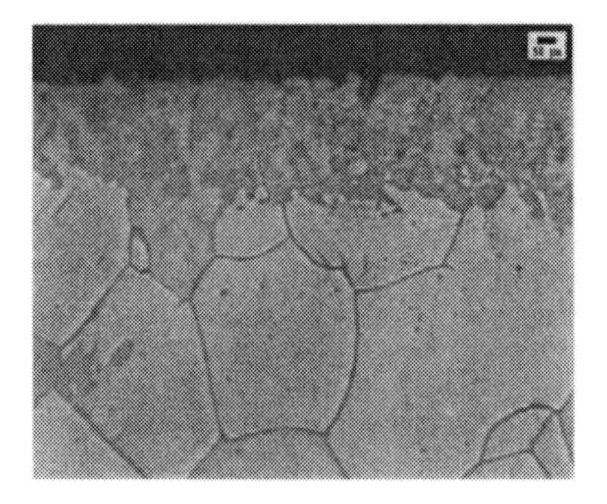

Figure 3. Microstructural alteration in the subsurface layer of Inconel 693.

3) Inconel MA758

Inconel MA758 is an "ODS mechanically alloyed superalloy." As a result, MA758 would be expected to display improved high-temperature grain boundary strength, as well as higher oxidation resistance and heat corrosion resistance. The half-down dimensional loss of the MA758 coupon is around 35 μm after the 7-day test (Figure 2). The internal penetration by the molten glass and the molten salt is about 100 μm for each region and 200 μm in the salt-vapor contact region. The MA758 coupon developed a different microstructure, characterized by numerous submicron-sized chromium oxide occlusions dispersed throughout the sub-surface area (about 30 μm deep) when in contact with molten glass (Figure 5a). The occlusions grew larger and deeper in the molten salt region (100 μm deep; Figure 5b) and in the salt-vapor region (200 μm). There is no sign of significant intergranular attack in this metal besides the isolated oxide occlusions. Overall, the corrosion damage to the microstructure of MA758 is significantly less than that observed for Inconel 690.

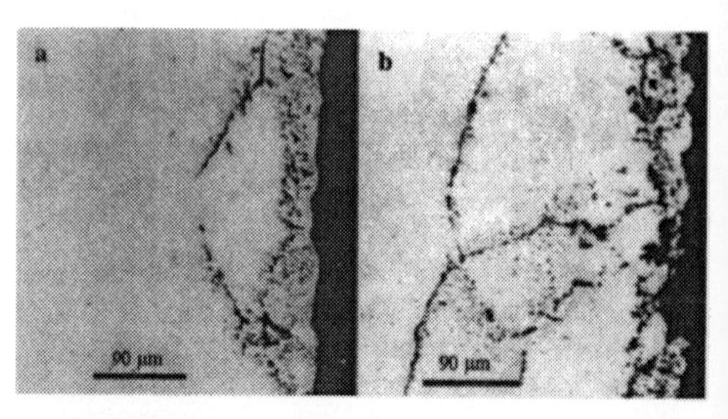

Figure 4. Aluminum oxide scale formed in the subsurface layer of Inconel 693: isolated intergranular attack in the glass contacted area (left); developed intergranular attack in molten salt contacted area (right).

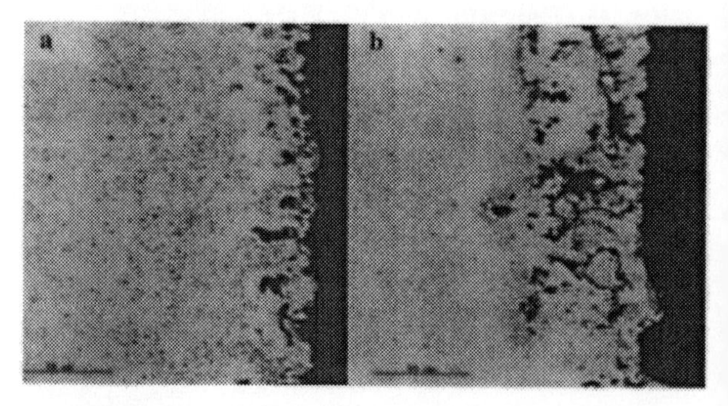

Figure 5. Internal corrosion damage of Inconel MA758 after 165 hour test: a) chromium oxide occlusions formed in the subsurface layer of metal contacted with molten glass; b) occlusions become larger and corrosion extends to a depth of 100 μm

4) Experimental Alloy 3001

Alloy 3001 is another ODS mechanical super-alloy with 12% Co (Table I). Alloy 3001 performed well in many aspects of the metal corrosion test. The subsurface layer of the metal affected by corrosion is about 50 μm deep. Within this layer are globular chromium oxide occlusions (Figures 6a-b) for the metal in the glass and the salt regions. The size of the chromium occlusions grows on moving closer to the metal surface (to 3-5 μm). Like Inconel MA 758, alloy 3001 shows no signs of extensive and interconnected grain boundary attack besides the development of the oxide occlusions in the areas in contact with the molten glass and the molten salt. However, the part of the coupon that was exposed to the salt vapor displayed some level of internal penetration along the grain boundaries to a depth of 200 μm. Overall, the corrosion damage of the microstructure of experimental alloy 3001 is significantly less than that observed on Inconel 690.

Significance of Internal Penetration to Component Failure

The results from the coupon tests indicate that Inconel 690 is clearly not the best available alloy in a sulfate- and chloride-rich environment. If, during melter operation, a transient molten salt phase is formed on the glass pool or Cl or S become locally enriched in the glass melt and the vapor phase close to the melt-cold cap region, increased structural damage to exposed components would be expected, including increased dimension loss and deeper intergranular attack (Figures 1 and 2). It is also clear from this study that addition of Al, as in Inconel 693 (Table I, Figure 3), and mechanical alloying coupled with grain-boundary-reinforcement (ODS), as in MA758 (Figure 5) and Alloy 3001 (Figure 6), significantly improve the resistance to grain boundary attack by molten glass or salt, thus reducing the depth of internal penetration and consequent structural degradation. Analysis of failed Inconel 690 components from joule-heated melters suggests that the metal had been deeply penetrated by the glass and salt species near the cold cap. Oxide, and occasionally sulfate and chloride, alteration products developed extensively along the grain boundaries, which would reduce the mechanical strength and facilitate further penetration and internal oxidation inside the alloy. Furthermore, the local oxygen fugacity, which controls the degree of Cr depletion from the alloy, would likely also depend on the depth of metal-metal grain boundary attack. The damaged and weakened grain

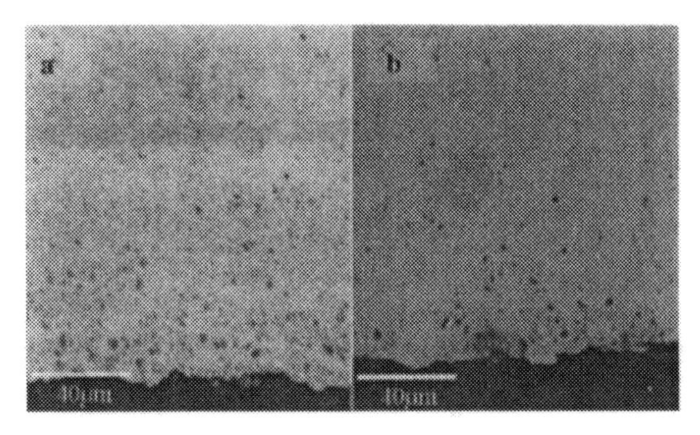

Figure 6. Minor corrosion damage of Alloy 3001 after 165-hour test: a) globular occlusions of chromium oxide formed in the molten glass contact area; b) zone of the occlusion formation in the molten salt contact area.

boundaries also likely act as channels to transport oxygen and other reactants, such as chloride and sulfate. The extent of this grain boundary damage can thus be analyzed by modeling the concentration profile of Cr inside a reacted alloy along the direction perpendicular to the metal-glass interface. In the extreme case, in which oxidation occurs only on the surface of the metal, the concentration profile of Cr should be adequately described by a diffusion equation. However, as the internal penetration and oxidation progress along the grain boundary, the oxidation of Cr starts to occur inside the alloy as well as on the surface, which then pushes in the boundary determined by volume diffusion deeper into the metal. The region in which Cr diffusion ceases to be the dominant process should then indicate the depth to which the grain boundaries have been significantly damaged. Based on such an analysis, the results for Alloy 3001 after the 3-day test show that volume diffusion is not the dominant process for the observed Cr depletion from the glass-melt interface up to 70 μm deep; beyond that, however, the Cr concentration profile follows that predicted by the diffusion equation. The Cr concentration profiles of the four alloys after 3- and 7-day tests for the glass-contact and salt-contact parts have been analyzed similarly. Figure 7 shows a typical concentration profile obtained by SEM/EDS. The internal penetration depths derived from the analysis are summarized in Figure 8. Consistent with the SEM results for the 7-day tests, Inconel 690 suffered considerably deeper grain boundary damage by the glass and salt as compared to the other three alloys. The internal penetration of the alloys MA758 and 3001 and the Inconel 693 are comparable in the both regions. It is important to note that the mere dimension loss measurement does not differentiate Inconel 690 significantly from the other three alloys (Figure 1), nor does the profile analysis of the 3-day test coupons (Figure 8). This also highlights the fact that the penetration depth is not linear in time, at least over a 7-day period, with the non-linearity varying from metal to metal. The internal penetration in Inconel 690 appeared to accelerate with time while the other three alloys displayed some levels of slowing.

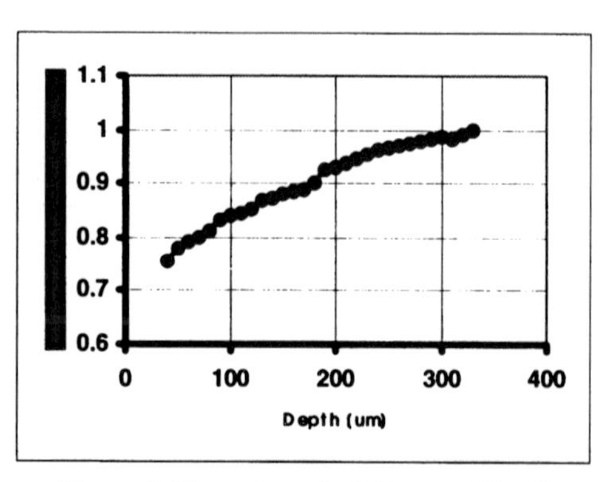

Figure 7. Chromium depletion profile of Alloy 3001 after 7-day test in contact with molten glass. Results from SEM/EDS line scans perpendicular to the metal-glass interface.

In conclusion, while Inconel 690 is the most commonly used alloy for joule-heated waste glass melters, in comparison to other alloys, it did not perform well in LAW glass melts rich in sulfate and chloride. The alloy coupons after 3- and 7-day crucible-scale tests were severely damaged, especially along the grain boundaries. The results suggest that, of the alternative metal materials tested in this work, the ones that will likely show superior performance to Inconel 690 are the two high-Cr mechanically alloyed superalloys, Inconel MA 758 and 3001, and an Al-added alloy, Inconel 693; this conclusion has since been substantiated in long-duration pilot melter tests.

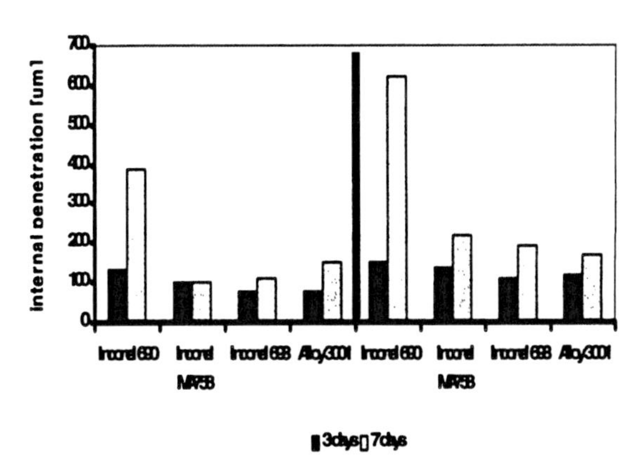

Figure 8. Depth of internal penetration for Inconels (690, 693, and MA758) and Alloy 3001 after 3 (solid) and 7 (hatched) day corrosion tests; the estimated uncertainty is 15μm.

ACKNOWLEDGEMENTS

The authors thank P. Branch for preparing test coupons; C. Mooers and A. Buechele for assistance with SEM/EDS analysis; and M.C. Paul for manuscript preparation. Special Metals Corporation provided coupons of many of the alloys used in this study. This work was supported in part by the Department of Energy Office of River Protection through Duratek, Inc.

REFERENCES

[1]I.S. Muller and I.L. Pegg, "Glass Formulation and Testing with TWRS LAW Simulants," VSL Final Report, January 1998.

[2]H. Gan, X.D. Lu, I. Vidensky, C. Paul, and I.L. Pegg, "Corrosion of K-3 Refractory and Metal Alloys in RPP-WTP LAW Glasses," Final Report, VSL-01R3540-1, March 2001.

[3]D.F. Bickford, "Selection of Melter Systems for the DOE/Industrial Center for Waste Vitrification Research," WSTC-TR-93-762, December 1993.

[4]W.N. Rankin, "Evaluation of Glass-Contact Materials for Waste Glass Melters," Advances in Ceramics **8** 559 (1985).

[5]D.F. Bickford, R.S. Ondrejcin, and L. Salley, "High-Temperature Materials for Radioactive Waste Incineration and Vitrification," Proceedings of the Third

International Symposium on Ceramics in Nuclear Waste Management II (American Ceramic Society, Chicago, April 1986).
[6]J.C. Marra, J.W. Congdon, R.F. Schumacher, A.L. Kielpinski, A.A. Ramsey, J. Etheridge, and R. Kirkland, "Corrosion Assessment of Refractory Materials for High Temperature Waste Vitrification," WSRC-MS-94-0571P, Proceedings of Symposium on the Corrosion of Materials by Molten Glass (American Ceramics Society, Cincinnati, May 1995).
[7]H. Gan, L. Fu, F. Perez-Cardenas, A.C. Buechele, R.K. Mohr, I.L. Pegg and P.B. Macedo, "Corrosion of Materials by Molten Glass," Ed. G.A. Pecoraro, J.C. Marra and J.T. Wenzel, Ceramic Transactions, **78** 372, Indianapolis, IN (1996).
[8]H. Gan, A.C. Buechele, C.-W. Kim, X. Huang, R.K. Mohr and I.L. Pegg, "Corrosion of Inconel-690 Electrodes in Waste Glass Melts," Mat. Res. Soc. Symp. Proc. **556** 278 (1998).
[9]G.Y. Lai, "High-Temperature Corrosion of Engineering Alloys," ASM International, (1990).
[10]"Inconel Alloy MA758", Publication #IAI-126, IncoAlloys International, Inc., 1993.

TECHNOLOGY ROADMAPPING FOCUSSES VITRIFICATION AT THE INEEL

John McCray, Chris Musick, Arlin Olson, and Keith Perry
Idaho National Engineering and Environmental Laboratory
P.O. Box 1625
Idaho Falls, ID 83415-5218

EXECUTIVE SUMMARY

Science and technology development programs have long struggled to provide timely, cost-effective, and useful solutions to real problems. The High Level Waste Program at the Idaho National Engineering and Environmental Laboratory finds itself with a unique opportunity to develop and apply technology for the disposition of 1.2 million gallons of sodium bearing waste (SBW). Although the National Environmental Policy Act process is not officially complete with regard to selecting a final technology for disposition of this waste, the HLW program has developed a science and technology roadmap to treat SBW via vitrification. This roadmap endeavors to identify the key uncertainties that require technical solution, develops a strategy for resolving these uncertainties, and provides guidelines for the timing of when these key deliverables will be provided. The HLW program has utilized the roadmap in preparing the detailed work plans that reflect the actual scope, schedule, and cost for the activities in the technology development program. Although the fruition of these work plans and the roadmap is in its earliest stages, preliminary feedback and actual results highlight the importance of having an umbrella methodology governing and integrating a vast amount of experimental activity.

BACKGROUND

The direct vitrification roadmap was developed on the heels of roadmap preparation for other candidate technologies to disposition SBW. These other roadmapping activities educated all of the roadmap developers, from the technologists and end users to the facilitators of the roadmap development. In order to maintain a level of independence in the roadmap development, facilitators who were familiar with the HLW program assisted the technologists in

To the extent authorized under the laws of the United States of America, all copyright interests in this publication are the property of The American Ceramic Society. Any duplication, reproduction, or republication of this publication or any part thereof, without the express written consent of The American Ceramic Society or fee paid to the Copyright Clearance Center, is prohibited.

identifying the needs and key uncertainties associated with direct vitrification of SBW.

After the needs and uncertainties were brainstormed, the facilitators provided a framework for the technologists and engineers to prioritize the activities. The needs and uncertainties were then broken down into three main risk categories titled high, medium, and low. A high uncertainty is defined as a potential negative result that could cause the failure of the vitrification process. A medium uncertainty could result in a high cost impact but not something that would cause the entire process to fail. A low uncertainty would affect the optimization of the process operation, but not result in a high cost relative to the overall project cost. This prioritization would later assist in the scheduling of the key work activities.

Using the list of uncertainties developed earlier, the technologists set out to develop work scope, cost, and schedule estimates towards resolution of these uncertainties. The INEEL also invited assistance from prominent vitrification experts in the Tanks Focus Area (TFA) in developing descriptions of necessary and sufficient work scope to resolve the uncertainties. In a relatively short and intense period, a work scope strategy was developed as well as schedule and cost impact to the program.

Prior to implementing these roadmap activities into detailed work plans, the roadmap was reviewed by well known scientists and engineers. The TFA Technical Advisory Group (TAG), a group independent from the aforementioned members of the team that assisted in the actual development of the roadmap, convened in Idaho to review the listed needs and uncertainties, as well as the strategy proposed to resolve those needs. The TAG review emphasized simplicity where possible, avoiding work scope that would be impossible to complete within the provided schedule windows. The roadmap developers initially planned to identify areas where improvements could be made that impacted efficiency or cost effectiveness of the facility. However the advisory group emphasized that the team should focus on creating a single robust technology, adding improvements could be added if they were indeed deemed cost effective. The modifications made to the roadmap, based on the TAG comments, completed the initial roadmapping phase and led to the development of detailed work plans for implementation.

Lower level application of the roadmap

While the roadmap is an excellent tool in focussing the project on the key uncertainties, the strategy described in the roadmap is at too high a level to be implemented on a day-to-day basis. The INEEL employs a detailed work planning procedure for all of its work scope, and the roadmap flowed into this process almost seamlessly. Individual areas of the vitrification program, like formulation development, pilot scale testing, and offgas system development, were broken down into individual tasks that could be carried out through the year. The key to keeping the roadmap continuity in this process was to first understand the key uncertainties that the tasks should be designed to address and, second, understand the interfaces between activities. The formulation, melt rate, and pilot scale testing parts of the program highlight these points well and will be used in the remainder of the paper as an example of how the roadmap is being implemented.

Understanding the logical flow of work

A project like vitrification of SBW is arranged in a Work Breakdown Structure (WBS). A WBS includes projects, control accounts, work packages, and tasks, and thus it is easy to lose sight of the big picture and integration requirements within this WBS. The roadmap kept these key aspects of the project at the forefront throughout the work planning process. The roadmap called very clearly for a glass formulation to be developed for the best known waste composition at the time to meet the critical path for pilot scale testing midway through the fiscal year. The roadmap also highlighted the need to understand the effects of reductant addition to this high-nitrate waste on melt rate and other glass product characteristics. At first, the desire on the part of the team to conduct pilot scale ($1/10^{th}$ scale of planned production scale) testing as soon as possible to understand the mass balance around the melter unit operation led to the melt rate tests using reductant addition and the pilot scale testing to occur in parallel. However, once the team returned to the roadmap to verify that their planning was fulfilling the resolution of key uncertainties, it was apparent that this was not the most fruitful ordering of activities. In fact, the melt rate testing was providing the information that was listed in the roadmap as a predecessor to being able to conduct a full pilot scale test. The schedules in the detailed work plan were changed to reflect this logic.

The predecessor-successor relationships identified in the roadmap not only identified key points of integration across work packages, but also identified the key deliverables that each work package must provide to be deemed a success. No good test or experiment can occur without a clear test plan that lays out the objectives of the test. Obviously then, a good test plan cannot even be outlined without these objectives being known and clearly stated. The roadmap methodology, that is the identification of the scope of work that is necessary and sufficient to resolve key uncertainties, fits hand and glove with the test plan and overall work plan development. The key uncertainties should need only slight modification, or perhaps restatement on a lower level, to become test objectives. As each test plan is developed to meet these objectives and, therefore, resolve key uncertainties, the chance for project success rises sharply. Then, as work scope yields data and key deliverables, the roadmap provides a clear framework for who needs those data, by when, and in what form to most benefit the project. This methodology is reflected in the way that the technology development detailed work plans have been and are being implemented.

REAL WORK AND REAL RESULTS

As stated previously, the roadmap is being followed through the implementation of detailed work plansMilestones have been met and key deliverables completed. Results thus far and future plans are described below.

Glass formulation development

The most accurate characterization of the SBW available was delivered to the glass formulation experts. This characterization of tank WM-180, a tank whose contents exist in the same form (concentrated) that will be processed in the actual facility, shows the highly acidic (1.1 M), high nitrate (~5 M) composition of this waste, as well as the presence of significant amounts of sulfate (~0.5 to 0.6 M, or the equivalent of ~1.1% of the glass on a dry oxide basis). The abbreviated composition is shown in table 1. This information was then used, through crucible testing and leveraging data gathered from other DOE projects, to develop a composition of glass forming chemicals and waste loading that would be used in further testing. The recommended oxide waste loading was 30%, and the glass former amounts listed in table 2 were recommended for this composition.

Table I. Characterization of WM-180 SBW

Element	M in Waste	Element	M in Waste
Acid	1.10E+00	Mercury	1.91E-03
Aluminum	6.28E-01	Molybdenum	1.82E-04
Arsenic	4.71E-04	Nickel	1.39E-03
Barium	5.26E-05	Potassium	1.85E-01
Beryllium	7.33E-06	Ruthenium	1.18E-04
Boron	1.16E-02	Sodium	1.94E+00
Cadmium	7.12E-04	Strontium	1.12E-04
Calcium	4.53E-02	Titanium	5.45E-05
Cerium	4.46E-05	Uranium	3.17E-04
Chromium	3.16E-03	Zinc	9.90E-04
Cobalt	1.82E-05	Zirconium	5.97E-05
Copper	6.58E-04	Chloride	2.84E-02
Gadolinium	1.67E-04	Fluoride	4.58E-02
Iron	2.05E-02	Iodide	1.27E-04
Lead	1.23E-03	Nitrate	5.11E+00
Lithium	3.20E-04	Phosphate	1.29E-02
Magnesium	1.14E-02	Sulfate	5.10E-02
Manganese	1.33E-02		

Table II. Glass forming chemical composition

Glass Forming Element	Weight Percent (Oxide Basis)
Boron	15%
Calcium	5%
Iron	10%
Lithium	5%
Silicon	65%

Melt rate and reductant selection studies

In addition to developing the initial glass formulation for this waste, chemists and engineers are testing various organic reductants to potentially increase melt rate and waste loading. Redox state is described as the ratio of ferrous (Fe^{2+}) to total iron in the glass. This ratio gives insight as to whether or not the proper amount of reductant has been added to the melter feed to optimize melt rate by enhancing denitration, and avoid potential process problems such as foaming, sulfide and noble metal formations, etc. The desired Fe^{2+}/Fe_{total} is between 0.05 and 0.3. A ratio of greater than 0.05 is desirable as it is statistically different from 0 given analytical uncertainty, and values of greater than 0.3 decrease the durability of the glass and/or precipitate noble metals.

As the sulfate concentration of the waste and its solubility in the glass are potentially the limiting factors of waste loading, sulfate control and volatilization (as SO_2) is highly desirable. As a part of reductant selection studies, sulfate minimization through redox control is also being investigated.

As stated above, the roadmap indicated that melt rate tests and reductant selection studies are a key part of the development process and had to be initiated as early as possible. In early February, a laboratory scale melter test was conducted at Pacific Northwest National Laboratory (PNNL) using the Research Scale Melter (RSM) melter. The main objectives of this test were to develop an indication of the sustainable melt rate for this waste and glass composition, understand the efficacy of sugar as a reductant, observe sulfate salt evolution, test the robustness of glass formulation while varying oxide waste loading, as well as to gather initial material balance data. These objectives were met over the 4 consecutive days that the test occurred.

The melt rate sustained during the RSM melter run was much higher on average than originally anticipated. Table 3 shows the initial data broken down into the relevant areas.

Table III. Melt rate data from RSM test conducted February 2001

	Average	**Low**	**High**
Feed Rate (L/hr)	2.5	1.5	3.5
Feed Rate (kg/hr)	3.3	2.1	4.8
Glass rate (kg/hr)	0.60	0.37	0.86
Melt rate (kg/m2/day)	791	487	1137

Initially, pre-test predictions had the melt rate hovering near what turned out to be the low value for the test. While these data are promising relative to melt rate, it is not yet fully understood which mechanism is driving these phenomena. One hypothesis points to the molten sulfate salts as a heat transfer mechanism on top of the melt, allowing significantly higher cold cap surface area contact.

An increase of reductant concentration was not only shown to increase melt rate and produce a desirable redox ratio, but there were also some indications that additional reductant was able to reduce the molten salt layer on the melt surface.

Towards the end of the run, sugar concentrations were varied from 135 g/L to 196 g/L. This variance of sugar concentration plotted versus redox ratio showed, in the RSM melter at 35% waste loading, that a sugar concentration of approximately 170 g/L was required to initiate the reduction of iron. Stated differently, the first 170 g/L of sugar chemically reacted with the nitrate contained in the waste, and iron reduction did not begin until the entire nitrate quantity was exhausted. The rapid steady state response of the RSM allowed for fairly quick changes in sugar concentration with startlingly fast chemical response times. Figure 1 shows some of the sugar concentration and redox ratio data plotted as a function of time.

Sulfate salt formation was observed during the RSM run. During crucible testing of the 30% waste loading glass formulation, little sulfate was evident. At higher waste loadings, however, more molten salt was evident once the cold cap was burned off for inspection. The salts also appeared in the glass crucibles that were poured at the highest waste loading of 35%. These results are not surprising as, at 30% waste loading, the amount of sulfate in the feed material would account for 1.07% of the glass on an oxide basis, whereas the glass formulation scientists were hoping for a 0.8% retention in the glass due to solubility. Indications were good, however, that running at the target 30% waste loading would keep molten salt down to an acceptably low level. The variation of the waste loading also demonstrated initially that the formulation was robust within a waste loading variation of 30% to 35%.

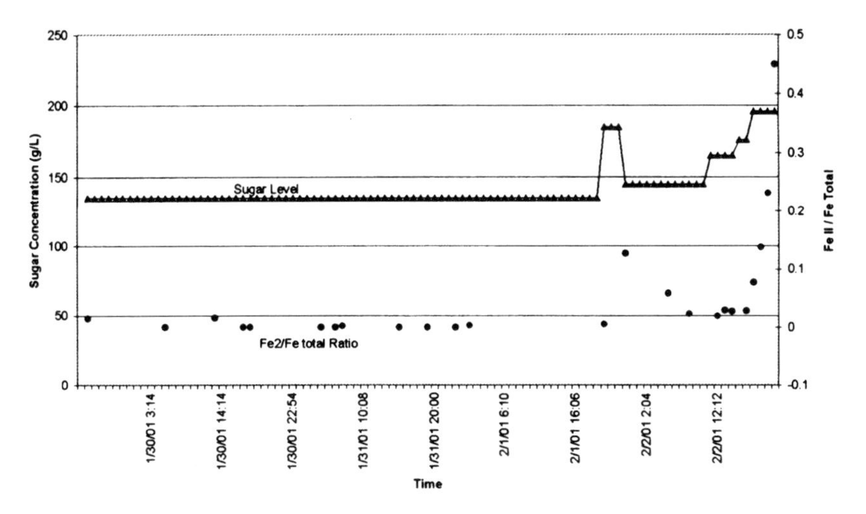

Figure 1, Sugar Concentration and Redox Ratio as a Function of Time

The RSM melter offgas system was sampled to gather some scoping material balance information. The results of these offgas measurements may not be entirely indicative of the type of offgas species concentrations that would be seen in the actual facility as the configuration of the melter is very different. The baseline melter configuration for the Idaho Waste Vitrification Facility includes plenum heat and the offgas port is much higher relative to the melt surface than the RSM melter. This suggests that the entrainment of particulate matter in the RSM offgas would probably be conservatively high relative to the actual facility. Also, initial continuous emissions monitoring (CEM) data show additional reduced forms of nitrates in the offgas, notably the presence of N_2O in addition to the expected NO_2, NO, and N_2. Complete offgas data will be available shortly and allow for detailed analysis as well as an attempt to close the material balance around the melter.

Pilot Scale Melter Testing at CETL

The 1/10th pilot scale melter at Clemson University, called the EV-16 melter, was used next to gather data for this effort. This melter was to be run twice in FY01 with two different waste compositions to provide an incremental scale increase from crucible to RSM to pilot scale. The test objectives for the first run were very similar to the RSM test, including gathering material balance data and indications of molten salt evolution. The scale of the EV-16 melter does not readily allow for compositional changes, so a baseline waste loading of 30% was to be maintained throughout the run. Sugar levels were adjusted in an attempt to attain a target value of 0.2 to 0.3 Fe^{2+} to total iron ratio that is desirable due to operational constraints of this melter and waste form. An offgas-sampling subcontract allowed for both CEM records as well as isokinetic sampling trains.

Many useful insights were gained during the test. The melt rate for the EV-16 run was, on average, lower than predicted using RSM run data. The lowest melt rate seen during the RSM run would correspond to, due to surface area differences (EV-16 has 11.5 times the surface area of the RSM), a feed rate of approximately 286 ml/min. So far, the average feed rate to sustain what was perceived as desirable cold cap coverage has averaged approximately 275 mL/min. At higher plenum and offgas temperatures, feed rates of up to 330 mL/min were observed. This correlation between melt rate and glass temperature will impact future glass formulations.

On a few occasions, the offgas system flow has seen unacceptably high pressure drops due to the buildup of material in the pipe bends of the system, both upstream and downstream of the film cooler. The material downstream of the film cooler is more salt-like, while the material upstream of the cooler resembled a solidified feed material. Figures 2 and 3 show pictures of the both the upstream and downstream film cooler flanges, respectively. These pictures were taken during a pause in melter system operation. Although the pictures are not crystal clear, one can see the solid material buildup on the upstream flange, while the downstream flange is coated with a light, powdery material. In fact, the actual

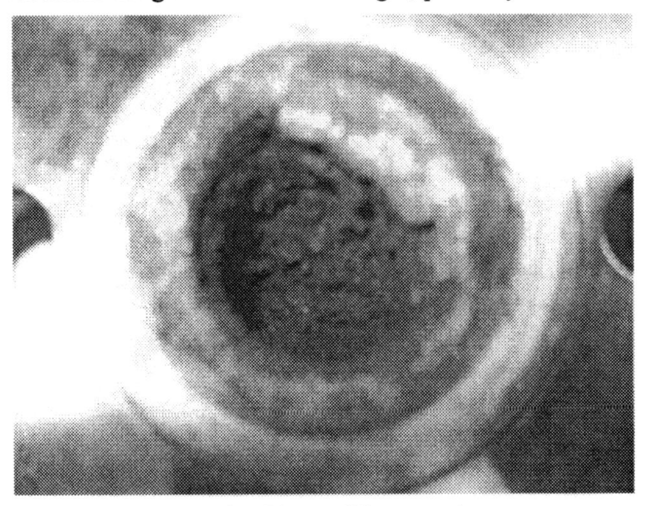

Figure 2, Film Cooler Flange (Upstream)

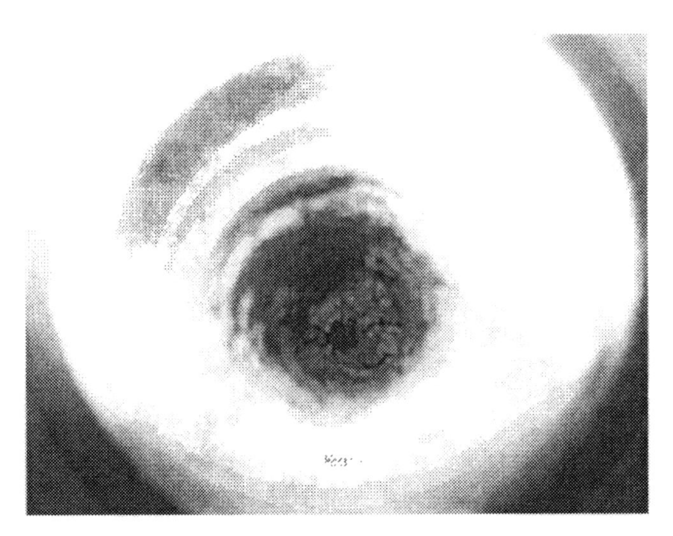

Figure 3, Film Cooler Flange (Downstream)

area where the air enters the cooler is visible because of black, flow-restricting material that has built up in the area immediately upstream. These materials have been mined out and saved for analysis. The composition of these materials will give great insight into whether a caustic or acidic scrub is warranted and whether or not to filter the flow of air for these solid materials.

Sulfate evolution has been higher than expected for the given waste loading. Salt pockets have been probed and shown to be up to 1" deep at times. When such salt pockets are observed, anthracite (69% carbon) grains have been introduced to the surface of the melt, with effective, observable results. The carbon particles seem to react with the sulfate as a slight foaming is witnessed over some of the melt surface. The particles that land in the glass are actually noted to be stable enough (i.e., they do not immediately combust) that if these same particles contact salt in short time (30 to 60 seconds), additional reactions are noted. It is not clear, especially with respect to stoichiometry, what the reaction mechanism is, nor if indeed all of the observable reactions are volatilization of sulfate. This phenomenon needs to be researched fully, although initial indications are promising. This information will help assess the viability of using only sugar as a reductant relative to the multipurpose potential of a sugar/carbon blend to both drive redox and sulfate elimination.

ONGOING AND FUTURE WORK

The work described above will be leveraged in additional testing in both FY01 and following years in agreement with the plan laid out in the roadmap. Additional lab and pilot scale tests will be conducted to choose reductant(s), continue material and energy balance development, understand the operational differences due to additional waste characterization, as well as test potential NO_x abatement technologies. A compositional variation study (CVS) is being conducted in parallel with these testing efforts. This study uses statistical analysis to make many (in this case 60) glasses of slightly varying composition to develop a compositional envelope for the waste and the resultant glass. Per the roadmap, the CVS data will ultimately be used in the development of waste qualification and certification documentation necessary to see the waste products disposed of in deep geologic isolation as presently prescribed.

The overall effort of developing vitrification as a method for treating and disposing of SBW is both vast and complex. The roadmap has allowed for a breakdown of this effort into manageable scopes of work, each designed to solve one or more specific needs necessary for demonstrating viability, limiting cost, and optimizing the operation of a vitrification facility in Idaho. The roadmap is often relied upon as a source of understanding for what each piece of work should accomplish, and this information has be translated into detailed work planning that can be tracked in the existing corporate structure. The effort of developing this roadmap was by no means insignificant, but the effort has already paid off in the integration of technology development activities and a solid understanding of the tasks necessary for successful completion of this mission.

GLASS FORMULATION FOR DIRECT VITRIFICATION OF INEEL CALCINE HLW

J. V. Crum and J. D. Vienna
Pacific Northwest National Laboratory
PO Box 999, K6-24
Richland, WA 99352

D. K. Peeler and I. A. Reamer
Westinghouse Savannah River Company
Savannah River Technology Center
Aiken, SC 29808

ABSTRACT

The Idaho National Engineering and Environmental Laboratory (INEEL) High-Level Waste Technology Development program has the goal of defining processes for immobilizing Idaho Nuclear Technology and Engineering Center (INTEC) high-level wastes (HLW) to a qualified waste form for disposal before year 2035. Approximately 4,400 m^3 of calcine are presently stored in stainless steel bins at INTEC. One option for treating the calcine waste is direct vitrification. In this option, the calcine is fed to the melter with minimal feed conditioning.

To demonstrate that direct vitrification is a feasible option for INEEL, glasses have been formulated for both a specific Zr-calcine simulant (Run 78 Calcine) and a Blend calcine. These waste compositions contained high concentrations of Al_2O_3, CaO, F, ZrO_2, and SO_3. Glass formulation efforts resulted in waste loadings for the Run 78 calcine and Blend calcine compositions of 38 and 40 mass%, respectively. At these loadings, all process and product performance criteria were met. Specific glass formulations for each waste stream were successfully processed in a pilot-scale melter at Clemson University.

INTRODUCTION

Approximately 4,400 m^3 of radioactive calcine exist that originated from reprocessing of spent nuclear fuel at the Idaho Chemical Processing Plant. This

To the extent authorized under the laws of the United States of America, all copyright interests in this publication are the property of The American Ceramic Society. Any duplication, reproduction, or republication of this publication or any part thereof, without the express written consent of The American Ceramic Society or fee paid to the Copyright Clearance Center, is prohibited.

waste must be immobalized into an acceptable waste form for disposal into the high level waste (HLW) repositiory. Vitrification is a demonstated and proven technology for treatment of HLW. One of the options being considered as a treatment process for immobilizing Idaho National Engineering and Environmental Laboratory (INEEL) HLW calcine is direct vitrification (with minimal feed conditioning). For the proposes of these studies, Run 78 calcine and Blend calcine[1] compositions were used as simulant waste forms.

Past formulation efforts with Run 78 calcine waste simulant have achieved up to 38 mass% waste loading in glass[2]. The goal of this work is to formulate frit compositons that accept higher than 38 mass% waste loading while satisfying glass processing and product constraints listed in Table 1. Two glass property constraints have changed for new glass formulation efforts. In prior work, T_M had only an upper limit of 1150°C and crystallinity after canister centerline cooling (CCC) heat treatment was limited to 2 mass%. Higher waste loading was achieved by removing the unnecessarily restrictive crystallinity constraint on CCC glass. In this work, T_M was restricted to 1125 ± 25°C, which eliminated the flexibility of formulating glass compositions that melt at lower temperatures and allow higher F concentrations in glass.

Table 1. Glass Property Constraints

Property	Property limit
Melting Temperature (T_M)	1125 ± 25°C
Viscosity (η) at T_M	$2 \leq \eta \leq 10$ Pa·s
Liquidus Temperature (T_L)	$T_L \leq T_M$-100°C
Quenched glass	Single phase
CCC glass	No crystallinity limit
PCT-A normalized releases -- r_B, r_{Li}, and r_{Na} for both quenched and CCC glass	$r_i \leq 1.00$ g/m^2

WASTE COMPOSITION

Originally, Run 78 calcine was used as the waste simulant for frit development. The simulant was changed to a Blend calcine once that composition became available. Table 2 shows the compositions of the two waste simulants. These waste simulants both have high concentrations of Al_2O_3, CaO, F, SO_3, and ZrO_2. The high F concentration in the wastes is expected to limit loading in glass because of the low F solubility in borosilicate glass. High concentrations of F in borosilicate glass promote amorphous and/or crystalline phase separation upon cooling and even at the T_M.

Table 2. Simulant Waste Compositions in Mass Percent

Component	Blend Calcine	Run 78 Calcine
Al_2O_3	32.06	24.71
B_2O_3	2.30	1.92
CaO	27.95	33.44
F	14.90	13.46
Na_2O	3.05	4.38
P_2O_5	0.43	3.36
ZrO_2	13.54	13.53
SO_3	2.22	2.26
Total(a)	96.45	97.06

(a) Only major waste components are listed. For full compositions see Crum et al. 2001[2].

RESULTS

Fluorine Solubility

Knowing the solubility limit of F in glass and how glass composition influences it is crucial for glass formulation efforts. For this reason, a model has been fit to available data to predict whether a glass will remain a single phase after quenching. The solubility of F was found to correlate with non-bridging oxygen (NBO) concentrations. Thus, a model was suggested of the form:

$$\frac{[F]}{[O]} = a + b \cdot \frac{([Na]+[Li]+[K]+2\cdot[Ca]-[Al])}{([Al]+[Si])}$$

where component concentrations are in mole fractions of elements in the glass. This model provides a reasonable qualitative gauge of F solubility after quenching.

Frit Development Work with Run 78 Calcine

Table 3 shows the compositions of the successful frits 5, DZr9, and DZr10 from previous work[3], along with new frits formulated to allow higher waste loading. Maximum loading (ML) frits ML102-ML108 were selected to evaluate the potential for increased waste loading. Table 4 shows the results of the initial testing. All of the quenched glasses were visually crystallized at 50mass% waste loading. Of the glasses, ML107-50 and ML-108-50 were the least crystallized after quenching. ML107 and ML108 were loaded with 45 mass% Run 78 calcine to produce a homogenous glass upon quenching.

Table 3. Frit Compositions

Frit	SiO_2	Na_2O	Li_2O	B_2O_3	Fe_2O_3	La_2O_3	ZrO_2	TiO_2
5	64.61	15.16	9.23	6.80	4.20	0.00	0.00	0.00
DZr9	59.30	14.60	10.00	10.50	0.00	4.00	1.60	0.00
DZr10	59.30	14.60	10.00	10.50	4.00	0.00	1.60	0.00
ML102	80.00	5.00	5.00	10.00	0.00	0.00	0.00	0.00
ML103	70.00	4.00	10.00	10.00	4.00	0.00	2.00	0.00
ML104	65.00	4.00	10.00	15.00	4.00	0.00	2.00	0.00
ML105	60.00	8.00	10.00	15.00	5.00	0.00	2.00	0.00
ML106	70.00	12.00	6.00	10.00	0.00	0.00	2.00	0.00
ML107	61.18	5.62	12.00	14.08	0.62	6.50	0.00	0.00
ML108	61.18	5.62	12.00	14.08	0.62	0.00	0.00	6.50

All of the glasses at 50 and 45 mass% waste loading were heat treated according to CCC profile[4]. Durability of both the quenched and CCC glasses were measured according to the standard 7-day Product Consistency Test (PCT-A)[5].

Table 4. Visual Observation and PCT Results of the as-Fabricated and CCC Heat-Treated Glasses

Glass ID	Observations	r_B, g/m^2	r_{Li}, g/m^2	r_{Na}, g/m^2
ML102-50-Q	Crystallized	0.07	0.43	-
ML102-50-CCC	Crystallized	0.27	0.17	0.04
ML103-50-Q	Crystallized	0.08	0.15	0.08
ML103-50-CCC	Crystallized	0.50	0.36	0.17
ML104-50-Q	Crystallized	0.09	0.15	0.09
ML104-50-CCC	Crystallized	0.11	0.17	0.09
ML105-50-Q	Crystallized	0.11	0.19	0.15
ML105-50-CCC	Crystallized	0.60	0.47	0.28
ML106-50-Q	Crystallized	0.11	0.20	0.14
ML106-50-CCC	Crystallized	0.26	0.46	0.15
ML107-45-Q	Homogenous	0.07	0.14	0.19
ML107-45-CCC	Crystallized	0.04	0.12	0.17
ML107-50-Q	Crystallized	0.06	0.12	0.17
ML107-50-CCC	Crystallized	0.14	0.21	0.21
ML108-45-Q	Homogenous	0.06	-	0.03
ML108-45-CCC	Crystallized	0.07	-	0.03
ML108-50-Q	Crystallized	0.06	0.18	0.06
ML108-50-CCC	Crystallized	0.53	1.29	0.19

Frits ML107 and ML108 both indicate that 8 mass% increases in waste loading are achievable when crystallinity constraints are not placed on CCC heat-treated glass. Although these glasses were all highly crystallized after CCC heat-treatment, PCT release was not strongly affected by the heat-treatment, with the exception of ML108-50.

Frit Development with Blend Calcine

Frit ML107 was selected for testing with the new Blend calcine waste composition because it out performed all of the other frits with Run 78 calcine. Results of ML107 testing with Blend calcine composition are shown in Table 5. The target loadings of Blend calcine were reduced from 45 mass% with Run 78 calcine to between 44 and 42 mass% to account for the increased F concentration in Blend calcine. Once PCT releases and F volatility were measured and found to be acceptable other glass properties of interest were measured on samples of ML107 frit with 40 mass% Blend calcine.

Table 5. Visual Observations, PCT, and Measured F Results for ML107- Glasses at 40, 42, and 44 mass% Blend Calcine Waste

Glass ID	Observations	r_B (g/m^2)	r_{Li} (g/m^2)	r_{Na} (g/m^2)	Target F	Meas. F
ML107-40 Q	Homogenous	0.12	0.19	0.13	5.96	5.44
ML107-42 Q	Homogenous	0.11	0.18	0.12	6.26	5.58
ML107-44 Q	Crystallized	0.10	0.18	0.11	6.65	5.99

Testing of Candidate Frit ML107

The ML107 glass with 40 mass% of Blend calcine was chosen for full characterization to allow for possible variations in waste composition and melter processing which might result in exceeding the homogeneity boundary between 42 and 44 mass% waste loading glasses. Table 6 shows the measured properties of glass with Frit 107 at 40 mass% waste loading. All targeted property values (Table 1) were met with the exception of the lower viscosity limit at T_M ($\eta \geq 2$ Pa·s). At $T_M = 1100°C$, the measured glass viscosity was 1.47 Pa·s. A decision was made to relax the lower viscosity constraint for the purposes of a pilot scale melter demonstration. Further investigation of the "true" limiting viscosity should be conducted before vitrification plant design begins.

Comparison of Glass Formulations for Run 78 Calcine and Blend Calcine

Glass formulation efforts have made steady progress from the first formulation (Frit 5) up to the present ML107 frit. Achievable waste loadings in glass have increased from 35 to 45 mass% of Run 78 calcine and up to 42 mass% of Blend calcine (even though F concentration in waste have also increased). Waste loading increases are the result increasing the alkalis and B_2O_3

concentration and decreasing the SiO_2 concentration in the frit, which increased F solubility and thus waste loading in glass.

Table 6. Characterization of Glass with Frit ML107 at 40 mass% Waste Loading of Blend Calcine Waste

Property or Test	Measured Value
Quenched glass	100% Amorphous
CCC heat treated glass	~ up to 2 mass% CaF_2
Phase separation temperature	773°C (CaF_2 crystallization)
T_L	~1004°C (Zr-containing phase)
Average r_B for quenched glass	0.03 g/m^2
Average r_{Li} for quenched glass	0.10 g/m^2
Average r_{Na} for quenched glass	0.11 g/m^2
Average r_B for CCC glass	0.05 g/m^2
Average r_{Li} for CCC glass	0.13 g/m^2
Average r_{Na} for CCC glass	0.10 g/m^2
Viscosity at 1100°C	1.47 Pa·s
Electrical conductivity at 1100°C	31.53 S/m
Glass transition temperature	419°C
Softening point	492°C
Density	2.72 g/cm^3
Devitrification at 750°C/48 h	Estimated 90 vol% crystallized
Remelt of devitrification sample at 1000°C/1 h	100% amorphous

Table 7 shows comparisons of some of the glass processing and product acceptance properties for each of the glasses characterized for scaled melter demonstrations. All of the glass properties shown have remained similar with the exception of T_L and viscosity. Liquidus temperature has increased considerably from the Run 78 calcine to the Blend calcine glass formulations. Viscosity for a given temperature, or T_M, is lower for Frit DZr9 with 38 mass% waste loading than Frit 5 with 35 mass% waste loading. This is due to the increased F concentration in the waste along with the increased concentrations of B_2O_3 and alkalis and reduced SiO_2 content in the frit for the DZr9-38 mass% Run 78 calcine glass. These changes in frit composition were necessary to increase the solubility of F in glass and thus increase waste loading. Frit ML107 was an attempt to raise F solubility and T_M up to 1100°C. The viscosity of ML107 at 40 mass% Blend calcine waste is between Frit 5 and DZr9 at their respective waste loadings. Based upon the available data, formulations of glasses with higher T_M will not lead to increased waste loadings.

Table 7. Frit Performance Comparisons

	Frit 5	DZr 9	DZr 10	ML107
B_2O_3	6.8	10.5	10.5	14.1
Fe_2O_3	4.2	0.0	4.0	0.6
La_2O_3	0.0	4.0	0.0	6.5
Li_2O	9.2	10.0	10.0	12.0
Na_2O	15.2	14.6	14.6	5.6
SiO_2	64.6	59.3	59.3	61.2
ZrO_2	0.0	1.6	1.6	0.0
F (in glass)	4.71	5.11	5.11	6.10
Waste	Run 78	Run 78	Run 78	Blend
Loading	35%	38%	38%	40%
r_B, g/m^2	0.098	0.199	0.208	0.121
r_{Li}, g/m^2	0.168	0.065	0.065	0.126
r_{Na}, g/m^2	0.033	0.418	0.339	0.188
T_M at $\eta = 2$ Pa·s	1095	1030	NA	1064
T_M at $\eta = 10$ Pa·s	954	896	NA	921
CCC crystal, vol%	0.5-5	1-2	3-6	1-2
T_L, °C	868	<900	<900	1004

CONCLUSIONS

The steady increases in waste loading of Run 78 calcine and Blend calcine are the result of increase F solubility in glass and relaxing of the constraint on mass% crystallinity in the final glass product. As waste loading in glass has increased viscosity (η) and T_L have been affected. Frit ML107 with 40 mass% Blend calcine waste shows an additional potential limit on waste loading. In an effort to accommodate higher concentrations of F in glass, T_M and T_L will likely converge resulting in an additional barrier. T_L will increase because, as waste loading increases, the concentration of ZrO_2 in glass will lead to higher T_L of Zr-containing phases such as zircon or baddeleyite. Conversely, viscosity will decrease because F strongly decreases viscosity. F-containing phases will continue to remain a problem upon cooling but Zr-containing phases, which generally crystallize slowly, will become an issue inside the melter at the melting temperature.

These formulation efforts have provided useful suggestions for scaled melter demonstrations. Additional data is needed to formulate an optimized glass composition for this waste type. Given that relaxing the constraint on mass%

crystallization in the final glass product has allowed increased waste loading, additional studies are required to address radionuclide partitioning.

ACKNOWLEDGEMENTS

The authors would like to acknowledge the following people: Lynette Jagoda, Mike Schweiger, Don Smith help completing the study and E. William Holtzscheiter for management and guidance
This study was funded by the Department of Energy's (DOE's) Office of Science and Technology, through the Tanks Focus Area, and office of Waste management through the Idaho National Environmental and Engineering Laboratory high-level waste program. This study was performed as a collaborative effort by PNNL (operated for DOE by Battelle under contract DE-AC06-76RLO 1830), SRTC (operated for DOE by Westinghouse Savannah River Company under contract DE-AC09-96SR18500), and INEEL (operated for DOE by Bechtel, Babcox and Wilcox Incorporated under contract DE-AC07-941D 13223).

REFERENCES

[1]Mohr, C. M., L. O. Nelson, and D. D. Taylor. 2000. *Optimization of Calcine Blending During Retrieval From Binsets,* INEEL/EXT-2000-00896, Idaho National Engineering and Environmental Laboratory, Idaho Falls, Idaho.

[2]Crum, J. V., J. D. Vienna, D. K. Peeler, and I. A. Reamer, 2001, *Formulation Efforts for Direct Vitrification of INEEL Blend Calcine Waste Simulant: Fiscal Year 2000*, PNNL-13483, Pacific Northwest National Laboratory, Richland, Washington.

[3]Musick, C.A., B.A. Scholes, R.D. Tillotson, D.M. Bennert, J.D. Vienna, J. V. Crum, D.K. Peeler, I.A. Reamer, D. F. Bickford, J. C. Marra, N.L. Waldo. 2000. *Technical Status Report: Vitrification Technology Development Using INEEL Run 78 Pilot Plant Calcine,* INEEL\EXT-2000-00110, Idaho National Engineering and Environmental Laboratory, Idaho Falls, Idaho.

[4]Marra, S. L. and C. M. Jantzen. 1993. *Characterization of Projected DWPF Glasses Heat Treated to Simulate Canister Centerline Cooling (U),* WSRC-TR-92-142, Rev. 1, Westinghouse Savannah River Company, Aiken, South Carolina.

[5]American Society for Testing and Materials (ASTM). 1998. "Standard Test Method for Determining Chemical Durability of Nuclear Waste Glasses: The Product Consistency Test (PCT)," ASTM-C-1285-97, in *Annual Book of ASTM Standards*, Vol. 12.01, ASTM, WEST Conshohocken, Pennsylvania.

A SNAPSHOT OF MELT RATE TESTING AND REDUCTANT SELECTION FOR THE INEEL SODIUM-BEARING WASTE VITRIFICATION PROGRAM

John A. McCray and Daniel L. Griffith
Bechtel BWXT Idaho, LLC
P. O. Box 1625
Idaho Falls, ID 83415-5218

ABSTRACT

Vitrification is being investigated as a treatment method for the remaining 1 million gallons of sodium-bearing waste (SBW) at the Idaho National Engineering and Environmental Laboratory (INEEL). SBW is an aqueous, acidic solution containing radionuclides, heavy metals, and high concentrations of dissolved nitrate salts. A Settlement Agreement between the Department of Energy (DOE) and the State of Idaho requires the tanks containing this waste to be taken out of service by December 31, 2012.

To minimize the number of potential unit operations required for SBW vitrification, direct liquid feeding has been suggested. To aid in the denitration of the waste, help control foaming in the melt, and potentially help prevent sulfate salt layer formation, the use of organic reductants is being evaluated. This paper describes the test methods used and results to date for reductant selection, waste loading maximization, and melt rate determination.

INTRODUCTION

Liquid acidic radioactive wastes generated and stored at the Idaho Nuclear Technology and Engineering Center (INTEC) of the INEEL, originated from spent nuclear fuel reprocessing, decontamination efforts, and other facility operations. These wastes were classified as either high-level waste (HLW) or SBW[i]. In the past, both waste types were calcined in the New Waste Calcining Facility (NWCF) to make a dry solid granular product that is both easier and safer to store.

---

[i] SBW is distinguished from HLW in that it was not a product of first-cycle reprocessing, and it has a relatively high concentration of sodium.

To the extent authorized under the laws of the United States of America, all copyright interests in this publication are the property of The American Ceramic Society. Any duplication, reproduction, or republication of this publication or any part thereof, without the express written consent of The American Ceramic Society or fee paid to the Copyright Clearance Center, is prohibited.

A Settlement Agreement between DOE and the State of Idaho requires that the INTEC tank farm facility (TFF) be taken out of service by December 31, 2012. This facility closure will require removal and treatment of the approximately one million gallons of SBW remaining in the TFF.

The stored HLW calcine does not meet long-term disposal requirements. As such, eventual further treatment will be required to convert it to an acceptable, leach-resistant form. Vitrification is included in most proposed INTEC HLW treatment options. Direct vitrification[ii] is being investigated as a treatment method for SBW immobilization.

SBW VITRIFICATION DEVELOPMENT

The SBW vitrification development program is comprised of multiple activities including waste characterization, development of glass formulations, the determination of waste loading limits, melt rate determinations, off-gas characterization and treatment, and the scale-up and verification of the vitrification process with various sized melters[1]. Current development efforts are concentrated on the SBW contained in tank WM-180. The chemical composition of the WM-180 waste (excluding the contained radioisotopes) is presented in Table I.

The status of the development work for WM-180 SBW vitrification is: characterization has been completed at INEEL[2], initial glass formulation development has been completed via crucible-melt work at both Pacific Northwest National Laboratory (PNNL) and Savannah River Site (SRS), melt rate and reductant-comparison crucible testing is in progress at INTEC, and an initial research-scale melter (RSM) run has been successfully completed[iii] at PNNL[3,4].

MELT RATE TESTING

The determination of melt rates, one of the major development activities and the primary topic of this paper, is actually a combination of activities and interfaces aimed at optimizing the SBW vitrification waste loading and process throughput. Two main sub-topics are: 1) reductant selection and reduction-oxidation (redox) potential control, and 2) elimination of sulfate salt desegregation in the glass melter.

[ii] Direct vitrification does not involve calcination or any other pre-treatment. The liquid waste is fed directly into the melter (with the appropriate glass formers and reductant)

[iii] The RSM-1 run, conducted in January/February 2001, used sucrose as a reductant

Table I. INTEC TFF Tank WM-180 SBW Composition

Element	Concentration (Moles/L)	Element	Concentration (Moles/L)
Aluminum	6.3E-01	Mercury	1.9E-3
Arsenic	2.4E-4	Molybdenum	1.8E-4
Barium	5.3E-05	Nickel	1.4E-3
Beryllium	7.3E-6	Potassium	1.9E-1
Boron	1.2E-2	Ruthenium	1.2E-4
Cadmium	7.1E-4	Sodium	1.9E+0
Calcium	4.5E-2	Strontium	1.1E-4
Cerium	4.5E-5	Titanium	5.5E-5
Cesium	1.65E-3	Uranium	3.2E-4
Chromium	3.2E-3	Zinc	9.9E-4
Cobalt	1.8E-5	Zirconium	6.0E-5
Copper	6.6E-4	Chloride	2.8E-2
Gadolinium	1.7E-4	Fluoride	3.4E-2
Iron	2.1E-2	Iodide	1.3E-4
Lead	1.2E-3	Nitrate	5.1E+0
Lithium	3.2E-4	Phosphate	1.3E-2
Magnesium	1.1E-2	Sulfate	5.1E-2
Manganese	1.3E-2	H^+	1.10

Reductant Selection and Redox Control

The need for the use of a chemical reductant was determined at the onset of SBW vitrification development. The redox state in a waste melter is identified as probably the most important process control parameter[5]. Redox state is generally determined by measuring the ratio of ferrous to total iron (Fe^{2+}/Fe_{total}) in a sample of the glass. An acceptable processing range expressed in terms of that ratio is between 0.01 (verification that some Fe^{2+} is present) and 0.3[iv].

Under highly oxidizing conditions (Fe^{2+}/Fe_{total} ratio < 0.01), liberation of O_2 from the thermal reduction of Fe in the melt can cause excessive foaming. This creates an insulating layer that reduces heat transfer to the cold cap, resulting in production slowdown or melter shutdown. Under strongly reducing conditions (Fe^{2+}/Fe_{total} ratio > 0.3) excessive electrode corrosion or the formation of elemental metals and sulfides can occur. The accumulation of metals and sulfides can eventually short-circuit the melter's electrodes.

[iv] The Fe^{2+}/Fe^{3+} ratio is also often used to identify the state of waste glass redox. In this case, the acceptable processing range is between 0.01 and 0.5

Reductant selection will be based on the results of crucible and melter tests. These tests include: 1) the development of redox curves (redox vs reductant concentration) for several reductants; 2) melt rate furnace testing with several reductants; 3) RSM reductant comparison testing; and 4) pilot-scale melter runs for verification.

An initial crucible test to determine an acceptable reductant concentration was performed for sucrose[v]. The results of this testing provided the starting sucrose concentration for the first RSM melter run (135 g/l). This melter operation was successful, demonstrating a very trouble-free flowsheet. Sucrose therefore is the "baseline" reductant for SBW vitrification.

Similar crucible tests were performed at INTEC to determine optimum reductant concentrations of sucrose, glycolic acid, corn starch, and powdered activated carbon. More volatile reductants[vi] were not tested.

The sample generation procedure used for the reductant optimization tests was significantly modified from that used for the initial sucrose concentration determination. The modified procedure incorporates pumping the slurry feed[vii] into a 500 ml crucible located inside a "dry-out" muffle furnace set at approximately 1000°C. The direct slurry feeding minimized the excessive foaming previously experienced, allowing for significantly larger samples to be produced. Following the feed addition, the crucible was removed from the first furnace, covered with a platinum lid, and immediately placed in a second furnace (rapid-temperature) at 1150°C for ½ hour. The redox state of each resulting glass was determined by dissolving a small sample in a mixture of HF and H_2SO_4, and measuring the concentrations of ferrous and ferric iron using a relatively simple spectrophotometric method[6]. Figures 1 through 4 plot the glass redox state vs the amounts of each reductant used per mole of nitrate in the feed. From the "breaks" in these graphs, an estimate of the "optimum" concentration of each reductant can be determined. The estimated optimum sucrose concentration closely matches that estimated from the RSM operation (~160 g/L).

---

[v] Testing was performed by John Vienna of PNNL. The "acceptable" reductant concentration is that which provides a glass with an Fe^{2+}/Fe_{total} ratio of approximately 0.05 to 0.2

[vi] i.e, formic acid

[vii] The "slurry feed" consisted of a continually mixed blend of SBW surrogate, glass formers, and reductant

Figure 1. Redox Vs. Sugar/Nitrate

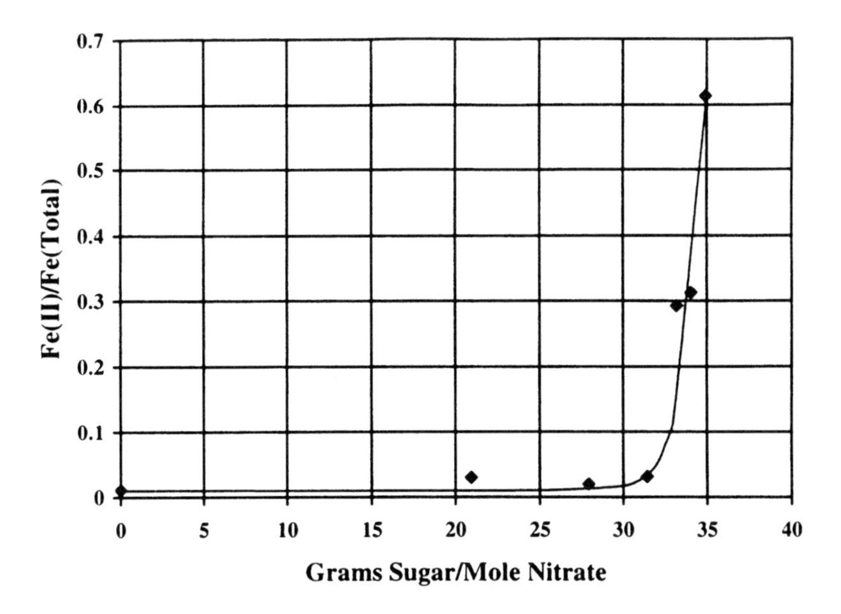

Figure 2. Redox Vs. Glycolic Acid/Nitrate

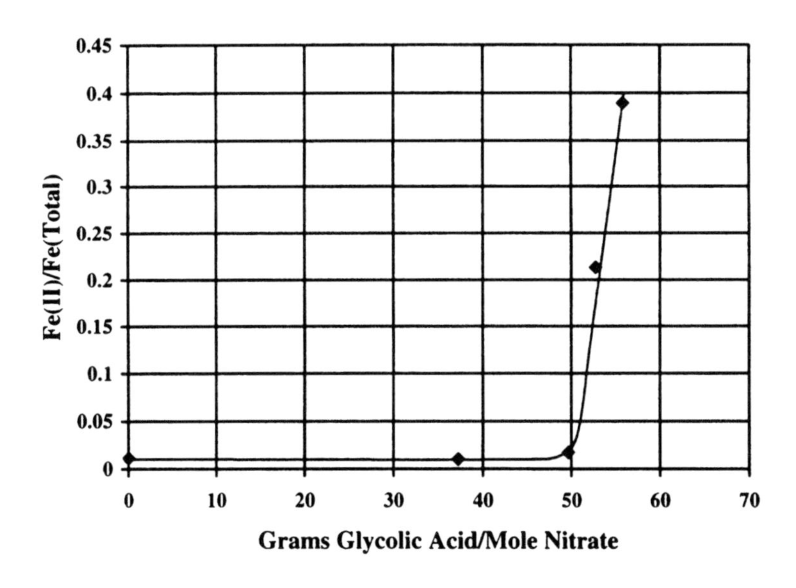

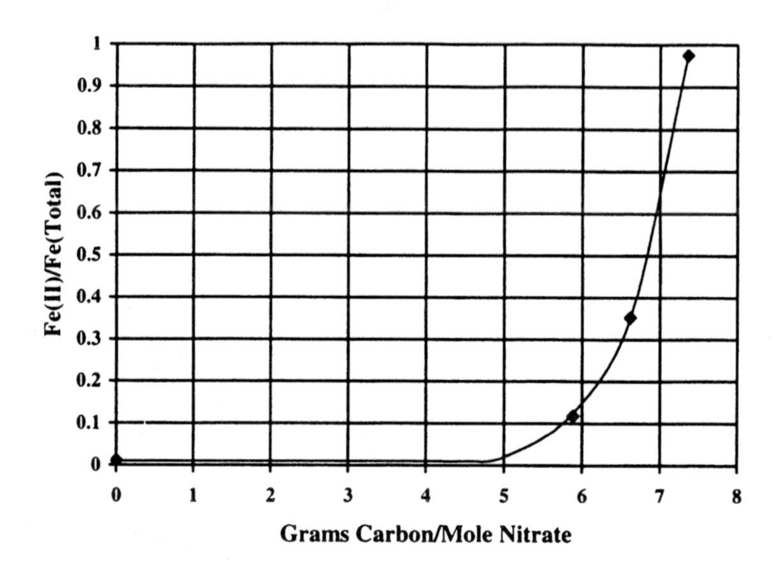
Figure 3. Redox Vs. Activate Carbon/Nitrate
Fe(II)/Fe(Total)
1
0.9
0.8
0.7
0.6
0.5
0.4
0.3
0.2
0.1
0
0
1
2
3
4
5
6
7
8
Grams Carbon/Mole Nitrate

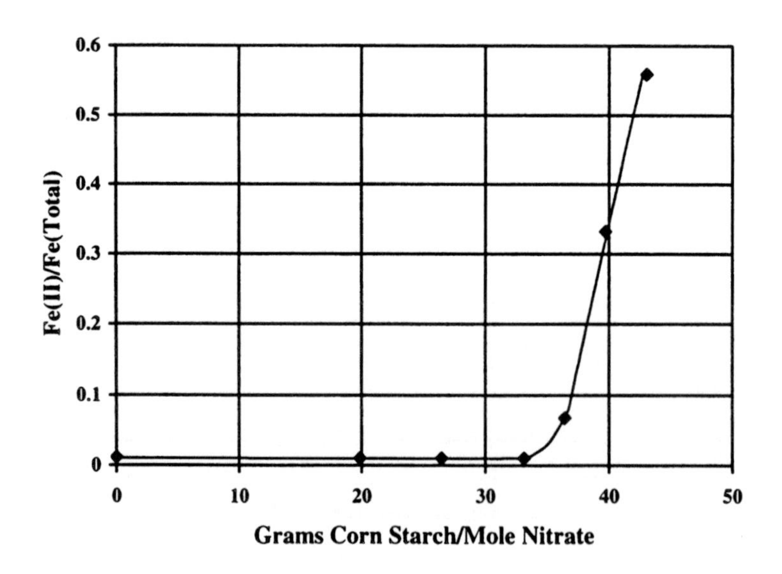
Figure 4. Redox Vs. Corn Starch/ Nitrate
Fe(II)/Fe(Total)
0.6
0.5
0.4
0.3
0.2
0.1
0
0
10
20
30
40
50
Grams Corn Starch/Mole Nitrate

With respect to oxidation of the reductants to carbon dioxide and water, the nominal equivalent weight of these reductants[viii] is 3 for activated carbon, 7.1 for sucrose and corn starch, and 12.7 for glycolic acid. The amounts actually needed to reach the "breaks" depicted in Figures 1-4 were ~6.2 g/mole nitrate for activated carbon; ~33 g/mole nitrate for sucrose, ~36 g/mole nitrate for corn starch, and ~51 g/mole nitrate for glycolic acid. In terms of the nominal number of equivalents of the reductant required per mole (of nitrate), these figures correspond to a 2.07 eq/mole for activated carbon, 4.65 eq/mole for sucrose, 5.07 eq/mole for corn starch, and 4.02 eq/mole for glycolic acid. The reason for the relatively high "efficiency" of activated carbon is that a higher percentage of the nitrate was thermally driven off in the form of NO_x (NO and NO_2), not reduced to N_2 and/or N_2O. This increased NO_x production and the fact that a greater temperature drop was noted in the muffle furnace during the drying-out step when using activated carbon, verifies that it is much less reactive than the other reductants.

Sulfate Salt Control

The high concentrations of sulfate salts in SBW raises concern about its effect on waste loading. Because the presence of a sulfate salt layer on the glass melt surface is deemed unacceptable, formulation development has been directed with avoidance of such a layer as a key criterion. Although a 30% waste loading glass formulation (based on metal oxide equivalents) has been developed for SBW, subsequent waste characterization refinement suggests that the sulfate concentration is likely to be about 40% higher than originally anticipated. Essentially all vitrification development work, to date, has been performed using SBW surrogate at the lower sulfate concentration.

Although the glass literature indicates that sulfate can be reduced to gaseous SO_2 in the melter, it is identified as solely a thermal process[7]. Conversely, sulfate reduction to a metal sulfide is referred to only as a chemical reaction. Although by no means conclusive at this point in time, some of the RSM melter run results

---

[viii] The nominal equivalent weight of an organic reductant is determined by dividing its molecular weight by the total charge in oxidation state undergone by its carbon moiety during the assumed reaction. The calculation of a reductant molecule's carbon oxidation state assumes that it is electrically neutral and that each of its oxygens has a charge of –2 and each of its hydrogens a charge of +1. For example, the mean oxidation state of each of the two carbon atoms in glycolic acid (C2H4O3, MW=76) is +1. Thus, 6 electrons must be taken from the respective two carbon atoms to convert them to carbon dioxide. Consequently, that reductant's equivalent weight = 76/6 or 12.7.

suggest that chemical reduction of sulfate to SO_2 may occur. If so, accurate redox control could allow for an increase in waste loading by driving sulfur to the offgas.

As a sulfate salt layer was present on the surface of virtually all SBW crucible samples, that approach is inappropriate for sulfate solubility determinations. During the RSM run, the waste loading was gradually increased, from 30 to 35 wt% (again, based on metal oxide equivalents), corresponding to final glass sulfur concentrations from 0.57 to 0.66 wt%. At all times during the run, at least some salt was present on the melt surface. Based on visual observations of sulfate salt layer volume changes, it was decided that a waste loading of 30-32 wt% would be the recommended maximum. However, the final several hours of operation were conducted using a much higher feed sulfur concentration (that corresponding to final glass sulfur concentrations of approximately 0.92 wt%), and a sucrose concentration increased to approximately 200 g/l[ix]. Although the sucrose concentration was determined to be too high (too reducing) as evidenced by a final Fe^{2+}/Fe_{total} ratio of 0.45[x], the fact that the amount of sulfate salt on the final glass surface was small suggests that some sulfate may have volatilized (as SO_2) due to chemical reduction. Further investigation into potential sulfur volatilization via redox control is planned.

CONCLUSIONS

Several conclusions can be drawn from the test results to date:

1. Activated carbon is likely not a preferred reductant for direct-feed SBW vitrification because its degree of reactivity is insufficient to reduce nitrate at lower temperaures.
2. The total absence of foaming or other operational problems during the RSM melter run leaves sucrose as the "baseline" reductant for SBW vitrification.
3. Because of the high nitrate concentration of SBW (typically 5 molar), a large amount of reductant will be required to eventually shift the redox state of the melt to within the acceptable range.
4. When the redox state of the melt finally begins changing, only a small amount of additional reductant will be required to attain an acceptable Fe^{2+}/Fe_{total} ratio in the melt.
5. Although not verified, the tests suggest that sulfur may be driven to the offgas if the redox state can be accurately controlled.

[ix] This is approximately 10% higher than the sucrose concentration just used, and 50% higher than the 135 g/l used at the beginning of the run

[x] The final glass redox was a result of only a few hours of operation at the above described conditions, and was not at equilibrium. This measured redox would have increased further had the feeding not been terminated.

6. Due to the extreme slopes of the redox curves, significant deviations in glass redox can result from minor differences in operational parameters or procedure.

REFERENCES

[1]A. L. Olson, et al, "Pre-Decisional Sodium Bearing Waste Technology Development Roadmap," INEEL/EXT-2000-01299, September 2000.

[2]T. G. Garn, internal INEEL memorandum to J. A. Rindfleisch, "Sampling and Characterization of WM-180," January 11, 2001.

[3]N. Soelberg, et al, "Test Plan for Vitrification Demonstration Tests of INEEL Sodium-Bearing Waste," PNNL/INEEL, January 2001.

[4]R. W. Goles, et al, "Test Summary Report INEEL Sodium Bearing Waste Vitrification Demonstration RSM-01-1," PNNL-13522, Pacific Northwest National Laboratory, Richland, WA (2001).

[5]V. Jain and Y. Pan, "Glass Melt Chemistry and Product Qualification," CNWRA 2000-05, September 2000.

[6]D. D. Siemer and J. A. McCray, "Real Time Determination of the Redox State of Glasses- Direct Potentiometry vs Chemical Analysis," in this volume.

[7]J. L. Ryan, "Redox Reactions and Foaming in Nuclear Waste Glass Melting," PNL-10510, Pacific Northwest Laboratory, Richland, WA (1995).

THE COLD CRUCIBLE MELTER: HIGH-PERFORMANCE WASTE VITRIFICATION

Antoine Jouan, Roger Boen
and Jacques Lacombe
CEA, Valrhô–Marcoule, BP 17171
30207 Bagnols-sur-Cèze
France

Richard Do Quang
COGEMA, 1 rue des Hérons
78182 Saint Quentin en Yvelines
France

Thierry Flament and Guillaume Mehlman
SGN, 1 rue des Hérons
78182 Saint Quentin en Yvelines
France

ABSTRACT

The idea of heating glass directly by Joule effect arose in response to one of the major concerns of the nuclear power industry, which is to limit the temperature of the containment materials in order to maximize their service life and minimize the production of secondary waste.

Glass is generally heated by Joule effect using electrodes in the furnace chamber or immersed directly in the melt. Although this classic design has been adopted in many countries to vitrify nuclear waste, its major disadvantage when the glass is melted inside a refractory vessel is the deterioration and ultimate disposal of the melter, which constitutes a cumbersome wasteform.

Joule heating can also be implemented through direct induction in the glass by means of a suitable electromagnetic field. In this case, the glass may also be contained in a cooled structure that is transparent to the field: this is the operating principle of the Cold Crucible Melter (CCM), which allows direct Joule heating of the glass within a solidified glass shell formed on contact with the cooled melter wall. The CEA and COGEMA have developed melters of this type to fabricate glasses and glass-ceramics, initially for the vitrification of high-level fission product solutions. This technique has progressively been applied to reactor waste, including solid and liquid materials, both organic and inorganic. It is capable of melting, refining and pouring glasses of all types at unlimited temperatures with virtually no melter corrosion. The devices now in operation can reach glass fabrication rates of

To the extent authorized under the laws of the United States of America, all copyright interests in this publication are the property of The American Ceramic Society. Any duplication, reproduction, or republication of this publication or any part thereof, without the express written consent of The American Ceramic Society or fee paid to the Copyright Clearance Center, is prohibited.

several hundred kilograms per hour, and there are no obvious technical obstacles to the design of units capable of melting one metric ton or more per hour.

Several test facilities are now available today at Marcoule, in France, to vitrify organic or inorganic solid and liquid waste. COGEMA is considering the use of this type of melter in one of its vitrification units at La Hague, and other industrial applications are planned in Europe and Asia. Cold crucible melters could unquestionably be used very competitively to vitrify the thousands of cubic meters of solutions and tons of fission product calcines now stored in the United States.

INTRODUCTION

The vitrification of fission product solutions—whether of defense or civilian origin—addresses two fundamental objectives: containment of the radioactive material in a stable and durable matrix, and volume reduction. This approach is widely acknowledged in the international scientific community, and has been implemented in the major industrialized nations.

Although various methods are used in different countries, there are two fundamental processes:

- The continuous process used in France and Great Britain was the first to be developed industrially (1978). It produces the final glass wasteform in two steps: the feed solution is first evaporated and calcined in a rotating kiln, then melted with primary glass frit at 1150°C in an induction-heated metal vessel and poured into a stainless steel canister[1].
- The single-step continuous process in which solution evaporation and glass melting occur within the same device, a ceramic melter heated by Joule effect using metal electrodes that limit the process temperature to 1150°C. This process was first implemented at Mol (Belgium) in 1985[2] and has been used in the USA since 1996 at the West Valley and Savannah River sites.

The 1150°C glass melting temperature limit, and the quest for more reliable and durable melting facilities spurred research to design melting devices capable of meeting specific requirements and of operating at higher temperatures. This effort led to the development of the Cold Crucible Melter in France.

FRENCH VITRIFICATION EXPERIENCE

France has acquired considerable experience in nuclear waste vitrification, as demonstrated by the performance of the vitrification facilities run by the French nuclear industrial operator, COGEMA. Beginning with the AVM, commissioned in 1978, the design of the COGEMA vitrification plants (R7 and T7 at La Hague) has been continually improved through innovation and through the incorporation of operating experience. The facilities now in operation present outstanding availability, safety and product safety records, having produced over 4000 metric tons of glass incorporating 3.5 billion curies.

France has maintained a research and development staff of about a hundred engineers and technicians in this area, jointly funded by the *Commissariat à l'Énergie Atomique* and by COGEMA, that has developed the cold crucible melter technology since the beginning of the 1980s.

THE COLD CRUCIBLE MELTER: A WARM HEART IN A COLD BODY

In the Cold Crucible Melter (CCM) concept, the glass is heated by Joule effect arising from currents induced directly in the material by an external inductor. Because the heat is not transmitted by conduction the crucible can be cooled—hence the name "cold crucible".

The technique is based on the use of a water-cooled structure that is transparent to the electric field produced by an induction coil surrounding it; this allows currents to be generated inside the material contained in the structure. The diagram in **Figure 1** shows a sectorized metal structure that allows currents to flow in each sector—notably on the inner face, creating a nonzero electric field that induces currents in the material and thus heats it by Joule effect. The process material is thus molten at the core, and solidified on contact with the cooled melter wall.

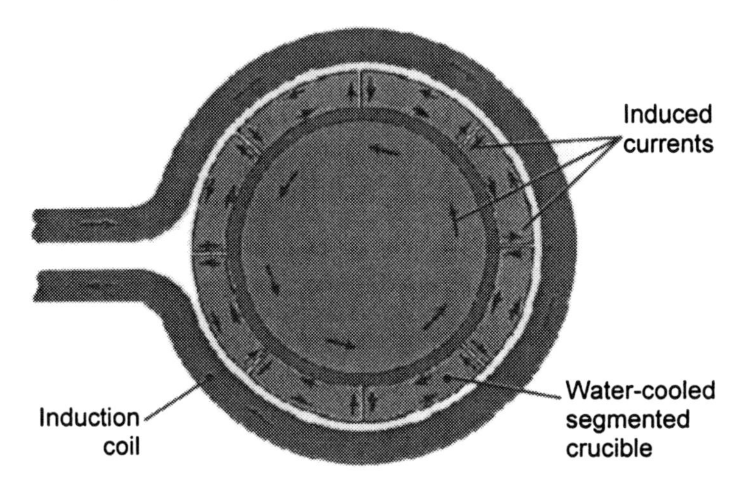

Figure 1. CCM operating principle

Schematically, the current penetration depth p depends primarily on the resistivity ρ of the material and on the current frequency F:

$$p = k\sqrt{\frac{\rho}{F}}$$

where k is a constant depending on the units used.

The penetration depth should be roughly equal to the crucible radius for maximum efficiency. For example, in the case of a glass melt with a resistivity ρ of

0.05 Ω·cm, p is equal to 0.2 m at a frequency of 300 kHz; these conditions are satisfactory for a melter 0.5 to 1 m in diameter.

Although the following discussion will focus on glass melting, it is important to note that the CCM technique is also capable of heating and melting metals or metal alloys of all types. Through the use of suitable operating frequencies and voltages, it is capable of inducing currents in gaseous plasmas to obtain veritable electric induction burner.

The salient feature of the process lies in the fact that all the components may be cooled. The molten material at the center of the crucible thus solidifies near the melter walls, which never exceed temperatures of about 200°C. This has a number of advantages:

- The crucible is protected against any risk of corrosion, ensuring a very long operating lifetime and allowing it to be used to process a wide range of materials containing even highly corrosive elements: glass with high P_2O_5 and even SO_3^- concentrations may be produced without difficulty. A glass melter has been used at Marcoule for more than 15 years, during which it has produced a broad spectrum of glass compositions and highly corrosive molten salts, and yet it still appears virtually new today.
- Extremely high—practically unlimited—core temperatures may be reached in the process material. Crucibles of this type have been used to melt uranium and zirconium oxides at temperatures exceeding 2500°C.
- A cold-crucible melter may be easily dismantled. Neither glass nor metal adheres to the cooled wall, which is therefore never subject to heavy high contamination.

The melter is highly compact, considering its ability to generate extremely high power densities in the glass—much higher than in an electrode furnace, where the power density is limited by electrode wear. Cold crucible melters are thus particularly well suited for obtaining high throughput in a small volume, and extend the range of potential compositions for solidifying fission product solutions or any other type of waste for which this type of treatment is applicable.

It is obvious, of course, that the presence of molten glass in contact with a water-cooled structure will have an impact on the power consumption of the process. This effect is mitigated, however, by the existence between the molten glass and the crucible of a solidified layer of glass, whose low thermal conductivity limits the heat flow toward the wall. The overall mean thermal balance can be expressed as follows, as percentages of the power supplied to the unit:

Dissipated by the current flowing in the inductor:	10%
Dissipated by the current flowing in the melter walls:	20%
Losses via heat flow to the melter wall:	40%
Dissipated in the glass melt:	30%

In other words, depending on the melter dimensions and design constraints (notably for nuclear applications), the power supply requirement is about 1.5 to

2 kWh per kilogram of glass, compared with 1 kWh in a conventional industrial melter.

COLD CRUCIBLE OPERATION: FROM STARTUP TO POURING

Induction heating is not possible directly in cold glass; Joule effect is observed only in the molten state. The glass must be at least partially melted before starting the process. Any method can be used for this purpose, from the simplest to the most sophisticated: preheating furnace, electrical resistors, burners, microwaves, metal susceptors, etc. The simplest method generally involves the use of a bar made of a reducing metal (Ti or Zr); when submitted to an electric field, the metal is heated and burns, and the energy released by the combustion reaction then heats and melts the glass.

If appropriate, the melt can also be stirred (by sparging or mechanically, using a water-cooled device), for example to prevent possible settling of insoluble platinum-group metals found in the fission product solutions. A water-cooled valve was developed to pour all or part of the glass from the melter into a canister.

THE COLD CRUCIBLE: AN OMNIVOROUS MELTER

The very concept of the cold crucible not only allows very high-temperature operation, but also ensures corrosion-free operation. The induction-heated CCM is compatible with numerous operating scenarios, and its capacity depends to a large extent on the selected conditions and on the materials in the feed stream.

The CCM is suitable for melting and pouring a very wide range of organic or inorganic solids and liquids, with or without suitable additives and for the complete destruction of organic materials.

Inorganic Materials

This category includes most fission product solutions. Numerous tests were carried out[3] in which the melter was supplied directly with the feed solution, together with glass frit (generally in solid form): some with simulated fission product solutions of the type generated in the La Hague plants and used to produce R7/T7 glass, and others with more specific high- and low-level liquid waste surrogates corresponding to the Idaho and Hanford compositions. All the tests showed that a liquid-fed melter could be operated at temperatures of 1300°C and higher, at rates well above 100 $L{\cdot}h^{-1}$ per square meter of melt surface area, producing in excess of 100 kg of glass per hour.

A cold crucible melter can also be supplied directly with solid inorganic materials, as demonstrated by CEA and COGEMA experience in this area with melting and pouring a wide variety of molten salts, glasses, glass-ceramics and ceramic materials, including basalt (melted at over 1500°C), Synroc (1600°C) and asbestos (1800°C).

Organic Material Destruction

Among the organic materials tested are ion exchange resins and plastics, both of which arise during the normal operation of nuclear power plants. Loading them on the surface of a glass melt in excess oxygen results in the combustion and complete destruction of the organic matter, which is converted to CO_2 and H_2O; the residual inorganic compounds are incorporated into the melt, the formulation of which is tailored to obtain the desired final composition. Here again, the CCM is particularly well suited for use with materials liable to contain highly corrosive materials.

PILOT–SCALE FACILITIES IMPLEMENTING CCM TECHNOLOGY

Several test platforms have been built in the Marcoule pilot facility since the 1980s in order to develop the Cold Crucible Melter Technology. These programs initially focused on processing HLW solutions from light water reactor fuel, producing simulated R7/T7 glass.

The oldest platform is a stand-alone CCM pilot that can be fed with simulated calcine and glass frit. The 550 mm diameter stainless steel melter can also be liquid fed. This platform, equipped with a 300 kW generator, has logged more than 5 000 hours of operation over a 15-year period. Approximately 50 metric tons of simulated HLW glasses have been produced. Since the beginning of the CCM programs, most of the process demonstrations have been performed on this platform with solid and liquid feed.

The second platform is a full-scale mockup of the R7/T7 vitrification process that can be equipped with a hot crucible melter or a 650 mm diameter CCM (with a 300 kW generator). In either case, the melter can be coupled to a calciner and fed with simulated calcine and glass frit. This platform is currently used to qualify the CCM process as it will be implemented in the R7 facility.

A third platform called EREBUS is dedicated to the more recent applications. It is equipped with a smaller 160 kW generator able to operate at frequencies ranging from 200 to 500 kHz. It can be operated with melters of various diameters up to 1 000 mm. The selection of the diameter will depend on the type of test to be performed, allowing for the limited available power. The feed systems allow simultaneous controlled feeding of solids (frit, powders) and liquids (surrogate solutions, sludges). The off-gas system is composed of a condenser, an acid recombining/washing column before the extraction device.

A fourth platform is used for the development of the combustion/vitrification process for organic compounds such as the previously mentioned ion exchange resins and plastics. The stainless steel CCM used on this platform has a diameter of 300 mm and is powered by a 240 kW generator. In order to finalize this development, a pilot industrial facility (equipped with a 550 mm crucible) has been built in collaboration with KEPCO/NETEC at Taejon in Korea[4].

THE ADVANCED COLD CRUCIBLE MELTER

Given the excellent results demonstrated by the CCM technology, COGEMA and the CEA have been developing an advanced cold crucible melter (ACCM) since 1995. The operating principles of the ACCM are similar to those of the CCM, but the design has been optimized to increase the capacity.

The ACCM retains all of the advantages of the CCM technology: flexibility with respect to waste composition, mechanical stirring of the melt, virtually unlimited equipment service life, modular design and reduced equipment size. A liquid-fed ACCM with a glass production capacity of 100 $kg \cdot h^{-1}$ would be about 2 meters in diameter and 2 meters high (taking into account the inductor, cooling systems and dome apparatus) and would weigh a few tons. These simple figures illustrate the substantial cost-savings associated with the ACCM technology in comparison with the LFCMs.

In support of the ACCM development program for commercial applications, COGEMA and the CEA have built a new one-half scale test platform at Marcoule. It comprises a CCM measuring 1.4 m in diameter (including the inductor) and a 650 kW induction generator. The design glass production capacity is 200 $kg \cdot h^{-1}$ with solid feed, or 50 $kg \cdot h^{-1}$ with liquid feed. The platform is scheduled to begin operation in the second quarter of 2001.

COMMERCIAL APPLICATIONS

The *Compagnie Française d'Électrothermie Industrielle* (CFEI) markets the CCM technology in non-nuclear fields. CCM technology-has-been used for non-nuclear applications since 1995, for example to produce high added-value glasses or enamels, with two melters. Because of the protection provided by the cold glass layer, glasses or enamels can be melted at high temperature with no pollution from the materials of the wall as in traditional glass melters. It is also possible to switch glass compositions in less than 8 hours since glass does not adhere to the cooled walls. In 1999, 500 tons of industrial glass of varying compositions were produced with a single 1200 mm diameter CCM. COGEMA is also providing the CCM Technology to international customers for nuclear applications.

COGEMA is considering the deployment of this technology at La Hague, on one existing vitrification line, to process specific corrosive, high-viscosity material in the near future. For this application, the advantage of high temperature has been fully used by raising the target processing temperature from 1150°C to around 1350°C, thus allowing the selection of a new matrix resulting in an overall glass volume reduction by a factor of 3. The process and its ancillary technologies (pouring valve, instrumentation, etc.) have been qualified on the corresponding full-scale platform at Marcoule.

A CCM coupled with a calciner has been proposed for the Hanford TWRS-P Phase IA HLW studies. A large demonstration program, including the production of about 3 metric tons of surrogate glass in a pilot unit, provided confidence in the process and pointed to some possible advantages of further extending the range of test conditions. More specifically, it was found that the technology showed a potential for significantly higher waste loading, thus reducing the volume of HLW glass for disposal.

This technology is also being supplied to foreign customers, in various configurations:

- CCM with direct liquid feeding to process legacy HLW in Italy[5];
- CCM with direct liquid or solid (resin) feeding for simulated reactor waste in Korea[4].

CONCLUSION

The CEA, COGEMA and its subsidiary SGN have incontestably demonstrated their expertise in the area of high-level waste vitrification for over twenty years. The CCM technique combines a range of favorable performance features—suitability for nuclear industrial environments, ease of maintenance, resistance to corrosion, high-temperature melting capability—that make it particularly well suited to meet the requirements emerging around the world for vitrification of nuclear waste of all types.

Compared with the other techniques available today, the CCM and its advanced ACCM concept capable of reaching high throughput capacities provides a cost-effective answer to the cleanup challenges facing the world today and tomorrow.

REFERENCES

[1]A. Jouan, T. Flament and H. Binninger, "Vitrification Experience in France, Development and Perspectives", *International Workshop on Glass as a Waste Form and Vitrification Technology*, Washington DC, May 13-15, 1996.

[2]G. Hohlen, E. Titmann, S. Weisenburger and H. Wiese, "Vitrification of High Level Radioactive Waste, Operation Experience with the Pamela Plant", *Waste Management 1986*, Tucson, Feb. 1986.

[3]C. Ladirat, C. Fillet, R. Do Quang, G. Mehlman, T. Flament, "The Cold Crucible Melter: a Key Technology for the DOE Cleanup Effort", *Waste Management 2001*, Tucson, Feb. 2001.

[4]Dr. Song, M. You, T. Flament, and C. Brunelot, "The Cold Crucible Melter Vitrification Pilot Plant: a Key Facility for the Vitrification of the Waste Produced in the Korean Power Plants", *Waste Management 2001*, Tucson, Feb. 2001.

[5]C. Calle and A. Luce, "CORA Projet: the Italian Way for Conditioning High and Low Level Liquid Radioactive Wastes", *ICEM 97*, Singapore, Oct 12-16, 1997.

MILLIMETER-WAVE MONITORING OF NUCLEAR WASTE GLASS MELTS – AN OVERVIEW

P. P. Woskov, J. S. Machuzak, and P. Thomas
Plasma Science and Fusion Center
Massachusetts Institute of Technology
Cambridge, MA 02139

S. K. Sundaram
Pacific Northwest National Laboratory
Richland, WA 99352

William E. Daniel, Jr.
Westinghouse Savannah River Company
Aiken, SC 29808

ABSTRACT

Molten glass characteristics of temperature, resistivity, and viscosity can be monitored reliably in the high temperature and chemically corrosive environment of nuclear waste glass melters using millimeter-wave sensor technology. Millimeter-waves are ideally suited for such measurements because they are long enough to penetrate optically unclear atmospheres, but short enough for spatially resolved measurements. Also efficient waveguide and optic components can be fabricated from refractory materials such as ceramics. Extensive testing has been carried out at a frequency of 137 GHz to temperatures up to 1500 °C. Performance of refractory waveguides at high temperature has been shown to be satisfactory. A novel new method for viscosity monitoring has also been tested with simulated nuclear waste glasses. It has been shown that a viscosity range of over 30 to 3000 Poise can be monitored with one instrument. Results of these laboratory tests and the potential of millimeter-wave sensors for on-line glass process monitoring are presented.

INTRODUCTION

The vitrification and long term storage of high level and low activity nuclear waste represents one of the major challenges of the U. S. DOE Environmental

To the extent authorized under the laws of the United States of America, all copyright interests in this publication are the property of The American Ceramic Society. Any duplication, reproduction, or republication of this publication or any part thereof, without the express written consent of The American Ceramic Society or fee paid to the Copyright Clearance Center, is prohibited.

Management's effort to clean up nuclear waste sites. Optimization of the vitrification process through improved process control could significantly reduce costs associated with this effort. Improved process control would result in increased glass manufacturing efficiencies, reduced storage volumes through increased waste loading, and reduced storage risk by insuring the long stability quality of the poured glass product.

Present nuclear waste melters such as the Defense Waste Processing Facility (DWPF) and planned melters for the Hanford site lack sophisticated diagnostics due to the hot, corrosive, and radioactive environments of these melters. New robust diagnostics are needed that can operate on-line in these environments. Parameters important for monitoring include temperature profiles of the glass melt, cold cap, plenum off gases, molten glass conductivity, density, and viscosity. With real time monitoring of these parameters it should be possible to implement feedback control of the vitrification process to achieve significant cost savings.

Much of the desired on-line monitoring capabilities can be realized through the use of millimeter-wave (MMW) technologies. Electromagnetic radiation in the 10 – 0.3 mm (30 – 1000 GHz) range of the spectrum is ideally suited for remote measurements in harsh, optically unclean and unstable processing environments. Millimeter waves are long enough to penetrate optical/infrared obscured viewing paths through dust, smoke, and debris, but short enough to provide spatially resolved point measurements for profile information. Another important advantage is the ability to fabricate efficient MMW melter viewing components from refractory materials. Application of MMW pyrometry to a dc arc furnace melter has been previously demonstrated [1].

MILLIMETER-WAVE SENSOR APPROACH

The millimeter-wave sensor approach is illustrated in Figure 1. A sensitive MMW receiver is used to view the molten gas via a quasi-optical transmission line. The receiver both receives signals from the glass and transmits a MMW probe signal to monitor reflections. The transmission line makes use of waveguides and can have one or more optical elements such as a mirror. The transmission line is highly efficient making possible remote placement of the MMW receiver outside of any biological shield around the melter facility. Profile measurements are made possible by appropriate design of the waveguide components inside the melter for rotation and/or translation [1].

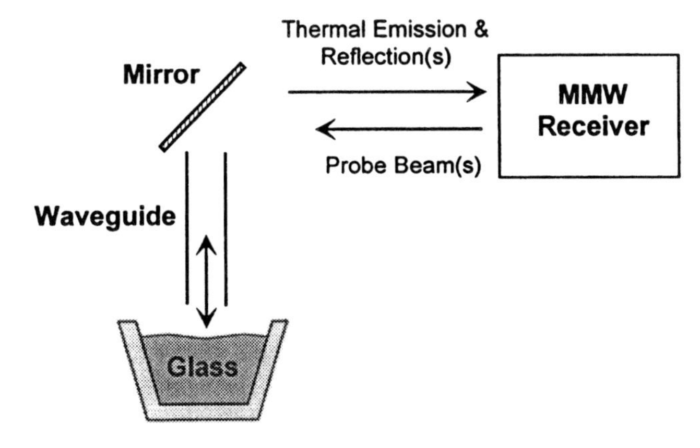

Figure 1. Millimeter-wave sensor configuration

The physical molten glass parameters that would be measured and the corresponding millimeter-wave phenomena that are exploited to achieve these measurements are listed in Table I. Temperature is determined by detecting the thermal emission from the viewed glass. The reflectivity of the glass, which can be used to determine the glass electrical resistively or conductivity [2] is obtained from measuring the millimeter-wave reflection amplitude. Glass density is determined by measuring reflection phase change in response to a pressure displacement. The reflection phase change gives the glass fluid displacement, which depends on density for a given pressure. Viscosity is then determined from the rate of phase change in response to a given pressure transient.

Table I. MMW Measurements

Parameter	Measured Effect
Temperature (T)	Thermal emission
Reflectivity (r)	Reflection amplitude
Density (ρ)	Reflection phase
Viscosity (η)	Reflection phase rate

BASIS FOR TEMPERATURE MEASUREMENT

The temperature of the viewed molten glass is given by the thermal emission signal. Unlike commercial pyrometers, which operate in the infrared portion of the electromagnetic spectrum, millimeter-wave thermal emission is linear with temperature. It is given in by the blackbody emission formula for a single polarization in this wavelength limit as [3]:

$$P = k_B T_{eff} d\nu \qquad (1)$$

where P is in Watts, k_B is Boltzmann's constant, T_{eff} is the absolute temperature in Kelvin that fills the receiver antenna, and $d\nu$ is the receiver bandwidth in Hertz. The receiver signal is linearly proportional to T_{eff}, which facilitates calibration and gives a MMW pyrometer much larger dynamic range than possible in the infrared.

The effective temperature that fills the receiver antenna depends on the efficiency of the transmission line and the emissivity of the sample of molten glass that is viewed. All sources and losses of thermal radiation need to be considered when determining T_{eff}. They are identified in Figure 2. T_{eff} is given by the formula:

$$T_{eff} = \varepsilon_{wg} T_{wg} + \tau_{wg} \varepsilon_s T_s + r_s \tau_{wg} \varepsilon_{wg} T_{wg} \qquad (2)$$

where the first term on the right is the thermal emission from the waveguide, the second term is the thermal emission from the glass, and the finial term is the reflected thermal emission from the waveguide, where ε_{wg} is the waveguide emissivity, ε_s is the sample glass emissivity, τ_{wg} is the waveguide transmission factor, r_s is the sample glass reflectivity, and T_{wg} and T_s are the temperatures of the waveguide and sample glass, respectively.

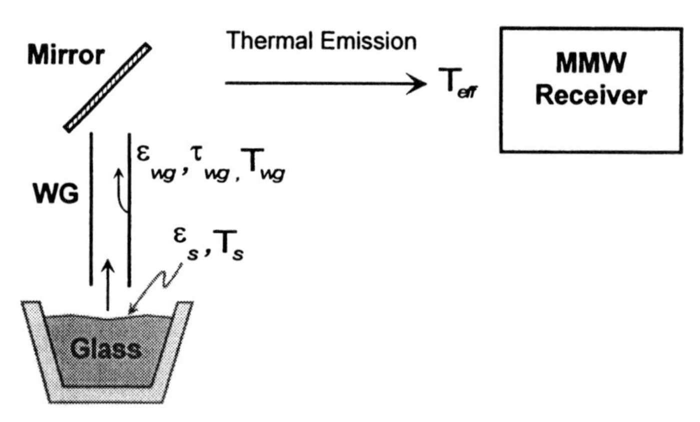

Figure 2. MMW receiver temperature measurement

It can be shown that the waveguide emissivity and transmission factor, and the sample glass emissivity and reflectivity are related by the expressions:

$$\varepsilon_{wg} = (1 - \tau_{wg}) \quad (3)$$

$$\varepsilon_s = (1 - r_s) \quad (4)$$

When the transmission line has no losses ($\tau_{wg} \approx 1$) and the sample glass is near a prefect blackbody ($\varepsilon_s \approx 1$), the receiver signal is representative of the glass temperature, $T_{eff} \approx T_s$. In other cases the other parameters in Equation 2 must be considered in determining temperature.

TECHNIQUE FOR REFLECTION MEASUREMENT

In most cases it is important to know the glass reflectivity (emissivity) not only to determine temperature accurately, but also to provide information of the electrical resistivity of the glass. A MMW reflection measurement is accomplished by redirecting a portion of the thermal signal back at the viewed sample. Using the broadband thermal signal as a probe of reflectivity avoids the standing wave effects of a coherent probe beam [1].

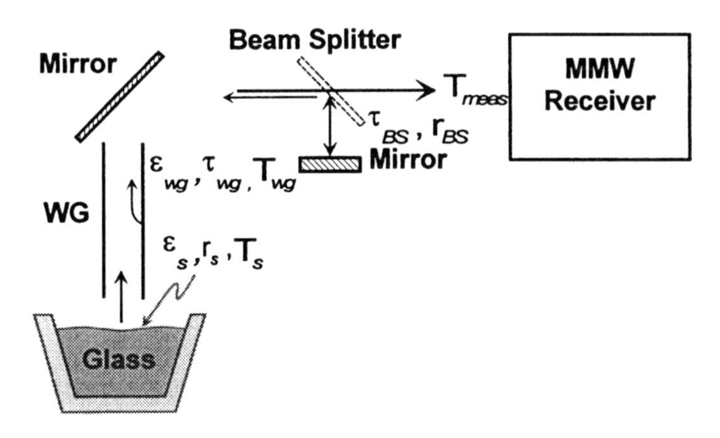

Figure 3. Thermal return reflection setup for reflectivity measurements

A thermal return reflection measurement is implemented by using a beam splitter in front of the receiver to divert some of the thermal signal from the melter to a side mirror as shown in Figure 3. When the side mirror is blocked to prevent reflection, the receiver signal is just the temperature signal, T_{eff}, reduced by the beam splitter transmission factor, τ_{bs}. When the side mirror is unblocked to return a signal to the sample the receiver signal increases by an amount dependent on the sample reflectivity. The view of the sample is aligned so that a signal incident from the waveguide is reflected back up the waveguide. The measured temperatures with the side mirror blocked and unblocked can be expressed as:

$$T_{meas} = \tau_{bs} T_{eff} \tag{5}$$

$$T'_{meas} = \tau_{bs} T_{eff} + r_{bs} T_{eff} \left(1 - r_s \tau_{bs}^2 \tau_{wg}^2\right)^{-1} \tag{6}$$

By taking these two temperature measurements one with the side mirror blocked and another with this mirror unblocked it is possible to solve for the viewed sample reflectivity, r_s. This is expressed as:

$$r_s = \frac{1}{\tau_{BS}^2 \tau_{wg}^2} \left(1 - \frac{T_{meas}}{T'_{meas}}\right) \tag{7}$$

Equation 7 with Equations 4 and 2 along with knowledge of the waveguide and beam splitter transmission factors makes possible the determination of both temperature and emissivity of the viewed molten glass.

BASIS FOR POSITION AND FLOW MEASUREMENTS

A MMW receiver used in practice for temperature measurements is a heterodyne receiver, that is, it uses a MMW local oscillator to frequency down shift the thermal radiation for detection with a conventional diode detector. Molten glass position and flow measurements are accomplished by using the leaked local oscillator signal from the antenna toward the viewed sample. This is a coherent signal and consequently the reflected phase can be determined. Reflected phase is an indicator of the distance to the reflecting surface from the receiver. The phase will cycle through 180-degree phase shifts for every ¼ wavelength shift in distance between the receiver and reflecting surface. The phase is expressed as:

$$\delta = 4\pi \frac{\Delta y}{\lambda} \tag{8}$$

where Δy if the molten glass displacement and λ is the local oscillator wavelength. A quadrature receiver or amplitude cycling from a single mixer receiver could be used to detect this phase.

For the position and flow measurements the waveguide is immersed into the molten glass and sealed with a window at the other end (as shown in Figure 4). This allows the waveguide to be pressurized to control the position and flow of the glass inside the waveguide. The absolute displacement of the glass divided into the displacement pressure in terms of water displacement gives the glass density directly. Viscosity is determined by inducing a sudden pressure transient in the waveguide and monitoring the velocity of the molten glass flow as manifested by the rate of change of the reflection phase.

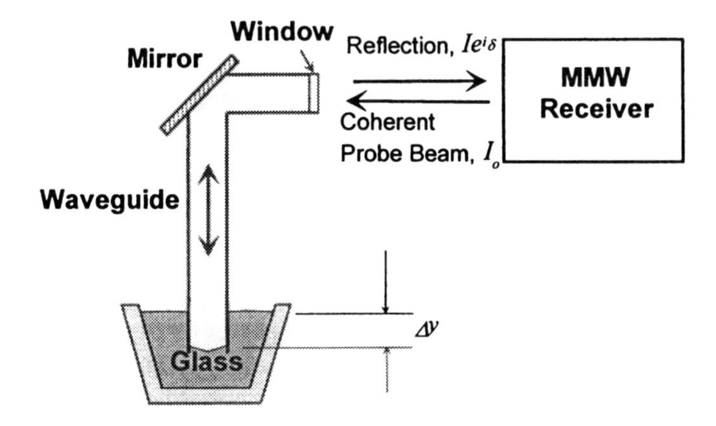

Figure 4. MMW position and flow measurement setup

REFRACTORY WAVEGUIDES

Efficient hollow millimeter-wave waveguides can be fabricated from refractory materials that can be used in high temperature environments. The transverse electric and magnetic field HE_{11} mode is the most efficient guided mode for electromagnetic propagation in hollow waveguides with internal diameter larger than a wavelength [4]. This mode is also an optimum for launching a diffraction-limited free space Gaussian beam, ideal for most diagnostic applications. The HE_{11} mode is a natural mode inside smooth walled dielectric tubes and can be achieved inside electrically conducting metallic tubes by circumferentially corrugating the inside wall with ¼ λ deep grooves at more than 2.5 grooves per λ of waveguide length. For typical metallic resistivities and dielectric indexes of refraction, smooth walled dielectric waveguides need to be much larger diameter to achieve the same transmission efficiencies as corrugated waveguides.

A number of different refractory waveguides were fabricated and tested at 137 GHz. Table II lists the types of waveguides tested, diameter, wall thickness, and measured transmission efficiency. Inconel 690 was corrugated with a 32/inch (1.26/mm) screw tap with V-shaped grooves. The measured losses of about 5.6% were higher than predicted by theory (< 1%) probably because the grooves did not machine well. After more than 30 hours of repeated use up to 1100 °C and then several hours up to 1180 °C the Inconel developed a black scaly oxide layer that increased transmission loses to 17%. Most of the oxidation occurred during the short period of use at the higher temperature.

Table II. Refractory Waveguides Tested at 137 GHz

Material	Inside Surface	Inside Diameter	Wall Thickness	Measured Transmission
Inconel 690	Corrugated	28.6 mm	N/A	94.6%
Inconel 690 *after use to 1180 °C*	"	"	"	83.0%
Vesuvius® Mullite	Smooth	41.3 mm	3.3 mm	97.6%
Coors® Mullite	Smooth	41.3 mm	6.4 mm	87.5%
Silicon Carbide	Smooth	41.3 mm	12.7 mm	79.5%

For higher temperature applications in oxidizing environments a ceramic, mullite ($3Al_2O_3{\cdot}2SiO_2$), and silicon carbide were evaluated. Mullite tubes were acquired from two vendors, Vesuvius McDaniel and Coors. The Vesuvius® mullite had additional impurities in it to aid in the manufacture of drawn tubing which made the material more opaque at 137 GHz, a desirable quality for the walls of a MMW waveguide. The Coors® mullite was a pure material with the waveguide manufactured by a powdered process. The refractive index for both types was determined to be about 2.5 at 137 GHz by interference measurements through flat plate samples.

The theoretically predicted transmission efficiency was 93% for these mullite waveguides. The measured transmission was higher for the Vesuvius® mullite and lower for the Coors® mullite suggesting that the walls were not thick enough to prevent interference form the outer waveguide surface in the Coors® sample. A silicon carbide waveguide from Carborundum of the same inner diameter as the mullite was also acquired. It was not a high resistivity dielectric and did not perform as well as the mullite.

LABORATORY TEST SETUP

Experimental millimeter-wave measurements of temperature, reflectivity, and glass flow were carried out on a crucible of molten glass inside an electrically heated furnace[1]. For these tests circular mullite waveguides were used, the 41.6 mm diameter Vesuvius® waveguide tested above was used for temperature and

[1] Deltech Model DT-31-RS-12

reflectivity measurements and a smaller 28.6 mm diameter Vesuvius guide was used for flow measurements.

The mullite waveguide was mated by a 90° steel miter bend to a circular corrugated brass waveguide of the same internal diameter. The refractory waveguide was inserted vertically into an electric furnace, as shown in Figure 5, where approximately two thirds of it was exposed to high temperatures. The brass part of the waveguide assembly, orientated horizontally above the furnace, facilitated the horizontal setup of the millimeter-wave receiver and antenna optics above and to the side of the furnace on an elevator platform. The elevator platform allowed raising and lowering the waveguide assembly with the electronics for waveguide immersion measurements into the glass.

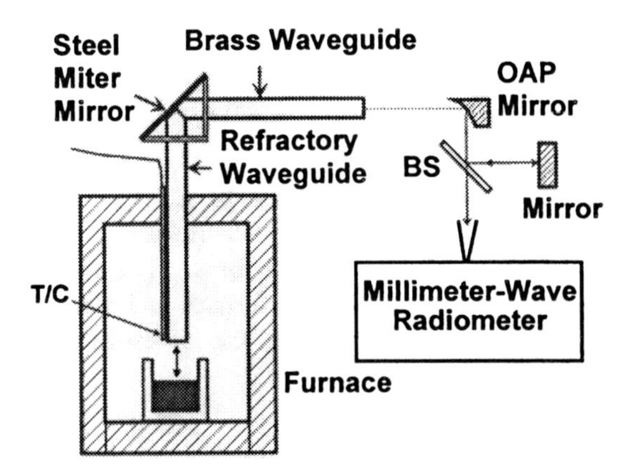

Figure 5. Experimental setup in the laboratory to test MMW senor methods

The receiver local oscillator frequency was 137 GHz with an intermediate frequency range of 0.4 – 1.5 GHz. The scalar horn antenna field-of-view was coupled to the waveguide assembly HE_{11} mode by an off-axis parabolic (OAP) mirror. A quartz beam splitter (BS) and a flat side mirror in the antenna field-of-view allowed the use of the furnace thermal emission as a probe of the emissivity of the viewed surface inside the furnace. An S-type thermocouple inside the furnace provided an independent measure of temperature.

EXPERIMENTAL TEMPERATURE AND EMISSIVITY MEASUREMENTS

The capability of millimeter-wave pyrometry to measure temperature and emissivity was tested by viewing various surfaces at high temperature inside the furnace. Figure 6 shows the results for viewing a ¼ inch (6.35 mm) thick plate of Coors® mullite resting on an alumina brick in the bottom of the furnace. The waveguide was lowered to within about 6 mm of the mullite surface to minimize diffraction of the return reflection.

The data is plotted as a function time in terms of the data acquisition count number and covers a period corresponding to several hours as the furnace was heated up, held at 1150 °C, and then cooled down. The top trace shows the thermocouple temperature and the middle trace shows the MMW pyrometer temperature, $T_{meas.}$ The pyrometer registers a lower temperature than the thermocouple because the waveguide has transmission losses and because the viewed mullite plate is not a perfect blackbody.

During the furnace temperature flat top at 1150 °C, the pyrometer side mirror was blocked and unblocked several times to provide several measurements of T'_{meas}. This part of the pyrometer temperature plot is shown expanded in the lower graph of Figure 6. The sharp dip in the pyrometer signal just after 10000 counts on the x-axis was due to blocking the antenna for calibration.

Using Equation 7 and the measured waveguide transmission given in Table 2, the reflectivity of the viewed mullite plate at 1150 °C was calculated to be $r_s = 0.10$. This corresponds to an emissivity of $\varepsilon_s = 0.90$. An alternative determination of emissivity can be made by using the measured thermocouple temperature for T_s in Equation 2 and solving for ε_s. The waveguide temperature can be approximated as a mean between the furnace and room temperatures. The emissivity computed in this way is plotted as the lowest trace in the upper graph of Figure 6.

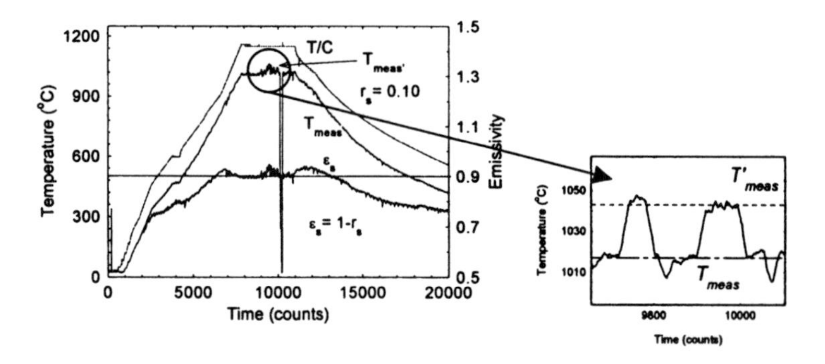

Figure 6. Measurements of temperature and emissivity of a mullite plate inside the furnace

At the temperature flat top the emissivity determined by the thermocouple and Equation 2 is in agreement with that calculated by Equation 7, as indicated by the straight line at $\varepsilon_s = 0.90$. During the furnace heat up and cool down the emissivity determined by Equation 2 is not reliable because of the different time responses of the thermocouple and MMW receiver. Also, the emissivity that is measured here cannot be used to determine the dielectric properties of mullite because the mullite plate is too thin in this case to be totally opaque to 137 GHz radiation.

The results of another test viewing an Inconel 690 plate inside the furnace are shown in Figure 7. As before the upper trace shows the thermocouple temperature and the lower trace is the measured millimeter-wave temperature. The millimeter-wave signal is very low because the Inconel plate is almost a perfect reflector. Applying the method of Equation 7 to the determination of emissivity results in a value of $\varepsilon_s = 0.04$ during the 1150 °C temperature flat top. This value is actually too high because Inconel does not change its resistivity with temperature enough to cause the emissivity to go measurably above zero at 137 GHz. The most likely reason that the calculated emissivity deviates from zero is because the waveguide transmission losses have increased at high temperature. An increase of about 2% in τ_{wg} would account for the observed millimeter-wave thermal emission.

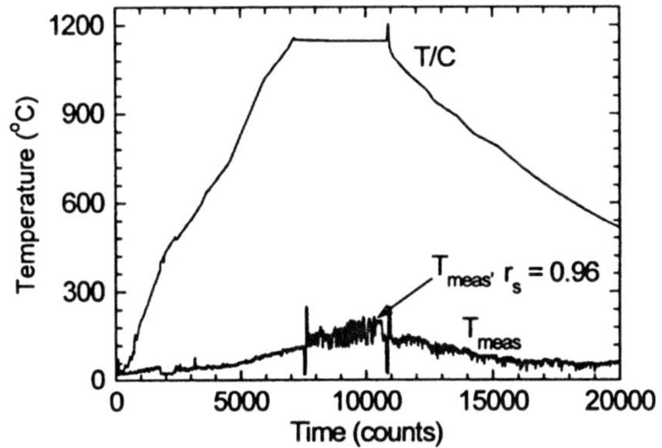

Figure 7. Temperature and emissivity measurements of an Inconel 690 plate

EXPERIMENTAL FLOW MEASUREMENTS

Molten glass flow measurements inside a sealed waveguide were carried out to test the feasibility of MMW viscosity measurements. After a few tests with the original 41.3 mm diameter mullite waveguide it was found that the glass flow rate was too fast to reliably measure viscosities below about 200 Poise due to the time response limits of the data acquisition system being used. The waveguide was changed to a 28.6 mm diameter mullite tube. With the smaller diameter waveguide viscosities down to about 20 Poise could be resolved.

The viscosities of the glass compositions used in these tests were obtained from a commercial laboratory[2]. MMW measured flow rates were then compared to these independently obtained viscosities. The flow rates were measured by immersing the waveguide into the glass and then pressurizing the waveguide to displace the molten glass a given distance. The pressure was then quickly released by triggering an exhaust solenoid valve and the time for the glass to flow back up the waveguide was recorded. This was done for more than one displacement and differences in flow time were used for comparison to viscosity to avoid surface tension effects. The furnace temperature was used to vary the glass viscosity.

A glass composition (Hanford #8) representing Hanford high-level waste (HLW) glass composition was chosen. The target composition (wt.%) is: SiO_2 = 45.14, B_2O_3 = 7.03, Na_2O = 20.00, Li_2O = 3.01, CaO = 1.17, MgO = 0.19, Fe_2O_3 = 14.56, Al_2O_3 = 6.02, ZrO_2 = 3.01, and others = 4.95. The results are shown in Figure 8. The known viscosity of the glass is plotted as a function of flow time difference between a 0.5 inch and 1.2 inch water pressure displacement. The points are experimental measurements and the curve is a straight line fit to the data on this log-log plot. The measured flow time tracks the viscosity perfectly over a 100:1 dynamic range of 20 to 2000 Poise. Work is currently in progress on viscosity measurements on other glasses and density measurements at high temperature are also being investigated.

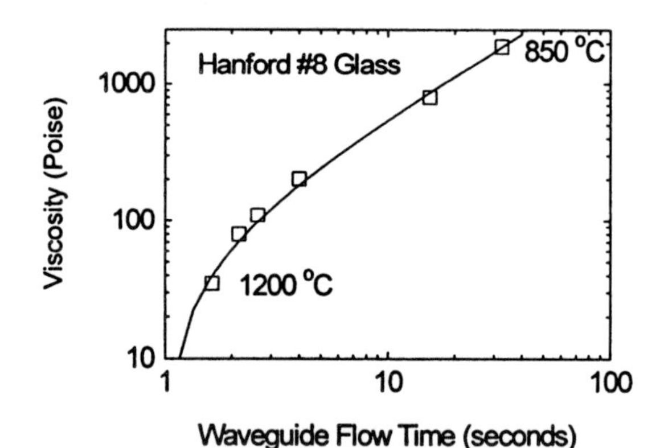

Figure 8. Viscosity versus measured MMW flow time inside a 28.6 mm diameter waveguide

[2] CELS – Corning Laboratory Services, Corning, NY 14831

SUMMARY

Significant new on-line monitoring capability for glass melters using robust millimeter-wave techniques have been described and tested in the laboratory. The capability to monitor temperature, reflectivity (for emissivity determination), density, and viscosity has been shown and the feasibility in the laboratory demonstrated. Refractory millimeter-wave guides have been developed and used routinely for efficiently interfacing sensitive millimeter-wave electronics with a high temperature furnace in an oxidizing environment. Implementing millimeter-wave diagnostics as described here on nuclear waste melters should make possible feedback control of the vitrification process. This would lead to reduced costs and risks in the waste clean up effort.

The monitoring techniques described here are not limited to in applicability only to waste glass melters. They can also be applied to other high temperature materials processes. It is expected there will be a wide interest in these monitoring techniques. Future developments of MMW technologies may also lead to new capabilities for high temperature materials research studies.

ACKNOWLEDGEMENT

This work was supported by the Environmental Management Science Program, U. S. Department of Energy under grant number DE-FG07-01ER62707.

REFERENCES

[1]P. P. Woskov, D. R. Cohn, D. Y. Rhee, and P. Thomas, "Active millimeter-wave pyrometer", *Rev. Sci. Instrum.*, Vol. 66, 4241-4248 (1995).

[2]S. Ramo, J. R. Whinnery, and T. Van Duzer; p. 291 in *Fields and Waves in Communication Electronics*, John Wiley & Sons, New York, 1984.

[3]M. A. Heald and C. B. Wharton; Section 7.2 in *Plasma Diagnostics with Microwaves*, John Wiley & Sons, 1965.

[4]J. L. Doane; Vol. 13 in *Infrared and Millimeter Waves*, Edited by K. Button, Academic Press, 1985.

COLD-CAP MONITORING USING MILLIMETER-WAVE TECHNOLOGY

S. K. Sundaram
Pacific Northwest National Laboratory
Richland, WA 99352

William E. Daniel, Jr.
Westinghouse Savannah River Company, Building 704-1T
Aiken, SC 29808

P. P. Woskov and J. S. Machuzak
Plasma Science and Fusion Center
Massachusetts Institute of Technology
Cambridge, MA 02139

ABSTRACT

Cold-cap, a porous solid crust of calcined slurry formed on the melt surface in slurry-fed joule-heated nuclear waste melter, plays a key role in stabilizing the process and increasing the production rate. It is difficult to model the complex chemical reactions taking place in the cold-cap. There is a need for a technology for on-line monitoring of the cold-cap. We have adapted millimeter wave (MMW) technology for this application. We used a novel waveguide system that consists of a smooth-walled mullite tube with a miter mirror of Inconel 690 attached at one end to guide the millimeter waves emerging from the cold-cap or melt efficiently to a MMW receiver that is located away from the melter. We demonstrated this technology in an engineering scale EnVitco melter (EV-16) at the Clemson Environmental Technology Laboratory (CETL) during a melter test campaign. During the test, glass composition containing high Ca-F-calcine waste was processed that is representative of the wastes at the Idaho site. The Tanks Focus Area (TFA) supported the melter tests. The results identified the hot and cold regions of the cold-cap or melt and their dynamics during processing in the melter. Implementation of this technology shows promise for better process control, potential feedback control, increased production, and reduced cost of disposal.

To the extent authorized under the laws of the United States of America, all copyright interests in this publication are the property of The American Ceramic Society. Any duplication, reproduction, or republication of this publication or any part thereof, without the express written consent of The American Ceramic Society or fee paid to the Copyright Clearance Center, is prohibited.

INTRODUCTION

Nuclear waste melters (such as the Defense Waste Processing Facility (DWPF)) are generally slurry-fed, joule-heated melters. Cold cap is a porous solid crust of calcined slurry that floats on the surface of the melt (similar to the batch blanket in commercial glass furnace). The chemical reactions that take place in the cold-cap are complex as there is typically a large number of different chemical species present in a typical waste melter feed. A 90% coverage of the melt surface by the cold-cap helps the melter operators in maintaining a stable melting process and therefore maximizing the production rate. This also helps continuous melter processing, by balancing the feed rate and the melting rate. Minimization of unwanted volatile emissions from the melt is another important function of a cold cap.

Several batch-modeling studies [1,2] have been conducted by commercial glass industry, wherein the dry batch is charged horizontally. Yasuda and Hrma [3] measured yield stress, viscosity, and flow distance down a hot inclined surface were measured on three (hydroxide, formate, and frit) simulated nuclear waste feeds at different temperature and oxide loading. The formate slurry exhibited higher yield stress, higher viscosity, and shorter flow distance than other feeds at temperatures at which steam evolution was not a dominating factor. When steam evolution became a dominating factor at elevated temperatures, the formate slurry exhibited a longer flow distance than the other feeds.

Demixing phenomena within the cold-cap can produce melting instability. Five demixing mechanisms reported in the literature [4-12] are: 1) evaporation of volatile components, 2) settling or buoyancy of solid particles, 3) floatation of solid particles, 4) segregation of melts of different densities, and 5) drainage of low-viscosity melts. Hrma and coworkers [13] examined the drainage of low-viscosity melts between refractory particles in a mixture of Na_2CO_3, Li_2CO_3, and SiO_2. Drainage was observed in the samples in which the molten carbonates occupied 50-100% of the space between silica particles. Choi [14] and Eyler et al [15] elucidated the complexities of modeling the cold-cap and integrating that into a melter model.

Existing waste glass melter technology lacks sophisticated diagnostics due to the hot, corrosive, and radioactive environments of these melters. There is no viable existing technology for on-line measurement of cold-cap temperature and temperature profile. New robust diagnostics using millimeter wave (MMW) technology has recently been overviewed [16]. We present the results of demonstration of MMW technology for non-contact cold-cap monitoring.

THEORITICAL BASIS

Millimeter-wave thermal emission is linear with temperature, given by the blackbody emission formula for a single polarization in this wavelength limit as $P = k_B T_{eff} d\nu$, where P is in Watts, k_B is Boltzmann's constant, T_{eff} is the absolute temperature in Kelvin that fills the receiver antenna, and $d\nu$ is the receiver bandwidth in Hertz. The receiver signal is linearly proportional to T_{eff}. The effective temperature that fills the receiver antenna depends on the efficiency of the transmission line and the emissivity of the cold-cap/glass melt that is viewed. All sources and losses of thermal radiation need to be considered when determining T_{eff}, which is given by the formula:

$$T_{eff} = \varepsilon_{wg} T_{wg} + \tau_{wg} \varepsilon_s T_s + r_s \tau_{wg} \varepsilon_{wg} T_{wg} \tag{1}$$

where the first term on the right is the thermal emission from the waveguide, the second term is the thermal emission from the glass, and the final term is the reflected thermal emission from the waveguide, where ε_{wg} is the waveguide emissivity, ε_s is the cold-cap/glass melt emissivity, τ_{wg} is the waveguide transmission factor, r_s is the cold-cap/glass melt reflectivity, and T_{wg} and T_s are the temperatures of the waveguide and cold-cap/glass melt, respectively. The basis has further been elaborated elsewhere [16].

MELTER DEMONSTRATION

Melter Description

EnVitco EV-16 melter at Clemson Environmental Technology Laboratory (CETL) at Clemson, South Carolina was used for the demonstration. A schematic of the melter system is illustrated in Figure 1. The EV-16 melter is a ceramic-

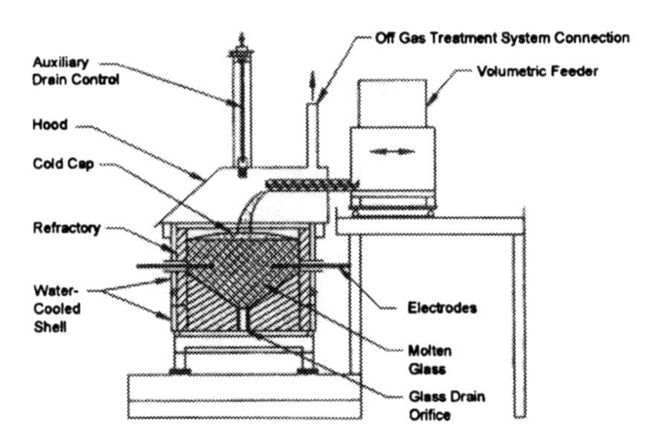

Figure 1. Schematic diagram of EV-16 Melter
(Off-gas system is not shown.)

lined, joule-heated, cold-top melter. The melter has four sidewalls and a bottom block of refractory. Melting volume available is about 18″ × 18″ × 14″. Heating is controlled by a total of four molybdenum electrodes, with one located on each side of the melter below the glass level. Maximum attainable temperature is 1500°C. Melting in the melter is controlled by the power consumption, not by the operating temperature. The power control consists of a Spang brand auto-tap transformer and control boards that control the power in two zones. The power set point for both zones is controlled by one potentiometer so that the power (typically 35 kW/zone or 70 kW total) in both zones should be equal. The glass melt is poured through a molybdenum tube mounted through the center of the bottom block of refractory. Flow through the drain is started using an electrode near the drain and stopped by a water-cooled, steel probe, which fits inside the drain orifice to stop the flow.

The EV-16 melter is equipped with a complete, self-contained off-gas treatment system (not shown in Figure 1). The system is driven by a positive-displacement blower that is capable of 40″ H_2O negative pressure, but most of this capacity is utilized as pressure drop across the various offgas treatment units. Typically, the melter operates at 0.5 – 1.0″ H_2O negative pressure at the plenum.

The melter test campaign was supported by the Tanks Focus Area (TFA) to support the Idaho vitrification program. We leveraged the melter campaign to demonstrate cold-cap monitoring, using MMW technology. Our demonstration was integrated with the objectives of the melter tests campaign.

Waveguide Selection and Setup

The test setup is shown in Figure 2. The MMW equipment was mounted on a sturdy platform on one side of the melter (circle on the right of Figure 2).

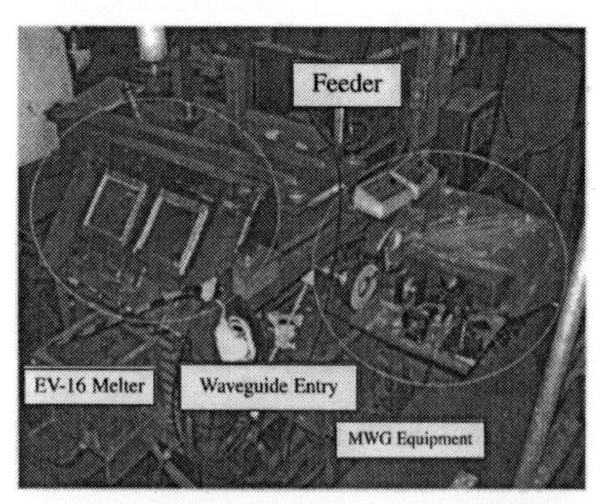

Figure 2. MMW test setup in EV-16 melter

A smooth-walled mullite tube was selected as the waveguide material based on our extensive testing [16]. The waveguide consisted of a mullite tube of about 1-5/8″ diameter that penetrates the melter plenum space with a 45°- Inconel 690 miter mirror sleeve attached to the one end inside the melter. The other end interfaces quasi-optically with the MMW receiver that is outside the melter through a graduated-flange that could be rotated manually 0-360°.

Feed and Glass Compositions

An auger volumetric feeder is used to feed the dry feed, prepared for the melter tests. The target feed and glass compositions are presented in Table I. The targeted waste has high Ca, F, and Zr.

Table I. Target feed/glass compositions

Surrogate calcine compounds as oxide products in glass	Surrogate calcine compounds as oxide products in glass (normalized)	Frit surrogate as oxide products in glass (mass % of frit fed to melter)	Glass composition as oxides (mass % @ 40% calcine surrogate)
Al_2O_3	31.81		12.72
B_2O_3	2.03	14.08	9.26
CaO	5.91		2.36
CaO (from CaF_2)	21.83		8.73
Cl	0.11		0.04
F	14.80		5.92
P_2O_5	1.36		0.54
CoO	0.04		0.02
Cs_2O	0.26		0.10
Fe_2O_3	0.54	0.62	0.59
K_2O	0.54		0.22
La_2O_3	---	6.50	3.90
Li_2O	---	12.00	7.20
MgO	0.82		0.33
MnO_2	0.03		0.01
Na_2O	3.03	5.62	4.59
NO_3	4.19		1.68
NiO	0.73		0.29
SiO_2	---	61.18	36.71
SnO_2	0.20		0.08
SO_3	2.20		0.88
SrO	0.34		0.14
ZrO_2	13.44		5.37
Total	100.00	100.00	100.00

Melter Start-up and Measurements

Figure 3 shows the melter start-up activities. Residual glass left in the melter was chiseled out and the waveguide was installed (Figure 3, top two images). Joule-heating in the EV-16 melter was established by heating the start-up frit, shown in Figure 3 (bottom left). Once the frit melted, the feeding was started (Figure 3 bottom right). Circles denote the waveguide.

The transverse electric and magnetic field HE_{11} mode is the most efficient guided mode for electromagnetic propagation in hollow waveguides with internal diameter larger than a wavelength [16]. The HE_{11} mode is a natural mode inside the smooth walled dielectric (mullite). The millimeter wave signals emerging from the cold-cap surface and guided by the waveguide was measured and converted to temperature on-line throughout the melter test campaign. This was compared to the reading from a melter thermocouple.

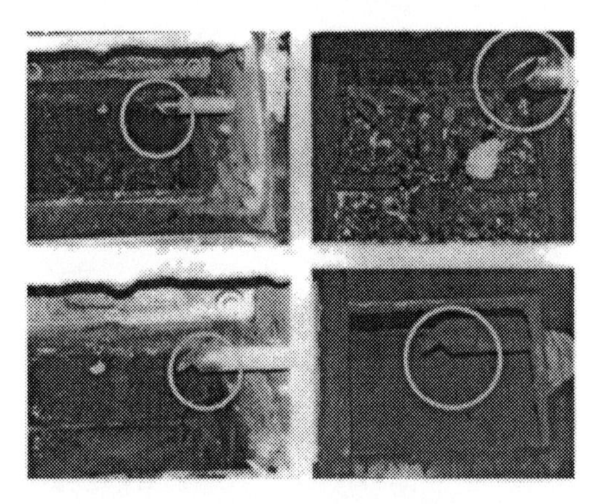

Figure 3. Views inside the melter startup (MMW waveguide circled)

Cold-cap coverage varied from 0 – 100% during the campaign. During selected steady-state segments of the campaign, the waveguide was manually rotated from 0° to 180° gradually at 5° intervals. Figure 4 shows the positions of the waveguide at 0, 90, and 180° positions (circles indicate waveguide). The plot on the right in Figure 4 shows the viewed spot size radius as a function of distance from the waveguide aperture. Over the 0-180° rotation range, most measurements occurred at a distance of 15-25 cm or a spatial resolution of about 3.6 cm (1.4"). This was repeated under different stages of cold-cap burning, starting from 100 % (complete coverage – cold-cap temperature) to 0% (no coverage – melt temperature). This was also repeated at different depths of penetration into the melter.

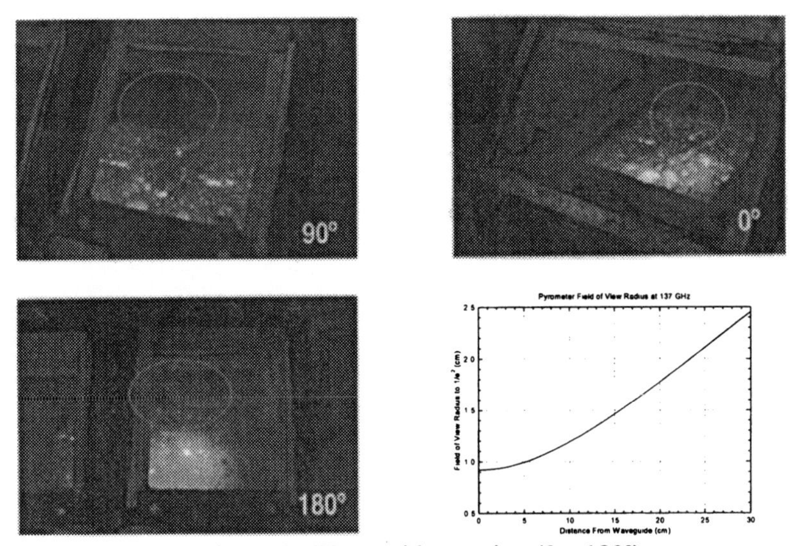

Figure 4. Waveguide rotation (0 – 180°)

DEMONSTRATION RESULTS

The measured surface temperature profiles provided a unique indication of the cold-cap status and the current operational process on going inside the melter. Figure 5 shows three representative profiles. The top trace was obtained with no cold-cap at the start of feeding. The lower trace with open circles was obtained when a cold-cap was present with continuous feed. The dip in temperature at about 110 ° corresponded to the position of a mound of feedstock on the cold-cap. The other lower trace (dashed curve) was obtained with a cold-cap when the melt was being poured from the melter. The dramatic dip in temperature at 90 ° occurs above the pour spout. The presence of the cold-cap and whether the melter was being fed or poured could be clearly determined from the magnitude and shape of the surface temperature profile.

Surface temperature contour plots were also generated by assigning color codes for different areas corresponding to the temperatures based on the field of view of the waveguide as it was rotated. A representative plot (in black and white) is shown in Figure 6 (left). The contour plot was overlaid on a photographic image of the melter top view (Figure 6 – right) to show the location of the region measured. It corresponded to a segment where the cold-cap was just burnt off and the melt was actively bubbling. The hottest region was at ≈ 933°C and the coldest region was at ≈ 683°C. The locations of colder regions close to the back wall of the melter corroborated well with the fact that the feeder was near that side of the melter.

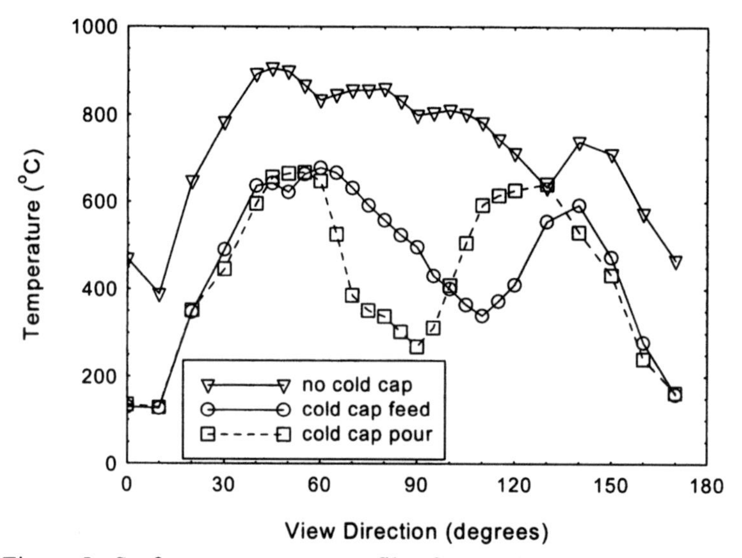

Figure 5. Surface temperature profiles front to back 5.5 inches (14 cm) from midplane

Figure 7 presents a 3-D view of the surface temperature profile of the melt after the cold-cap was completely burnt off. The hearth ranged approximately from 39 to 151° of the view angle. The temperature ranged from about 400°C to 950°C. The temperature profile was found to be similar across the hearth. The profile showed clearly hot (around 40° view angle) and cold (around 120° view angle) regions. Colder edges were seen on either side of the hearth.

The data demonstrate the usefulness of the MMW technology to monitor the cold-cap. Evaluation of the real-time temperature profiles generated during processing would indicate the dynamics of the cold-cap. In addition, the temperature profile could also be correlated to the off-gas species evolved during processing, thus indicating optimum process parameters to minimize the evolution of undesirable species.

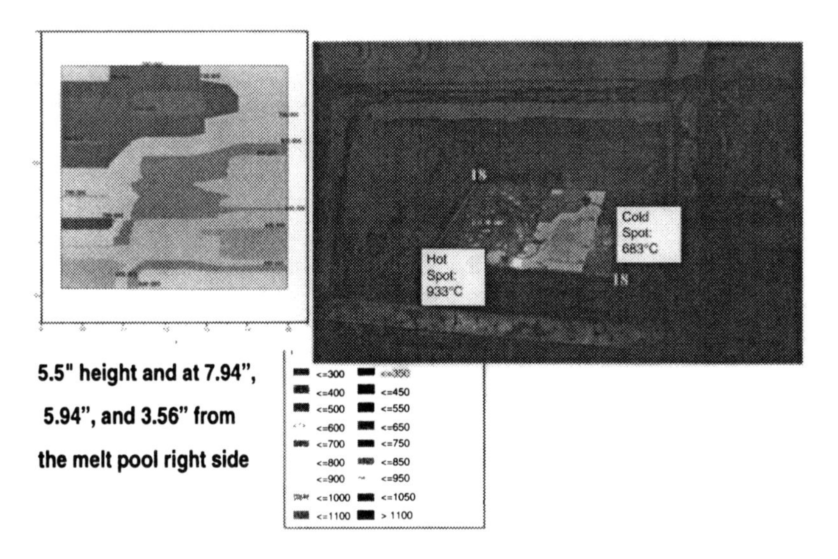

Figure 6. Surface temperature contours (without cold cap)

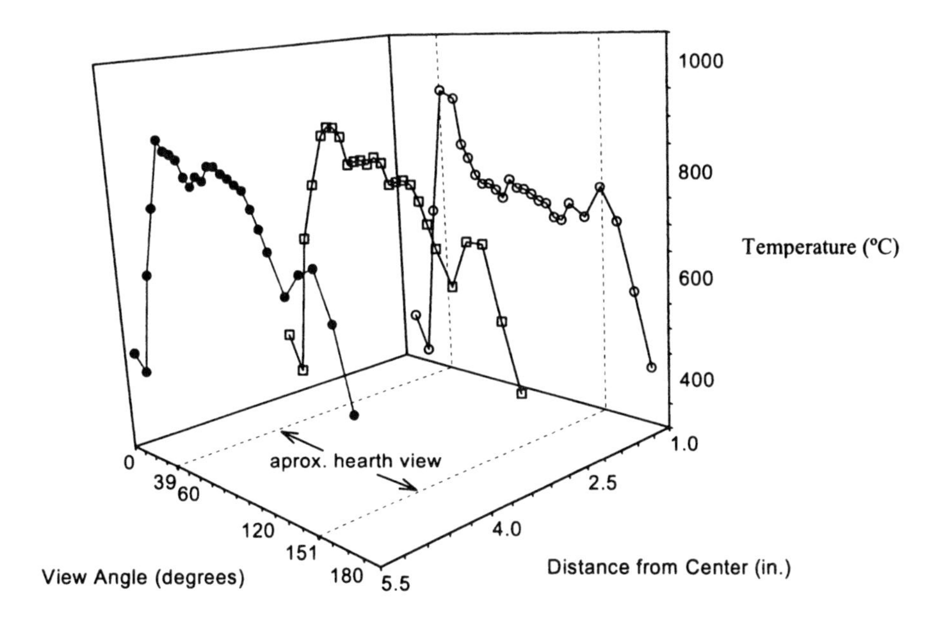

Figure 7. Melt surface temperature profile (without cold cap)

SUMMARY

We have demonstrated application of MMW technology to measure and profile cold-cap temperature in a melter, that is processing glass compositions designed for vitrification of Ca-F-rich calcine waste from Idaho site. We designed a novel waveguide for this application. The technology clearly identified the hot and cold regions of the cold-cap or melt surfaces and their dynamics during processing. Implementation of this technology would result in better process control, potential feedback control, increased production, and reduced cost.

ACKNOWLEDGEMENT

This work was supported by the Environmental Management Science Program, U. S. Department of Energy under grant number DE-FG07-01ER62707. We acknowledge Dr. E. W. (Bill) Holtzscheiter, Technology Integration Manager (TIM) - Immobilization, TFA, for supporting the melter tests campaign.

REFERENCES

[1]H. Mase and K. Oda, "Mathematical Model at Glass Tank Treemace with Batch Melting Process," J. *Non-Crystalline Solids*, Vol. 38-39, (1980).

[2]A. Ungan and R. Viskanta, "Melting Behavior of Continuously Charged Loose Batch Blankets in Glass Melting Furnaces," *Glastechn. Ber.*, Vol. 59 (10), (1986)

[3]D. D. Yasuda and P. Hrma, "The Effect of Slurry Rheology on Melter Cold Cap Formation"; pp. 349-359 in *Nuclear Waste Management IV*, Ceramic Transactions, Vol. 23, Edited by G. G. Wicks, D. F. Bickford, and L. R. Bunnell. The American Ceramic Society, Inc., Westerville, Ohio, 1991.

[4]R. E. Davis, "Batch Blanket Chemistry and Air Quality Emissions Associated with Cold-Top Electric Furnaces," *Ceram. Eng. Sci. Proc.*, **7**[3-4] 460-466 (1986).

[5]W. T. Cobb and P. Hrma, "Behavior of RuO_2 in a Glass Melt," pp. 233-237 in *Nuclear Waste Management IV*, Ceramic Transactions, Vol. 23, Edited by G. G. Wicks, D. F. Bickford, and L. R. Bunnell. The American Ceramic Society, Inc., Westerville, Ohio, 1991.

[6]C. Kroger and F. Marwan, "On the Rate of Reactions Leading to Glass Melts. VII, The Dependence of Melting Time of Glass Batches on the Nature and Amount of Products Produced by Batch Reactions," *Glasstech. Ber.*, **29** 275-288 (1956).

[7]U. Vierneusel, H. Goerk, and K. H. Schuller, "Suppression of Segregation in Crucible Melts," *Glasstech. Ber.*, **54**[10] 332-337 (1981).

[8]A. Dietzel and O. W. Florke, "Effect of Sulfate during Melting Process," *Glasstech. Ber.*, **32** 181-185 (1959).

[9]L. Nemec and J. Zluticky, "Possibilities of Photographic Recording of Some Processes Occurring in Molten Glass," *Sklar Keram.*, **29**[12] 353-358 (1979).

[10]H. Mitamura, T. Murakami, T. Banba, Y. Kiriyama, H. Kamizono, M. Kumata, and S. Tahiro, "Segregation of Elements of the Platinum Group in a Simulated High-Level Waste Glass," *Nuclear and Chemical Waste Management*, **4** 245-251 (1983).

[11]J. C. Hayes, "Laboratory Methods to Simulate Glass Melting Process"; pp. 14.1-14.8 in *Advances in Fusion of Glass* The American Ceramic Society, Inc., Westerville, Ohio, 1988.

[12]P. Hrma, "Melting of Foaming Batches: Nuclear Waste Glass," *Glasstech. Ber.*, **63K** 360-369 (1990).

[13]P. Hrma, C. E. Goles, and D. D. Yasuda, "Drainage of Primary Melt in a Glass Batch"; pp. 361-367 in *Nuclear Waste Management IV*, Ceramic Transactions, Vol. 23, Edited by G. G. Wicks, D. F. Bickford, and L. R. Bunnell. The American Ceramic Society, Inc., Westerville, Ohio, 1991.

[14]I. G. Choi, "Mathematical Modeling of Radioactive Waste Glass Melter"; pp. 385-394 in *Nuclear Waste Management IV*, Ceramic Transactions, Vol. 23, Edited by G. G. Wicks, D. F. Bickford, and L. R. Bunnell. The American Ceramic Society, Inc., Westerville, Ohio, 1991.

[15]L. L. Eyler, M. L. Elliott, D. L. Lessor, and P. S. Lowery, "Computer Modeling of Ceramic Melters to Asses Impacts of Process and Design Variables on Performance"; pp. 395-407 in *Nuclear Waste Management IV*, Ceramic Transactions, Vol. 23, Edited by G. G. Wicks, D. F. Bickford, and L. R. Bunnell. The American Ceramic Society, Inc., Westerville, Ohio, 1991.

[16]P. P. Woskov, J. S. Machuzak, and P. Thomas, S. K. Sundaram, and William E. Daniel, Jr., "Millimeter-Wave Monitoring of Nuclear Waste Glass Melts - An Overview," *this Volume.*

FURNACE SYSTEM DEVELOPMENT FOR THE PLUTONIUM IMMOBILIZATION PROGRAM

A.D. Cozzi, K. C. Neikirk, D. T. Herman and J.C. Marra
Westinghouse Savannah River Company
Aiken, SC 29808

T. Pruett and J. Harden
Clemson Environmental Technologies Laboratory
Anderson, SC 29625

ABSTRACT

A furnace system is being developed to meet the requirements for disposition of approximately 13 metric tons of excess weapons useable plutonium. The proposed immobilization form is a titanate based ceramic consisting primarily of a pyrochlore phase. Actinides (or surrogates) are mixed with the ceramic precursors, cold-pressed, and then densified via a reactive sintering process.

A number of items must be considered to specify the design, operation and maintenance of a furnace system for the glovebox environment. Furthermore, throughput requirements drive certain design features and dictate the number of furnaces required for the immobilization plant. A test program is ongoing to evaluate these design considerations resulting in the specification of a furnace system for plant operations.

INTRODUCTION

Immobilization of radioactive waste in a polycrystalline ceramic form was demonstrated in the early 1970's by McCarthy and Davidson[1]. Evaluation of the immobilization of a surrogate waste from commercial nuclear power generation resulted in the formation of more than twenty phases. Jantzen et al.[2] described the concept of tailoring the composition of the non-waste additives to produce specific crystalline phases. This process was outlined for two typical commercial nuclear wastes and for three types of the Savannah River Defense Waste.

Several methods of fabrication of the ceramic form have been investigated[3,4]. A pyrochlore based ceramic form has been selected for the immobilization of plutonium. The reactive sintering process used to form the pyrochlore and enhanced feed preparation abilities made possible the production of a ceramic with less than ten percent porosity via a uniaxial cold press and sinter method [5]. Thus, a furnace must be chosen to meet the criteria set forth to support production of plutonium bearing ceramics in a glovebox environment.

To the extent authorized under the laws of the United States of America, all copyright interests in this publication are the property of The American Ceramic Society. Any duplication, reproduction, or republication of this publication or any part thereof, without the express written consent of The American Ceramic Society or fee paid to the Copyright Clearance Center, is prohibited.

SPECIFICATIONS

A furnace development team was tasked with specifying a furnace capable of meeting the needs of the glovebox facility in terms of operation and maintenance while maintaining the functionality to produce repository acceptable ceramic pucks. The necessary and the desirable functions of the furnace were determined for the two inputs either by interviewing personnel recognized with glovebox operations or through brainstorming sessions with personnel responsible for development of the ceramic process. The furnace team assembled a requisition to be presented to a furnace vendor. Essential performance requirements specified for the furnace included:

- The furnace must follow the temperature profile specified by the baseline ceramic process with a load of 54 pucks on the tray assembly.
- The furnace set point temperature shall be stabile at the hold temperature.
- The furnace temperature uniformity within the useable furnace volume shall be +/-10°C or better during ramp up.
- The furnace insulation package shall be designed to maximize the durability of the insulation.

Several of the furnace components were deemed important enough to require an increased degree of specificity. Table I is the furnace components that were identified for increased consideration.

Table I. Furnace components identified for a higher level of specification.

Furnace Component	**Key Property**
Furnace Shell	Support cooling water jacket
Furnace Elements	Straightforward to replace in glovebox enclosure
Thermocouples	Backup Tc's recessed into insulation
Insulation	Endure the lifetime of the furnace (10y)
Control System	Interface with data acquisition system to log conditions
Door/Loading System	Interface with trolley used to load trays on furnace door
Purge Air System	Air must not impinge on trays
Offgas System	Chimney plug controlled by the furnace setpoint

The trays for the furnace were specified by a subset of the furnace team. A standard design (4-walled) for stacking trays was modified to interface with the remote puck-loading robot. A second design (pagoda) requiring less handling was also investigated. Figure 1 is a schematic of the two tray designs investigated. Several materials were investigated in an attempt to reduce the mass of the furnace trays and minimize ceramic puck – furnace tray interactions. The tray materials for evaluation included hardened alumina-rich fiberboard; yttria stabilized zirconia (YSZ) reticulated ceramic; and dense, coarse-grained alumina. The reactivity of the ceramic form with non-zirconia based tray materials requires that a layer of zirconia separate the plutonium ceramic from the tray.

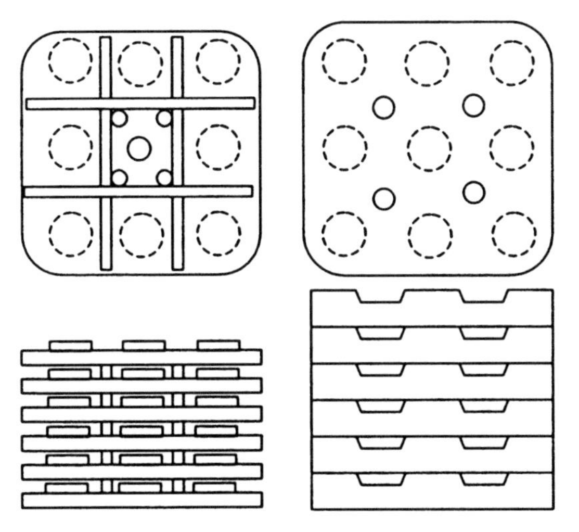

Pagoda 8 pucks/tray   4-walled 9 pucks/tray

Figure 1. Schematic of trays tested. Ceramic pucks are represented by dashed lines.

A custom furnace*, specified to meet the aforementioned criteria as closely as possible, was installed at the Clemson Environmental Technology Center (CETL). The furnace was integrated with a trace water cooling system, a forced air annular space cooling system, a purge air system and an offgas system. A data acquisition system (DAS) was added to monitor a number of elements including flow rates and temperatures of cooling water and air (purge, offgas, and dilution), furnace power and furnace temperature. The shell is a six millimeter steel jacket with a copper cooling trace welded to the sides. Twelve thermocouples (process variable (PV), over-temperature controller and ten undesignated) are inserted through two opposite sides. Figure 2 is the furnace with panels removed to view cooling traces and thermocouples. The insulation is comprised of three layers of progressively denser fiberboard totaling five inches in thickness. The furnace control system is coordinated through a programmable logic controller (PLC). The PLC provides safety interlocks to prohibit hazardous operation of the furnace as well as controlling the door/loading system that delivers the furnace door into the furnace. The PLC also controls the position (opened or closed) of the chimney plug using feedback from a programmed setpoint temperature.

* CM Furnaces, Inc. Bloomfield, NJ

Figure 2. Photo of furnace with panels removed.

FURNACE RUNS

A series of furnace runs were developed to evaluate the ability of the furnace to meet the operational specifications listed in Table I. Specifications were tested in order beginning with the tracking of the heating profile and progressing through integrated operations. Further testing was performed in an iterative manner to ascertain the thermal response of the furnace under various air flow conditions. Furnace tray materials and designs were also evaluated during this testing.

Initial Baseline Tests

To establish a "worst case" cooling scenario for the empty furnace, the chimney remained closed during the cooling portion of the sintering cycle and no air was introduced into the furnace cavity. The top side-panel of the furnace was removed to allow convection to cool the aluminum straps connecting the heating elements (as seen in Figure 2). Figure 3 is the plot of the furnace temperature annotated with significant events. Initial tests were also used to screen furnace tray materials and to provide temperature data for a thermal model of the furnace. Table II is the baseline sintering schedule provided by the ceramic process team.

Table II. Heating schedule for the initial furnace test.

Segment	Ramp (°C/min)	Temperature (°C)	Dwell (hr)	Chimney
1	3	300	0	Open
2	0	300	2	Open
3	5	800	0	Open
4	5	1350	0	Closed
5	0	1350	4	Closed
6	5	RT	0	Closed

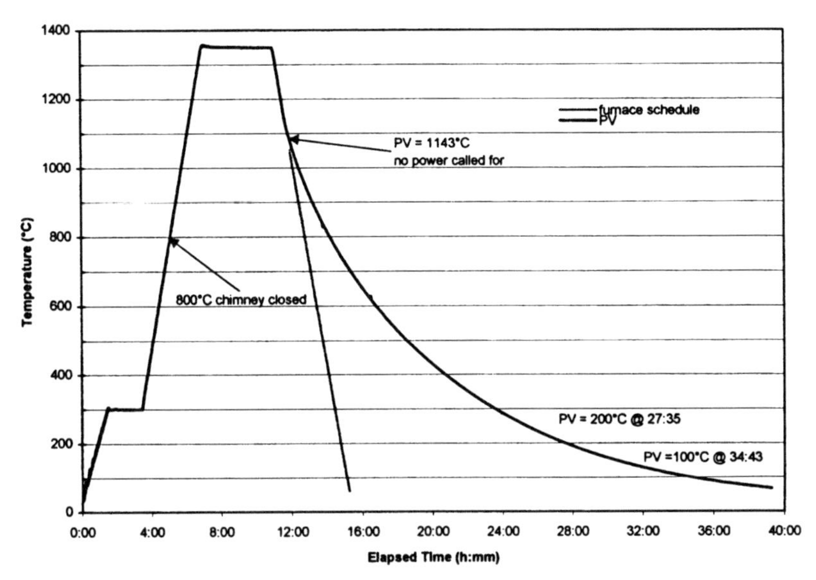

Figure 3. Initial furnace run with shell cooling only.

Full Thermal Load Tests with Sintered Pucks

After the baseline testing was completed, tests using a full thermal load of previously sintered pucks were completed using various air flow conditions. These tests were designed to assess the effect of air flow conditions on furnace temperature, power requirements and sintering cycle time. Several candidate tray materials were used during the full thermal load tests to assess the tray materials effect on temperature cycle as well as assess the robustness of the tray designs and materials of construction. Pucks that had been sintered previously were used in lieu of green pucks to represent the thermal load in the furnace. The use of previously sintered pucks allowed for the testing of various tray materials without jeopardizing the furnace due to the reactivity of green pucks. The reuse of pucks significantly reduced the raw materials costs associated with the initial testing. Green pucks will be used in future testing after the sintering cycle has been optimized.

Table III compares the performance of the tray materials and designs. The trays made from the hardened fiberboard slumped slightly during the first cycle and much more significantly during the second cycle. The mass of the pucks was approximately 4.5 kg per tray and contributed to the slumping experienced by the trays. The YSZ trays were fabricated to fit together snugly. It is suspected that during the heating portion of the cycle, thermal expansion of the vertical support rods initiated failure of the plates. This was apparent in all of the four trays that failed. The "heavy" alumina 4-wall trays received were thicker than specified. Inspection of the furnace after the third cycle, in which three trays exhibited cracks, revealed a breach of the furnace insulation. The presence of a breach allowed more air than expected to enter the furnace during the cool down portion of the cycle and thermally stressed the trays. Additional testing (five cycles to date) with coarse alumina trays of the proper thickness has shown no visible degradation.

Table III. Performance of tray materials during several sintering cycles.

Material	Design	Cycles to Failure	Failure Mode
Hardened fiberboard	4-wall	2	High temperature creep in the center of the tray
Hardened fiberboard	Pagoda	Not tested	N/A
YSZ reticulated ceramic	Pagoda	1	Cracks originating from vertical supports
Dense coarse alumina (heavy)	4-wall	3	Cracks originating from wall-base interface*
Dense coarse alumina (light)	4-wall	>5	No failure detected as yet

*Three bottom trays cracked.

Full Thermal Load Tests with Green Pucks

Subsequent furnace tests were run using the dense coarse alumina (light) trays with a YSZ insert. A limited number of green pucks were used to evaluate the temperature uniformity of the furnace cavity. Based on results from previous tests, purge air was introduced during heating (<700°C) to burn out binder and during cooling (<800°C) to accelerate cooling. Uniformity of the furnace cavity was determined by the measured shrinkage of green pucks placed in various positions on all of the trays. The variance in shrinkage among pucks on different trays is due to the vertical temperature gradient within the furnace. The gradient is a result of the rate in which cooling air is introduced into the furnace cavity and the leakage of air through the furnace door. Table IV is a sample of the variation in shrinkage measured among the trays for two furnace runs made with seemingly identical heating and cooling conditions.

Further analysis of the data in Run #2 indicated that the shell cooling air velocity was sufficient to place the furnace cavity under a slight negative pressure when all jacket panels were in place. The jacket air cooling drew hot air from the furnace cavity out through the element and thermocouple penetrations. The air was replaced by air drawn in through the door and thus, past the trays. This air flow condition lead to increased furnace power consumption and decreased cycle time. Confirmation of this phenomenon was confirmed by several piece of evidence. The evidence included the considerable increase in power needed by the furnace to maintain the 1350°C hold temperature and the significantly faster cool down of the furnace due to the draw of ambient air into the furnace. In Run #5, side panels were removed from the furnace (Figure 1) to minimize this effect. The removal of a jacket panel substantially reduced the velocity of cooling air passing over the furnace. These changes lead to better control of the furnace temperature gradient and thus improved control over product shrinkage. Furnace observations and configuration changes resulted in a significant improvement in the control of product shrinkage throughout the furnace.

Table IV. Puck shrinkage for various sintering conditions.

Tray Position	Run #2	Run #5
Top - 6	23.7%	24.6%
5	24.5%	24.6%
4	24.4%	24.5%
3	23.6%	24.5%
2	22.5%	24.6%
Bottom -1	20.2%*	24.2%
Cooling air (cfm)	5 purge 10 leak	5 purge 10 leak
Time to 100°C (h:mm)	24:08	29:29

*Shrinkage must be greater than 21.5% to fit in can.

CONCLUSIONS

A full-scale sintering furnace was installed at the CETL. A series of experiments were performed to evaluate the essential performance requirements specified for the furnace. Initial furnace testing without air flow demonstrated that:

- The furnace can follow the heating temperature profile specified by the baseline ceramic process with a load of 54 pucks on the tray assembly.
- The furnace set point temperature is stable at the hold temperature, Figure 3.
- The furnace temperature uniformity within the useable furnace volume is less than +/-10°C during ramp up as determined by the small variation in puck shrinkage.
- The furnace insulation package has exhibited no discernable wear during the initial runs.

Three tray materials and two tray designed were evaluated. The coarse alumina 4-wall tray has proved to be the most reliable to date. Potential savings in time from using the lower mass pagoda design are offset by the decreased puck load (54 on 4-wall stack, 48 on pagoda).

The source of cooling air must be carefully controlled to ensure that it is delivered through the purge ports. Negative pressures in the furnace cavity can not only effect the temperature uniformity within the furnace cavity, but would also present a potentially uncontrolled operating condition. In a glovebox environment, air flow from the furnace cavity containing plutonium into the cooling air system is undesirable and represents a contamination concern.

FUTURE WORK

The following items need to be addressed in future testing:

- Perform a series of experiments to evaluate the extent of the effect of purge air and cooling air on the sintering cycle time and gradients within the furnace. The goal of the testing is the optimization of the ceramic processing conditions.
- Operate and maintain the furnace under glovebox like conditions to provide input into the next generation furnace regarding maintenance and recovery from off-normal operations.
- Demonstrate the uniform sintering of a full load (54) of green pucks. Testing to date has been limited to six pucks (one per tray) per furnace run.
- Continue evaluation of tray materials. Alumina trays (which showed much promise) require a chemical barrier to minimize ceramic puck – tray interactions.

ACKNOWLEDGEMENTS

The authors would like to thank J.W. Congdon, K.M. Marshall and C. Rathz for providing all the green and sintered pucks for the furnace runs. The information contained in this paper was developed during the course of work under Contract No. DE-AC09-96SR18500 with the U.S. Department of Energy.

REFERENCES

[1] G.J. McCarthy and M.T. Davidson, "Ceramic Nuclear Waste Forms: I. Crystal Chemistry and Phase Formation," Bull. Am. Ceram. Soc. Vol. 54 pp. 782-786 (1975).

[2] C.M. Jantzen, J. Flinthoff, P.E.D. Morgan, A.B. Harker and D.R. Clarke, "Ceramic Nuclear Waste Forms," in Proceedings of the International Seminar on Chemistry and Process Engineering for High-Level Liquid Waste Solidification, Vol. 2 pp. 693-706 (1981).

[3] C.L. Hoenig and H.T. Larker, "Large-Scale Densification of a Nuclear Waste Ceramic by Hot Isostatic Pressing," Bull. Am. Ceram. Soc., Vol. 62 pp. 1389-1390 (1983).

[4] R.B. Rozsa and C.L. Hoenig, "Synroc Processing Options," Lawrence Livermore National Laboratory Report UCRL-53187 (1981).

[5] W. Brummond, G. Armantrout and P. Maddux, "Ceramic Process Equipment for the Immobilization of Plutonium," in Third Topical Meeting of DOE Spent Nuclear Fuel and Fissile Material Management, pp. 380-384 American Nuclear Society (1998).

PLUTONIUM IMMOBILIZATION PROJECT PHASE 2 COLD POUR TEST (U)

Mike E. Smith and E. Lee Hamilton
Westinghouse Savannah River Site
Savannah River Technology Center
Aiken, SC 29808

ABSTRACT

The Plutonium Immobilization Project is concerned with the disposition of excess weapons useable plutonium. The project will utilize the can-in-canister approach that involves placing plutonium containing ceramic pucks in sealed cans that are then placed in Defense Waste Processing Facility canisters. These canisters are subsequently filled with high-level radioactive waste glass. This process puts the plutonium in a stable form and makes it unattractive for reuse. Cold (non-radioactive) tests were performed to develop and verify the baseline design for the canister and internal hardware. This paper describes the second phase of these tests and their results.

INTRODUCTION

The Plutonium Immobilization Project (PIP) is a program funded by the U.S. Department of Energy to develop technology to dispose of excess weapons useable plutonium. Lawrence Livermore National Laboratory (LLNL) is the lead laboratory for the program with the Savannah River Site (SRS) partnering on key technical and engineering aspects of the program. In the two-part can-in-canister (CIC) approach, plutonium is immobilized at a nominal 9.5 weight percent concentration within titanate-based ceramic forms (pucks). The pucks are sealed in stainless steel cans and then loaded into long cylindrical magazines. These magazines are latched to racks inside Defense Waste Processing Facility (DWPF) canisters that will be filled with high-level waste glass at the DWPF. The ceramic form and the radiation barrier provided by the glass makes the plutonium unattractive for reuse. Presently the DWPF pours glass into empty canisters. The addition of a stainless steel rack, magazines, cans, and ceramic pucks to the canisters introduces a new set of design and operational challenges.

Remote operation considerations, high metal temperatures (structural integrity), minimizing thermal mass, and glass voiding were the primary CIC hardware issues. Demonstration of the effectiveness of the design in full scale testing was deemed necessary. These tests were done using non-radioactive glass. The scoping Phase 1 tests consisted of test pours and thermal modeling to evaluate preliminary hardware designs (cans, magazines, and racks), as well as identify any changes that were required to establish a firm baseline design. The results of Phase 1 are documented elsewhere[1]. The Phase 2 tests incorporated changes based on Phase 1 results and were used to verify the adequacy of the baseline design for the start of Title 1 plant design. Unlike Phase 1, Phase 2 will be used to demonstrate compliance with specific requirements identified in the Plutonium Immobilization Product Specifications (PIPS) and hence was performed in accordance with applicable repository quality assurance requirements (i.e. DOE-RW-0333P).

To the extent authorized under the laws of the United States of America, all copyright interests in this publication are the property of The American Ceramic Society. Any duplication, reproduction, or republication of this publication or any part thereof, without the express written consent of The American Ceramic Society or fee paid to the Copyright Clearance Center, is prohibited.

PHASE 2 TEST PLAN STRATEGY

The test parameters controlled were the pour rate, glass composition, glass stream temperature, glass stream fall height, hardware configurations, and glass fill height. The two main items to be investigated in the Phase 2 test were the extent of glass void formation and the degree of structural deformation of the CIC hardware in the DWPF canister. Two canisters (one low pour rate and one nominal pour rate) with installed hardware were determined to be adequate for this assessment. Two additional PIP canisters were to be filled with glass and stored for future proliferation resistance testing. A high viscosity DWPF surrogate glass (about 90 poise at 1150 °C) was used to fill all four canisters. This viscosity was determined to be the highest possible viscosity glass that would be fed to the DWPF Melter during CIC operations at the DWPF. SRS personnel performed the Phase 2 test at the Clemson Environmental Technologies Laboratory (CETL) with the aid of CETL personnel. The melter used was the Full-Scale Stirred Melter.

CAN-IN-CANISTER HARDWARE CONFIGURATION

Can-in-canister hardware consists of cans, magazines, and a rack (see Figure 1). In the PIP facility, pucks will be placed inside stainless steel cans remotely. An automated welding process will seal the can tops and then four cans are remotely placed inside each magazine. A total of seven magazines will be loaded through the neck of the DWPF canister using a telescopic robot. The canister will then taken to DWPF and filled with high-level radioactive waste glass. These steps present unique challenges ranging from remote handling issues to high temperature concerns. These factors directed the hardware development and selection for cold pour testing.

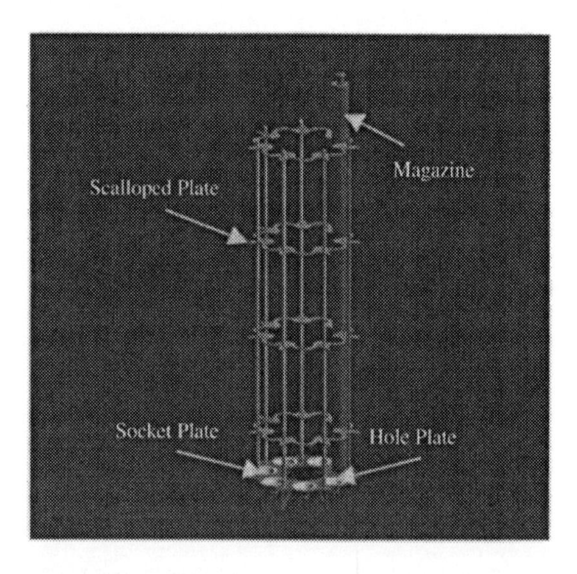

Figure 1. Simulation of Can-in-Canister

PHASE 2 COLD POUR TESTS

General Test Information

Four canisters were filled with high viscosity DWPF surrogate glass. Filled first were the two proliferation canisters (PR1 and PR2). The targeted nominal DWPF pour rate for the proliferation canisters was 82 to 109 kg/hr. These canisters had 4 magazines (16 cans) of Harbison-Walker Aurex 95 ceramic pucks and 3 magazines (12 cans) of surrogate plutonium ceramic pucks supplied by LLNL (designated as PR pucks). Not enough pucks from LLNL were available for the test and therefore another material was chosen for some of the cans. The Aurex 95 ceramic was picked because it has similar thermal properties to those projected for the actual plutonium containing ceramic pucks. The magazines were arranged so that the contents of each alternated between Aurex 95 ceramic and PR pucks.

The low pour rate and instrumented canisters were both filled with six magazines (24 cans) of Aurex 95 ceramic logs (4 per can) and one magazine (4 cans) of surrogate plutonium ceramic pucks. These surrogate plutonium ceramic pucks were fabricated by LLNL (designated EM/RW pucks) in accordance with applicable repository quality assurance requirements. The instrumented canister had thermocouples installed on the surface of the canister, the surface of cans, on the base plate, and inside the canister at various canister heights. It also had a camera installed on the top of the canister so that the inside of the canister could be viewed during glass pouring. The targeted pour rate for the low pour canister was 45 kg/hr. The targeted pour rate for the instrumented canister was between 82 to 109 kg/hr. Both of these canisters were sectioned after being filled to determine if glass voiding and/or hardware deformation had occurred. For all of the nominal pour rate canisters, a minimum pour rate of 68 kg/hr was required by the Experimental Test Plan. Table I summarizes the overall test configurations.

Table I. Phase 2 Canister Test Configurations

Canister	Purpose	Targeted Pour Rate	Puck Material	Instrumented (Yes/No)
Low Pour Rate	Destructive analysis	Low (45 kg/hr)	EM/RW pucks & Aurex 95 logs	No
Nominal Instrumented	Destructive analysis & thermal information	Nominal (82 to 109 kg/hr)	EM/RW pucks & Aurex 95 logs	Yes
Proliferation (PR1 and PR2)	Proliferation tests	Nominal (82 to 109 kg/hr)	PR pucks & Aurex 95 pucks	No

As with Phase 1, the main test parameters controlled were the 1) pour rate, 2) glass composition, 3) glass stream temperature, 4) glass stream fall height, 5) glass fill height, and 6) hardware configurations. By controlling these parameters, the important pour conditions (similar to those expected at the DWPF during can-in-canister glass filling) can be achieved. Below is a discussion of these various test parameters.

Pour rate - Feeding the Stirred Melter with glass at the desired rate controlled the pour rate as the melter was operated in an overflow pour mode. Each canister was placed underneath the superheater section of the Stirred Melter and glass was poured by means of the superheater pour valve. For pouring, the canisters were placed on a calibrated scale with insulation between the scale and the canister to protect the load cell. The pour rate was monitored by tracking the change in the weight of the canister over time.

Glass composition – The glass fed into the melter was Frit Phase 2 glass. SRS personnel developed the glass composition used for this test. The glass had a predicted viscosity of about 85 poise at 1150 °C.

Glass stream temperature – Unlike the DWPF Melter, there was no riser/pour spout with the Stirred Melter for this run. The glass stream temperature was determined by a thermocouple that is located in the superheater zone 3 glass pool just before it exits the pour valve and flows into the canister. The targeted temperature range was 1050 to 1150°C. In addition, periodic glass stream temperature readings were made just above the top of the canister by an optical pyrometer. These pyrometer readings were compared to those taken during DWPF pilot scale melter runs in the 1980s at various pour rates. This glass stream data was used as a "sanity check" to determine if pour temperatures were indeed similar to the DWPF Melter.

Glass stream fall height – The fall height from the bottom of the pour valve to the bottom of the canister was set up to be about the same (within 15 centimeters) as in the DWPF Melter.

Glass fill height – The minimum glass fill height specified for all canisters was 229 centimeters. During glass pouring the fill height was estimated by observing the height of the oxide layer that formed on the canister surface. This was compared to the calculated height of glass per the measured weight of glass poured. After the canisters were filled, they were removed and the actual glass height was then determined by inserting a tape measure into the top of the canister until it touched the top of the glass.

Hardware configuration – Each canister had the same hardware configuration. The hardware was chosen based on the Phase 1 results.

Proliferation 1 (Pr1) Canister Test

The filling of the Proliferation 1 (PR1) canister (DWPF canister S00301) resulted in a final glass weight of 1456 kilograms. The overall calculated pour rate was about 69 kg/hr. The pour rate was fairly consistent throughout the pour except for one pour stoppage. The lower than desired pour rate was due to the inability of the Stirred Melter superheater to pour at higher rates with this higher viscosity glass. After the canister was removed, the actual glass height was measured to be 246 centimeters. The calculated weight (assuming 6.17 kilograms of glass per one centimeter of glass in the canister) of glass poured into the canister was about 1518 kilograms. This was close to the actual measured weight of 1456 kilograms and implied that there was no significant glass voiding in the canister. The measured temperature of the superheater zone 3 glass just before the pour valve ranged from 1084 to 1109°C during the filling of this canister. In addition, the pour stream temperatures as measured by an optical pyrometer during the pour was 1070°C. Previous pilot scale work at SRS indicated that the glass stream temperature at a pour rate of 68 kg/hr is about 1040°C. Therefore, this pour stream appears to have been thermally similar to a DWPF pour at this fill rate.

Proliferation 2 (Pr2) Canister Test

The filling of the Proliferation 2 canister (DWPF canister S00302) resulted in a final glass weight of 1384 kilograms. The overall calculated pour rate was 69 kg/hr. After the pour the actual canister glass height was measured to be 232 centimeters. Therefore the calculated glass weight poured into the canister was about 1432 kilograms. This was close to the actual measured weight and again implied that there was no significant glass voiding in the canister. The measured temperature of the superheater zone 3 glass just before the pour valve ranged from about 1099 to 1110°C during the filling of this canister. In addition, the pour stream temperature as measured by optical pyrometer was 1050° C.

Low Pour Rate Canister Test

The filling of the Low Pour Rate canister (DWPF canister S00307) resulted in a final glass weight of 1453 kilograms. The overall calculated pour rate was 48 kg/hr. After the canister was removed, the actual glass height was measured to be 244 centimeters. Therefore the calculated glass weight poured into the canister was 1503 kilograms. This was close to the actual measured weight and again implied that there was no significant glass voiding in the canister. The measured temperature of the superheater zone 3 glass just before the pour valve ranged from 1075 to 1101°C during the filling of this canister. Four optical pyrometer temperature readings were taken on the pour stream and they ranged from 1020 to 1028°C. This agrees well with previous SRS pilot melter work.

There was an unusual observation made during the filling of the low pour rate canister. About twelve hours into the pour the canister glass weight was 567 kilograms. The oxide layer on the outside of the canister was, however, only 66 centimeters high. This glass height implied that only about 407 kilograms of glass had been poured. These two observations indicated that coning (the formation of a stalagmite) was occurring in the canister due to the low pour rate. After about nineteen hours of pouring the oxide layer was about 102 centimeters high but a column of oxide layer on one side of the canister was about 191 centimeters tall (see Figure 2). By the end of the glass pour, however, the oxide layer was at a fairly uniform height. Apparently a "stalagmite" had formed. As the stalagmite grew it finally reached a height where the glass then was hot enough to spill over the stalagmite. It then began to fill the canister at the lower glass height. An inspection of the inside of the canister after pouring showed a flat glass layer at the top of the canister. This confirmed the oxide layer observation and along with the final weight versus glass height data (and subsequent sectioning of the canister) showed that no major glass voiding had occurred. This pour test did show that pouring with a high viscosity glass at a very low pour rate (45 kg/hr) in a CIC canister was pushing the acceptable operational limits with regard to glass voids and coning.

Figure 2. Low Pour Rate Canister with Oxide Layer Column

Instrumented Canister Test

The filling of the Instrumented canister (DWPF canister S00304) resulted in a final glass weight of 1446 kilograms. The overall calculated pour rate was 72 kg/hr. After the pour the actual canister glass height was measured to be 236 centimeters. Therefore the calculated glass weight poured into the canister was about 1456 kilograms. This was close to the actual measured weight and again implied that there was no significant glass voiding in the canister. The measured temperature of the superheater zone 3 glass just before the pour valve ranged from about 1076 to 1085°C during the filling of this canister. In addition, three pour stream temperatures as measured by an optical pyrometer during the pour were 1028, 1059, and 1031° C. The instrumented canister had 30 thermocouples and a camera that viewed the inside of the canister during glass pouring. Details of the temperature data are given in the next section. The camera showed that the glass flowed from the centerline of the canister to the outside of the magazines. Sometimes the glass would flow around one magazine and then around the outside of several adjacent magazines before returning back to the centerline of the canister at a different location. This observation indicated that glass voiding was not occurring in this canister. This agreed with the comparison of the calculated and measured glass poured weights for this canister and gave yet more confidence that problematic glass voiding had not occurred.

Infrared Camera Test

An infrared camera was installed to measure the canister glass level during Phase 2 glass pouring due to concerns that the CIC hardware would hinder canister level measurements made from the camera at the DWPF. By viewing the image, the glass level in the canister can be approximated. The weight of the glass poured at the time that the image was taken was then measured and compared to the approximated weight. An example of this is given here. The approximated glass level from the image (see Figure 3) was 218 centimeters. The calculated glass level per the measured glass weight of 1241 kilograms at that time was about 208 centimeters (within 10 centimeters of the observed image level). The "accuracy" of the existing DWPF infrared camera technique is about +/- ten centimeters. The above data indicates that an infrared camera can be used to determine glass level at the DWPF when CIC hardware is used.

Figure 3. Infrared Image of Canister S00301 at 13:42 on 8/25/00

INSTRUMENTED CANISTER TEMPERATURE RESULTS

The Instrumented canister was configured with 30 thermocouples to determine the thermal profile of the canister during glass pouring with the can-in-canister hardware. One thermocouple was welded to the bottom of the outside of the canister. Six thermocouples were welded to the outside of the canister at heights of 71, 137, and 213 centimeters (two located 180° apart at each height). Nine thermocouples were inserted through the canister to a radial location of 7.6 centimeters from canister centerline and at canister heights of 31, 61, 99, 112, 130, 163, 196, 213 and 231 centimeters. Five more thermocouples were inserted through the canister to a radial location of 26.7 centimeters from canister centerline and at canister heights of 31, 61, 112, 163, and 213 centimeters. Two thermocouples were welded to the radial outside and inside of three different cans on magazine 1 (EM/RW puck cans) at heights of 61, 112, and 213 centimeters. Thermocouples were also welded to the base plate and one to the canister throat.

Figure 4 plots the temperatures of the glass inside the canister 7.6 centimeters from the canister centerline at various glass heights. The maximum glass temperatures occur as the glass level reaches the various thermocouples. The maximum glass temperature measured was about 1040 °C. The cool down rates for the glass agreed quite well with past tests at SRS. These cool down rates indicate that the glass in CIC canisters will cool fast enough so that devitrification (crystallization) of DWPF type glass at the centerline of the canister will not occur. This information is important because the centerline is the canister location where devitrification is most likely to occur. In addition, data from the can thermocouples showed that the maximum observed can temperatures (900 °C) were well below that in which the stainless steel would be expected to fail due to weakening at elevated temperatures. In summary, the heatup and cool down rates of the CIC hardware and glass were found to be acceptable for DWPF operations.

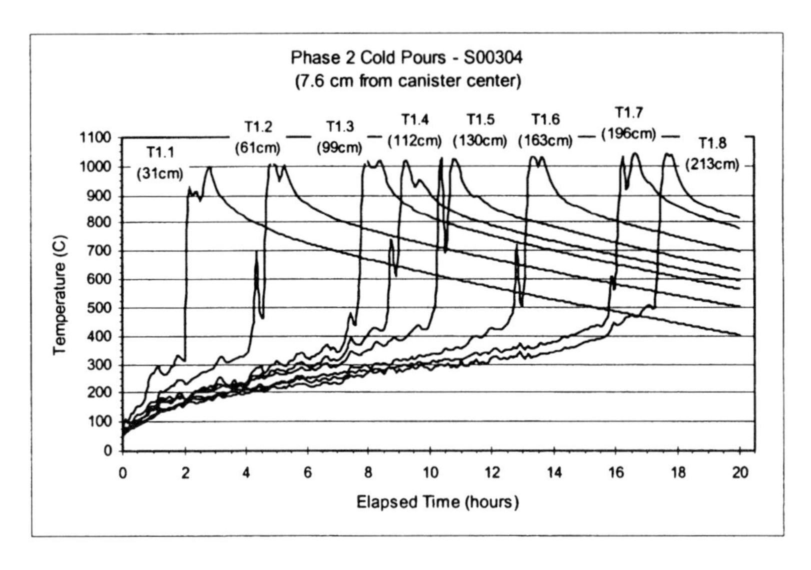

Figure 4. Canister S00304 Glass Temperatures 7.6 Centimeters from Canister Centerline during Pouring

CANISTER SECTIONING

After the low pour rate and instrumented canisters were filled with glass at the CETL, they were sectioned at 4 canister heights at the CETL. A vendor contracted by the CETL used a diamond wire saw technique similar to that used to cut the Phase 1 canisters. After cutting the canisters, they were to be inspected by SRTC personnel to determine the extent of glass voiding and hardware integrity. General observations of the cross-sectioned canisters are as follows:

1. At the 14.6 centimeter cuts, no glass voids were seen in the canister region. As was expected, only the interior of the magazine cones did not flood with glass.
2. There was no appreciable wetting of the stainless steel surfaces by the glass.
3. There was one small void in the low pour rate canister at the 58 centimeter cut that was about 33 centimeters long and 7.6 centimeters by 7.6 centimeters wide.
4. There were no other voids observed. There was good glass flow into the region between the cans and the magazines. Cans were tightly held in place and could not be removed by hand.
5. There was less glass cracking near the centerline of the canister (same as current DWPF canisters).

The small void found was similar in size to voids that have been observed in DWPF pour testing without the presence of CIC hardware. Therefore from an operational perspective, the voiding is considered insignificant. The good glass flow between the cans and magazines, especially at the lower canister heights, was also significant in light of the low pour rate and high viscosity glass used for this canister. After the canisters were sectioned and visually inspected, the CIC hardware was measured to determine if it had experienced any plastic deformation during the test. The conclusion was that the CIC hardware did not experience measurable plastic deformation during the pour or as the glass cooled.

CONCLUSIONS

The Can-in-Canister System (7 magazine/28 cans) was tested during the Phase 2 tests at the CETL by SRS personnel. Four canisters with the hardware chosen from Phase 1 were filled under various DWPF pour conditions. Comparisons of predicted glass pour weights with measured glass weights in the canisters, the video-taping of the inside of one canister during glass filling, and subsequent cross sectioning of the low pour rate and instrumented canisters, along with subsequent inspections of these two canisters showed little (if any) glass voiding or hardware deformation. Subsequent measurements of the hardware after the initial visual inspections verified that the hardware did not experience measurable plastic deformation during the test. In addition, temperature data from the instrumented canister indicate that the cans will not reach temperatures that could result in the rupture of the cans that house the plutonium containing ceramic pucks. The data also indicate that temperatures during glass pouring and cool down will not adversely impact DWPF operations or glass quality.

Therefore, the CIC System was proven to be a viable option for the disposition of excess weapons useable plutonium. The Phase 2 test was performed to demonstrate compliance with certain requirements specified in the Plutonium Immobilization Product Specifications (PIPS). Two of the canisters were filled in Phase 2 for possible testing related to the proliferation resistance of the CIC form.

REFERENCES

1. "Phase 1 Can-in-Canister Cold Pour Tests for the Plutonium Immobilization Project (U)", WSRC-TR-99-00337, M. E. Smith and Gregory Hovis, October 30, 1999.

REAL-TIME DETERMINATION OF THE REDOX STATE OF GLASSES – DIRECT POTENTIOMETRY VS CHEMICAL ANALYSIS

Darryl D. Siemer* and John A. McCray
Bechtel BWXT Idaho, LLC
P.O. Box 1625
Idaho Falls, ID 83415-7111

ABSTRACT

This paper describes recent INEEL research directed to the development of a practical means of obtaining real or near-real time feedback on the redox state of simulated radioactive waste (radwaste) glasses. Three basic approaches were investigated. The first is a streamlined version of the conventional spectrophotometric method which determines the ratio of ferrous to total iron in aqueous solutions prepared from the glasses. The second approach involves potentiometric measurement of Fe^{++}/Fe^{+++} ratios in similar solutions. The third monitors the voltage between a platinum sensing surface exposed to the molten glass and another exposed to a redox-buffered reference glass.

INTRODUCTION

The oxygen fugacity (pO_2) of a radwaste-type glass melter affects both its productivity and the durability (leach resistance) of its product[1]. If the system becomes too oxidizing, a gross degree of foaming may be encountered; if it becomes too reducing, elemental metals and sulfides may precipitate out and short out the heating electrodes. Unfortunately, this parameter is rather difficult to control for these reasons:

1) the feed streams are chemically and physically heterogeneous
2) these streams usually contain substantial concentrations of both oxidants (e.g., nitrate) and reductants (e.g., organic materials)
3) access to the internals of the melters is difficult - they are designed to be operated based upon detailed prior knowledge of their characteristics and what goes into them, not real-time feedback.

To the extent authorized under the laws of the United States of America, all copyright interests in this publication are the property of The American Ceramic Society. Any duplication, reproduction, or republication of this publication or any part thereof, without the express written consent of The American Ceramic Society or fee paid to the Copyright Clearance Center, is prohibited.

This study of how to go about obtaining more timely redox feedback represents a part of the Idaho National Engineering and Environmental Laboratory's (INEEL) effort to evaluate the applicability of vitrification to "sodium bearing waste" (SBW).

EXPERIMENTAL

Equipment: SPECTRONIC Model D1 spectrophotometer with test tube-type cuvettes, multichannel peristaltic pump (INSTATEC Model 78005-10) digital multimeters (DVMs), a muffle furnace, miscellaneous glass/plastic lab ware, top-loading analytical balances, micropipets, alumina crucibles, rods, and tubes, platinum crucibles and wire, etc.

"Redox Reagent": Dissolve 0.5 grams 1,10 (ortho) phenanthroline in 80 grams of glacial acetic acid and slowly add it to a solution of 80 grams of NaOH and 50 grams of boric acid (H_3BO_3) in ~800 cc of water - dilute to 1 liter.

Spectrophotometric Approach

The relative amount of each member of a reversible redox couple (e.g., ferric and ferrous ions) in a molten glass do not change much when it is rapidly cooled[i]. This means that the ratio of their concentrations in the "frozen" glass accurately reflects the redox state of the melter. Since radwaste-type glasses generally contain a good deal of iron (several percent) and appreciable amounts of each of its common oxidation states (Fe^{+++} & Fe^{++}) co-exist at desirable melter pO_2 levels, it has become the industry-standard "redox indicator"[ii]. While this "redox ratio" can, in principle, be determined in ways that do not require dissolution of the sample[iii], it is quicker/simpler/cheaper to perform the analyses on solutions. This is generally accomplished by dissolving powdered glass in a mixture of HF and H_2SO_4 followed by sequential spectrophotometric determination of first, ferrous and then, total iron[2]. Since the details of how this is done vary considerably, one goal of this project was to develop a version that addresses genuinely important issues while being simple enough to be readily adapted to remote use on radioactive samples. Key issues include:

1) dissolution must not affect the redox state of the iron
2) the subsequent analysis must be able to measure small amounts of ferrous ion in the presence of a large excess of ferric iron and vice versa.

---

[i] The main reason for this is that the high viscosity of molten glasses causes the mass transfer rate (diffusion) of species which could affect their redox state during cool-down (e.g.; elemental oxygen) to be extremely slow[2].

[ii] Most authorities now agree that a *conservative* estimate of the acceptable range of pO2 levels corresponds to Fe^{++}/Fe^{+++} ratios from 0.1 to 0.5. This corresponds to a Fe^{++}/Fe_{total} of 0.09 to 0.33.

[iii] E.g., Mössbauer spectrometry, magnetic susceptibility, laser-excited fluorimetry, & near IR/visible absorption spectrometry

Most of the differences noted in published protocols have to do with the extent of precaution exercised to prevent air-oxidation of ferrous iron during dissolution. These include: 1) doing the dissolution under an inert gas blanket[2-3]; 2) adding vanadate ion to "protect" the ferrous iron[4]; 3) doing it in the dark to forestall UV-induced oxidation; and, 4) minimizing the effects of all potential bias sources by doing both the dissolution and subsequent analyses quickly[5].

A good deal of repetitious experimentation indicated that the last approach best suits our needs. That effort indicated that most of the *potential* sources of analytical bias mentioned in the published protocols do not, in actual practice, cause difficulties. For example, several hours spent comparing the apparent Fe^{++}/Fe_{total} ratios of sub-samples of a single batch of a rather oxidized glass (i.e., there wasn't much ferrous iron in it – Fe^{++}/Fe_{total} = 2%) dissolved either under carbon dioxide or open to the atmosphere, indicated that a significant degree of air-oxidation of ferrous iron did not occur. This study also suggested that many of the analytical protocols are unnecessarily complicated, tedious, and, therefore, slow. For example, since the ferrous o-phenanthroline complex forms quite rapidly and exhibits the same molar extinction coefficient over a six-unit pH range (3 to 9), there is no compelling reason for these methods to insist that the pH of the final solution be adjusted to within +/-0.1 of any particular value or that the absorbance responses be noted only at specified times. The following is the version developed for INEEL's glass research group.

1) Grind the sample to a flour-like powder and place[iv] several (typically 2 to 10) milligrams of it into a conical-bottom, 50 cc plastic centrifuge tube.
2) Add 3-5 drops of concentrated H_2SO_4 and then 3-5 drops of concentrated (19 M) HF. Shake the tube for 2-3 seconds (radwaste-type glasses immediately dissolve)
3) Add approximately 10 mL of "Redox Reagent" plus enough water to bring the total volume to 25 mL (the tube's calibrations are accurate enough).
4) Pour some of the solution into a cuvette and immediately read its absorbance (A1) in the spectrophotometer at ~510 nm.
5) Pour the solution back into the centrifuge tube, add 50-100 milligrams of ascorbic acid, shake to mix, refill the cuvette and reread the absorbance (A2)[v].

[iv] If only the Fe^{++}/Fe_{total} ratio is needed, it is not necessary to weigh the sample – its weight is required only if the absolute iron concentration is of interest.

[v] If either of the absorbance responses is so high that compliance with Beer's law cannot be assumed (absorbances up to 1.5 are fine with a good-quality spectrophotometer), the analyst may

Calculations:

$$Fe^{++}/Fe_{total} = A1/A2$$
$$Fe^{++}/Fe^{+++} = A1/(A2-A1)$$
and
$$\text{Wt \% total iron}^{vi} = 12.4/\text{milligrams of sample}$$

The virtues of this method - it is simple, generates very little waste, and requires only a tiny sample – mean that it would be relatively easy to "remote". These characteristics facilitated discovery of a characteristic of DOE's HLW durability benchmark (EA glass) relevant to its applicability as a redox standard; i.e., that the redox state of individual granules vary significantly and in systematic fashion – the bigger chunks are more oxidized. The evidence is as follows: 1) the relative standard deviations (RSD) of replicate determinations done on individual granules of the glass was much greater than those seen when many granules were composited before analysis - typically 50% vs. 6% RSD; 2) when the as-received glass shards were screened prior to testing, the coarsest size cut (>20 mesh) consistently exhibited 40% lower Fe^{++}/Fe_{total} ratios than did the finest size cut (<40 mesh); 3) an individual, very large (over 1 cm long), shard reproducibly exhibited a Fe^{++}/Fe_{total} ratio of 0.035 – only about one fifth of the material's nominal redox ratio, ~ 0.18[6]. This phenomenon was not unique to EA glass – redox state heterogeneity was noted in several lab-prepared glasses.

Solution Potentiometry

In principle, the Fe^{++}/Fe^{+++} ratio of aqueous solutions can be determined with the same oxidation/reduction ("ORP") probes used as detectors for redox-type titrations. Our initial attempts at applying a conventional platinum wire-based ORP electrode to HF/H_2SO_4 solutions were not very promising because the results were irreproducible. However, a few more hours of "trial and error" testing indicated that this approach could be made to work if the surface area of the sensing electrode were increased, all "free" HF complexed with boric acid, and some chloride added to the solutions. Figure 1 is a plot of "Redox reagent"-determined Fe^{++}/Fe^{+++} ratios vs. potentiometric responses observed for five batches of a single glass former/simulant mix melted with differing proportions of sugar. The potentiometric measurements involved dissolution of several milligrams of glass in a 3 drops each of HF/H_2SO_4 followed by the addition of ~ 5 cc of water, one drop of concentrated HCl, and about 1 cm^3 of boric acid (enough to saturate the solution). Most of the resulting solution is transferred to a platinum crucible which serves as both the sensor & container, a homemade

---

either repeat the procedure using a smaller sample or dilute the solution, reread the absorbance, and correct for the dilution factor.

[vi] This constant (12.4) assumes a 1cm path cuvette.

Ag/AgCl/0.1 M Cl^- reference electrode is inserted, and the voltage between it and the crucible is measured with a DVM.

Figure 1: Redox Reagent-determined Fe^{++}/Fe^{+++} ratios vs. solution potentiometric responses

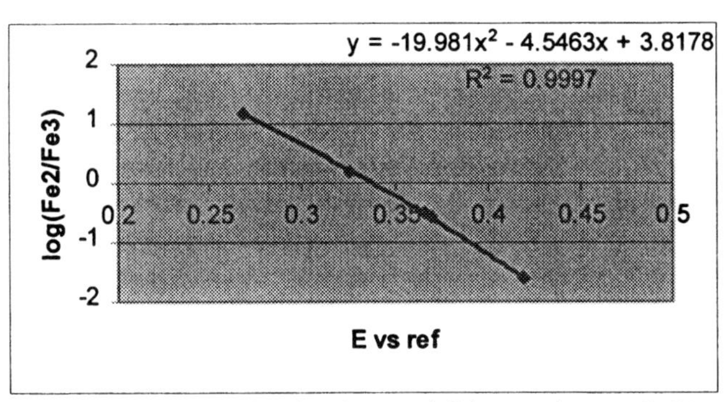

A linear least-squares fit of this data exhibits a slope of 55 mv/decade (59 mv/decade is "ideal") and a correlation coefficient of 0.997.

Direct Potentiomety[vii]

The fundamental weakness of any solution-based method is that other multivalent species may alter the measured Fe^{++}/Fe^{+++} ratio and thereby interfere. For example, while the potential of the V^{IV}/V^{V} redox couple is almost coincident with that of Fe^{++}/Fe^{+++} under melter conditions[Ref.1, Fig.3], it is quite different in most aqueous solutions. In practice this means that V^{V} (vanadate) oxidizes ferrous ion during the HF/H_2SO_4 dissolution of glass. While the subsequent addition of a ferrous ion selective complexing agent (e.g., o-phenanthroline or "Ferrozine") tends to reverse that reaction (i.e., the V^{IV} co-produced by the oxidation now serves to reduce Fe^{III}), there will still be a net interference if the glass contained vanadyl (V^{IV}) ion in addition to vanadate. If the sample contains more vanadium than iron, this interference can totally invalidate the test.

Consequently, in principle at least, the best way to determine the redox state of a radwaste melter is via in-situ "direct" potentiometry. This is done by measuring the voltage between reference and sensing electrodes immersed in the molten glass. The state-of-the-art approach to doing it utilizes two inert (usually Pt) sensing surfaces separated by a yttria-stabilized ZrO_2 "membrane" which is selectively conductive to oxide ion[7]. The voltage between the Pt surface contacted by the melt and the one separated from it by the ZrO_2 membrane and surrounded by a reference gas = $(RT/4F)\log_{10} [p(O_2)_{ref}/p(O_2)_{sample}]$

[vii] The reverse reaction is rather slow which probably accounts for the fact that SRS's current method (which adds V^{v} prior to dissolution) specifies a fifteen minute-delay between chromogen addition and the first absorbance measurement[4].

where:

$p(O_2)_{ref}$ is typically 0.21 atm.- i.e., air at one atmosphere)

R = gas constant, T =absolute temperature, & F = Faraday's constant.

ZrO_2 membrane-based oxygen sensors have been commercially available for about two decades and there is no compelling reason to believe that they could not be used in radwaste-type glass melters. They have not been embraced by the US radwaste glass industry because its melters are not designed to accommodate them and they are considered to be both too expensive and too fragile.

Space limitations permit only a cursory description of our own efforts to develop a practical direct potentiometric sensor. The first system examined utilized a "redox glass" reference electrode. It was constructed from a piece of double bore, alumina thermocouple (TC) insulation, one hole of which was plugged at the "bottom" end with a coarse-grade alumina cement "frit" and then partially filled with a low-melting borosilicate "eutectic glass" powder (nominal mp = 500°C) doped with a 10:1 mix of Fe^{++}/Fe^{+++} oxides. While the "reference glass" was being melted by inserting the frit-plugged end of the TC tube into a glass-working torch, a stiff platinum "reference wire" was threaded into it from the top. The "sensing electrode" consisted of several turns of platinum wire fed down through another length of TC insulation and wrapped around its lower end.

The nominal composition of a typical simulant glass ("RSM-1 glass") formulation examined in this study would be: 8.4% Al_2O_3, 10.6% B_2O_3, 4.2% CaO, 7.4 % Fe_2O_3, 3.5% Li_2O, 2.3% K_2O, 15.7% Na_2O, 1.1% SO_3, and 45.5% SiO_2. 2-4 gram portions of the sample glass were melted in alumina crucibles placed directly under the furnace's top access port[viii] adjacent to its feedback control TC. The electrodes were introduced through the port and poked into the melt. A general-purpose[ix] digital multimeter measured the voltage between the Pt wires. Figure 2 is typical calibration graph (response curve).

The first step in generating such a plot involves reduction of the sample by stirring some powdered charcoal into it. After copious gas evolution ceases, the electrodes are inserted into the melt and allowed to equilibrate for several minutes[x]. EMF readings and tiny samples of the glass are then taken at regular

[viii] The THERMOLYNE model 4900 muffle furnace used for this work features a 12.5 cm-wide, 10 cm-high, and 14 cm-deep chamber with a 2.5 cm diameter, access port in the middle of its top.

[ix] Electrometer-type voltmeters like those required for pH measurements are not necessary. An advantage of doing potentiometry in molten glasses is that the "exchange currents" are high.

[x] At 1150°C with these high-iron glasses, this system takes about five minutes to come to within 2-3 millivolts of steady-state.

intervals as the melt gradually air-oxidizes back to an equilibrium state[xi] (which can take from one to five hours depending upon the temperature and whether or not the glass is stirred).

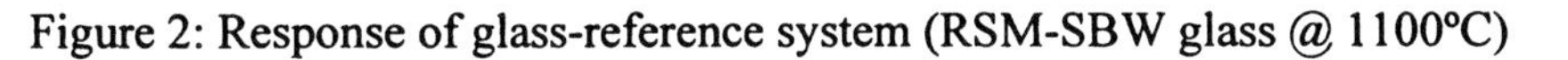
Figure 2: Response of glass-reference system (RSM-SBW glass @ 1100°C)

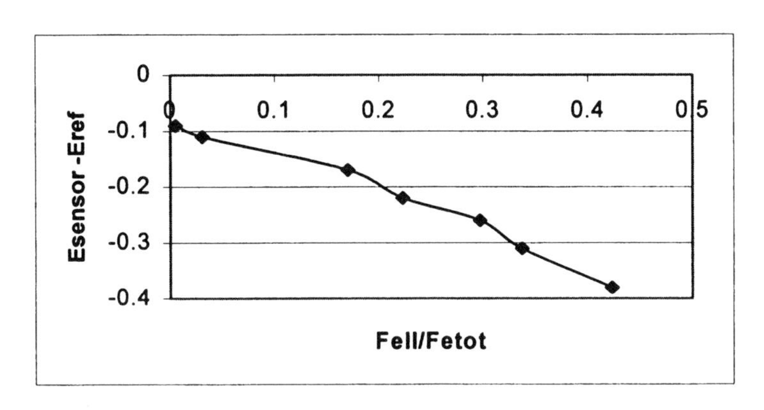

Several conclusions may be drawn from this figure. First, the detector's response to changes in the Fe^{2}/Fe_{tot} ratio are in the right direction; i.e., as the glass surrounding the "sensing" wire becomes more reduced ("electron rich"), its potential relative to the "reference" wire becomes more negative. Second, its response is of the correct order of magnitude i.e., a plot of voltage vs. $\log_{10}[Fe^{++}/Fe^{+++}]$ of this data exhibits a slope of ~0.36 volt/decade vs. the 0.27 expected for a rigorously "Nernstian" system at 1100°C. Third, the fact that the data points do not fall on a smooth curve, means that it exhibits "noise" and/or "drift"

Some of the reasons for its imperfections include: 1) neither electrode is shielded so that there are no chemical or thermal gradients along its sensing surface; 2) the Fe^{++}/Fe^{+++} ratio of the glass contained within the reference electrode slowly changes[xii] which causes its potential to drift.

An excellent review[8-9] of direct potentiometric sensors written two decades ago provided us with the key to a system which may be able to solve the problem. An "ideal" version of it would consist of two identical Pt wires sealed into the tips of two zirconia (which is more durable than alumina) shield tubes so that only a short length of each would be exposed to molten glass. The "sensing electrode"

[xi] To take a sample, the sensing probe is pulled up out of the furnace and a gob of molten glass pinched off its tip with pliers.

[xii] oxygen slowly diffuses through both the cement-plug "frit" and the wall of the alumina tube

would be immersed directly in the melt. The "reference electrode" would be inserted into a larger i.d. shield tube which would permit a "reference gas" (e.g., air) to be pumped downwards through it. The tip of the reference wire would be bent sideways to contact the inner wall of the outer shield tube immediately above the bottom. A small flow of reference gas introduced into the top of the shield tube with a positive displacement-type pump would serve to prevent molten glass from backing up into it and maintain the glass which contacts the tip of the "reference" wire at a constant redox state. The sensing electrode would be situated some distance away from the reference electrode to prevent the latter's reference gas from influencing the pO_2 of the glass it senses.

Figure 3 depicts the response of a non-ideal[xiii] version of this system to the same glass used to generate the data in Figure 2. 21 cc/minute of air was pumped into the melt through the reference tube. The "exchange currents" which served to "poise" the potentials of the electrode surfaces were ~ 0.1 ma [xiv].

Figure 3: Response of the pumped-air reference electrode system to (RSM-1 SBW glass @ 1100°C)

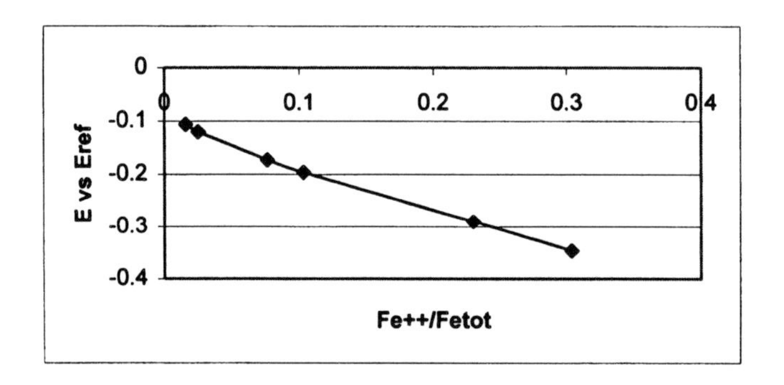

Note that this system responds much like the one discussed previously but doesn't seem to be as "noisy".

CONCLUSIONS

The full "acceptable" range of pO_2 concentrations in radwaste glass melters encompasses 7 orders of magnitude regardless of temperature[1]. This translates to

[xiii] "not ideal" because neither of the Pt wires were sheathed in a way which restricted the sensing surface to a region with uniform conditions.

[xiv] Exchange currents were determined by using a second DVM to measure the voltage imposed by the first when it was switched from its "volt" to "ohm" range ($I_{exch.} = V_{imp.}/ohms$).

a 7/4 or 1.75 order-of-magnitude range of Fe^{++}/Fe^{+++} ratios which, at 1100°C, corresponds to a potentiometric response range of 477 millivolts (1.75*RT/F). Another fact that should be kept in mind is that the mean residence time of solid-forming materials in typical radwaste melters is over one day. Taken together these suggest that "slow" response and/or a few millivolts worth of drift or offset would not seriously compromise the value of a potentiometric sensor as a process feedback transducer. The second system described in this paper possesses the same fundamental virtue that the commercial ZrO_2 membrane-based systems do; i.e., that the potential of its reference electrode can be reproducibly fixed at any convenient point by simply pumping a suitable reference gas (e.g., air or CO_2) into it. Most importantly, such a system should be cheap to make (or replace), easy to use, and physically rugged enough (no "membranes" to corrode) to use in a real melter. Of course, this conclusion must be considered as "tentative" at this point because it is largely based upon the results of scoping-type laboratory experiments done with rather simple glasses. During the next year we hope to be able to test it with a much wider array of glasses and install a ruggedized version in a pilot-scale melter.

REFERENCES

[1] H. D. Schreiber, C. W. Schreiber, M. W. Reithmiller, and J. S. Downey, "The Effect of Temperature on the Redox Constraints for the Processing of High-Level Nuclear Waste Into a Glass Waste Form", *Mat. Res. Soc. Symp. Proc.*, (176) 1990, 419-426.

[2]H. D. Schreiber, S. J. Kozak, A. L. Fritchman, D. S. Goldman, and H. A. Schaeffer, "Redox Kinetics and Oxygen Diffusion in a Borosilicate Glass Melt", *Physics and Chemistry of Glasses*, 27(4), 1986, pp. 152 –177 (fig. 3).

[3]O. Corumluoglu and E. Guadagnino, (Compilers), "Determination of Ferrous Iron and Total Iron in Glass by a Colorimetric Method - Report of ICG/TC", *Glass Technology*, 1999, 40(1), pp. 24-28.

[4]"Determining Fe^{+2}/Fe^{+3} and Fe^{+2}/Fe_{total} Using the HP8452A Diode Array Spectrometer", Procedure 1.8 Rev. 1, SRTC Mobile Laboratory Procedure (the Savannah River Site's current method)

[5]"Iron II and Total Iron Ratio", PNNL Procedure APSL0-2, 15 pp.

[6]C. M. Jantzen, et. al., "Characterization of the Defense Waste Processing Facility (DWPF) Environmental Assessment (EA) Glass Standard Reference Material", doc. WSRC-TR-92-346, Rev 1, June 1, 1993.

[7]H. Muller-Simon and K. W. Mergler, "Electrochemical Measurements of Oxygen Activity in Glass Melting Furnaces" *Glastech. Ber.*, 61 (1988), pp. 293-299.

[8]T. Tran and M. Brungs, "Application of Oxygen Electrodes in Glass Melts. Part 1. Oxygen Reference Electrode", *Physics and Chemistry of Glasses*, Vol. 21 (4), 1980, pp. 133 –140.

[9]T. Tran and M. Brungs, "Application of Oxygen Electrodes in Glass Melts. Part 2. Oxygen Probes for the Measurement of Oxygen Potential in Sodium Disilicate Glass", *Physics and Chemistry of Glasses*, Vol. 21 (5), 1980, pp. 178-183.

Crystallization in Nuclear Waste Forms

CRYSTALLIZATION IN HIGH-LEVEL WASTE GLASSES

Pavel Hrma
Pacific Northwest National Laboratory
K6-24, P.O. Box 999
Richland, WA 99352

ABSTRACT

This review outlines important aspects of crystallization in HLW glasses, such as equilibrium, nucleation, growth, and dissolution. The impact of crystallization on continuous melters and the chemical durability of high-level waste glass are briefly discussed.

INTRODUCTION

The U.S. Department of Energy must dispose of large quantities of high-level waste (HLW) at several facilities. For example, approximately 55-million gallons of HLW are being temporarily stored at the Hanford Site in Washington State. Present plans are to convert this waste to borosilicate glass for permanent disposal. Most U.S. HLWs are mixtures of predominantly refractory components with limited solubilities in borosilicate glass. Vitrification of these wastes and the slow cooling of HLW glass in canisters provide an opportunity for a variety of crystalline forms to precipitate. Similar to natural minerals, crystalline phases from HLW glass are typically solid solutions and may contain a large number of components in proportions in which they do not usually occur in nature.

Crystallization in a continuous melter may produce an undesirable sludge on the melter bottom, and precipitation of crystals in a HLW glass canister may have a detrimental impact on glass quality. These undesirable effects limit the level of waste loading in glass [1] and have in turn serious economic consequences.

This paper reviews some results of crystallization studies. It covers phase equilibria, crystal nucleation, growth, and dissolution. The impacts of crystallization on continuous melters and HLW glass quality are discussed.

PHASE EQUILIBRIA IN HLW GLASS

Crystallization from HLW glasses has been studied since the 1970s. Early studies focused on identifying primary crystalline phases and phases that precipitate during slow (canister centerline) cooling or the storage of hot glass [2-5]. These studies were motivated by the need to characterize the immobilized waste product and to determine the impact of crystallization on its behavior.

To the extent authorized under the laws of the United States of America, all copyright interests in this publication are the property of The American Ceramic Society. Any duplication, reproduction, or republication of this publication or any part thereof, without the express written consent of The American Ceramic Society or fee paid to the Copyright Clearance Center, is prohibited.

Commercial HLWs with high Mo content precipitate molybdates or form a separated modyblate phase that crystallizes on cooling [2,3,6]; chromates and CeO_2 are also common [2,6]. Over 10 different crystalline phases, mainly spinel and zirconium-containing minerals, and various silicates were identified in heat-treated Hanford glasses [7]. Glasses rich in fluorine precipitate fluorite, fluorapatite, and fluorosiliactes [8,9]. Crystalline phases were also investigated in glass for Am and Cm immobilization [10,11] and in low-activity waste glass [12].

Though pure compounds, such as RuO_2, ZrO_2, or $ZrSiO_4$, occur in HLW glasses, crystalline phases are typically solid solutions.. Spinel is a solid solution of trevorite with magnetite and nichromite [13], clinopyroxene is a solid solution of acmite with diopside and hedenbergide [7], and nepheline contains albite and iron silicate [14]. The composition of these solid solutions changes with time and the temperature of crystallization. For example, the concentration of chromium in spinel increases as the heat-treatment temperature approaches liquidus temperature (T_L) [13].

The presence of crystals in the melter became an issue only when the defense HLWs, which generate low radiolytic heat, were considered for vitrification by continuous melting. With exception of noble metals, crystallization does not occur at the processing-temperature interval of commercial HLW glasses because their waste loading is restricted by radiolytic heat to a low level. To protect continuous melters for defense HLW glass from crystal settling and accumulation, attention was focused on the T_L as a function of glass composition [7,15-21]. The aim was to reduce T_L below a prescribed level by optimizing glass composition. A large database of glass properties including T_L was systematically developed for the Hanford composition region [22]. The equilibrium fraction of crystalline phases below T_L was also reported [15,23-25] to assess the temperature at which the crystallinity fraction was at a low level that the melter could possibly tolerate. Most equilibrium data have been generated on spinel, zirconium-containing phases (baddelyite, zircon, and a host of alkali zirconium silicates), nepheline, and clinopyroxene (such as acmite).

Spinel is a ubiquitous crystalline phase in glasses containing Fe with Ni, Cr, Mn, and Zn [13,20,21,26]. Typical spinel crystals are shown in Figure 1. The liquidus surface in the spinel primary phase field is nearly planar in HLW glasses, so the T_L within this field is a linear function of glass composition, i.e., partial T_Ls are nearly constant (Figure 2). Hence,

$$T_L = \sum_{i=1}^{N} T_{Li} x_i \tag{1}$$

where T_{Li} is the i-th component partial liquidus temperature, x_i is the i-th component mass or mol fraction, and N is the number of components. It has been demonstrated for spinel that T_{Li}s of most components are closely related to the ionic potentials of metallic ions (Figure 3) [21].

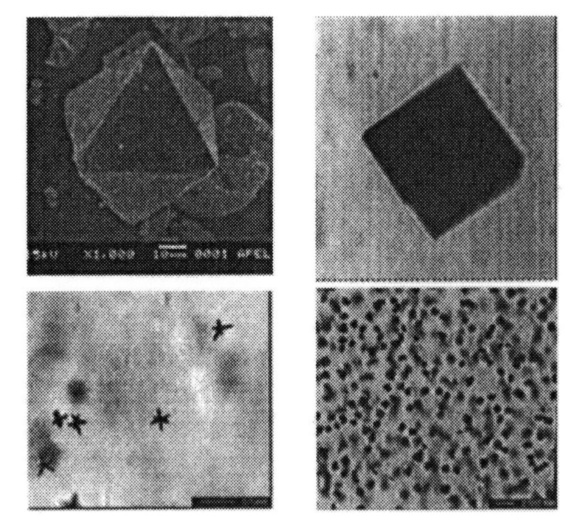

Figure 1. Spinel crystals in HLW glass (clockwise from top left): A scanning electron microscopy (SEM) image of an octahedral crystal isolated from glass by acid digestion; a large (approximately 20 μm) spinel crystal in transmitted light microscopy; a typical light microscopy image of HLW glass with spinel; star-like spinel crystals that typically grow in glass at temperatures below 850°C.

As Figure 2 shows, for glasses with a spinel primary phase, the T_L is most increased by Cr and Ni and mostly decreased by alkali oxides [7,20,21,24]. The T_L associated with the spinel primary phase is sensitive to p_{O2} [26-28]; in MS-7 glass, the maximum T_L was measured at $p_{O2} = 2.0\times10^{-4}$ Pa [28] (Figure 4); the T_L sharply decreased in a more reduced or more oxidized glass (see also [29]).

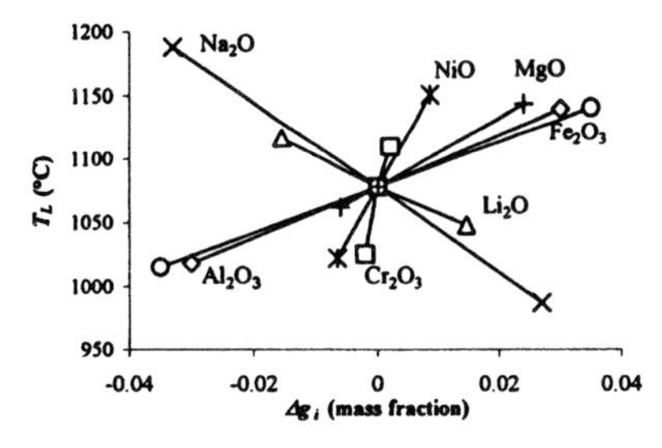

Figure 2. T_L as a function of mass fraction of a component added to (or removed from) MS-7 glass (baseline) [24]; $\Delta g_i = g_i(\text{glass}) - g_i(\text{baseline})$; g_i is the i-th component mass fraction

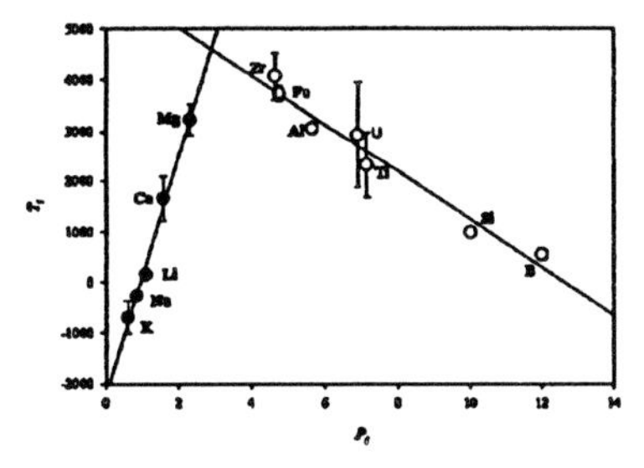

Figure 3. Partial liquidus temperatures of element versus ionic potential [21]

Zirconium-containing phases tend to become primary in HLW glasses with more than 7 mass% ZrO_2 [16,18]. Which phase precipitates can be estimated from a submixture rule [30]: if $Na_2O/(Na_2O+ZrO_2+SiO_2) > 0.22$ (the formula stands for the mass fraction of an oxide), alkali zirconium silicate, such as parakeldyshite, is likely to form; if $ZrO_2/(Na_2O+ZrO_2+SiO_2) > 0.22$, baddeleyite can be expected; otherwise zircon forms as a most common crystalline phase. The maximum solubility of ZrO_2 at 1050°C is roughly 16 mass% ZrO_2 at the triple point [31]. Micrographs of typical zirconium containing crystals are in Figure 5.

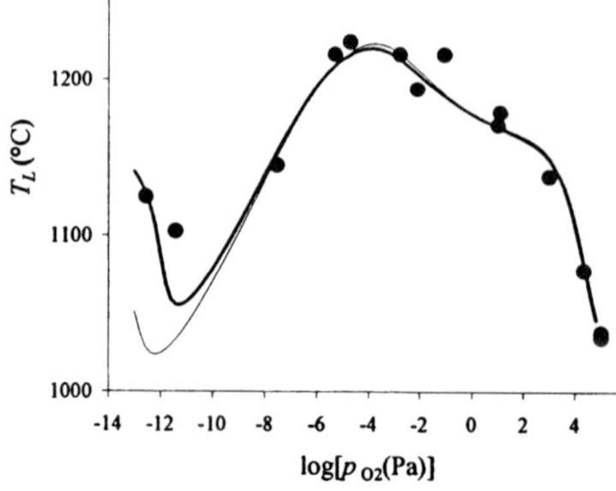

Figure 4. T_L of MS-7 glass as a function of p_{O2} [28]

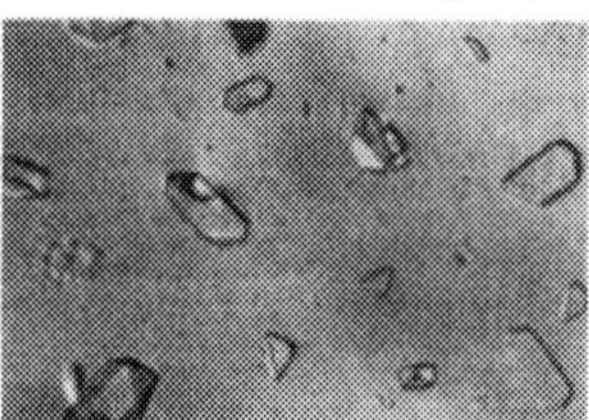
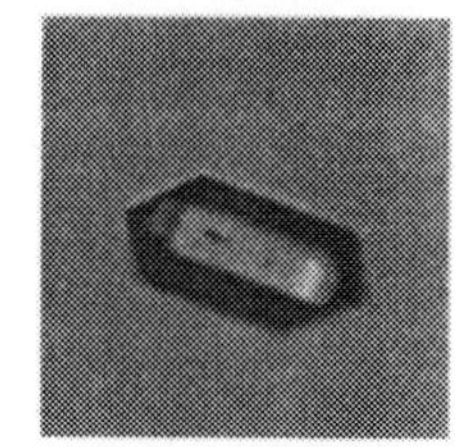
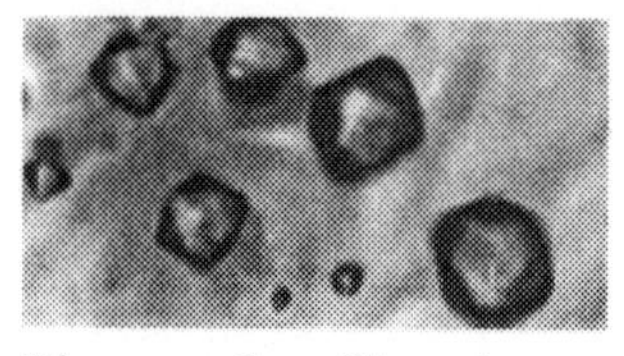

Figure 5. Zirconium-containing crystals (clockwise from top left): parakeldyshite; zircon; baddeleyite; baddeleyite dendrite

Nepheline, usually with Fe and excess Si, precipitates from low-Si glasses, in which $SiO_2/(Na_2O+Al_2O_3+SiO_2) < 0.62$ [14]. Nepheline crystals can have various forms, from individual crystals (Figure 6) to densely grown dendrites. More than 50 mass% nepheline can precipitate from a HLW glass [14,32].

The equilibrium fraction of a single crystalline phase below T_L has been measured for spinel [23,33] and nepheline [14]; it can be represented by the ideal solution equation that corresponds to the line fitted to data in Figure 7, the parameters of which depend on glass composition [24]. A nearly linear temperature dependence of the equilibrium fraction of spinel below T_L was also reported [27,34]. The linear dependence (and a linearized form of C vs. T function) is compatible with Equation (1).

The spinel equilibrium fraction in HLW glass rarely exceeds 3 mass%. The spinel distribution in a HLW glass canister was evaluated by Schweiger et al. [35]. Spinel crystals were organized in layers [35,36] due to the Liesegang effect.

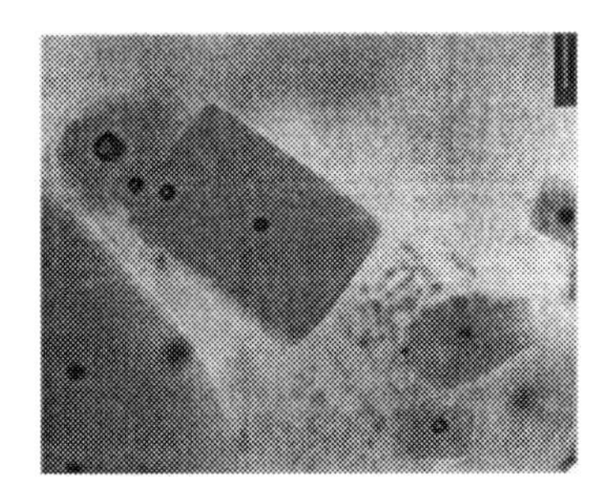

Figure 6. Nepheline crystals in NP-Al-1 glass heat-treated at 1150°C

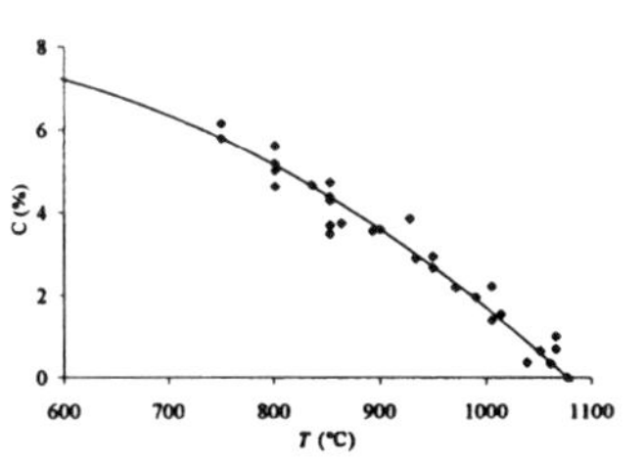

Figure 7 Equilibrium mass fraction of spinel in MS-7 glass as a function of temperature

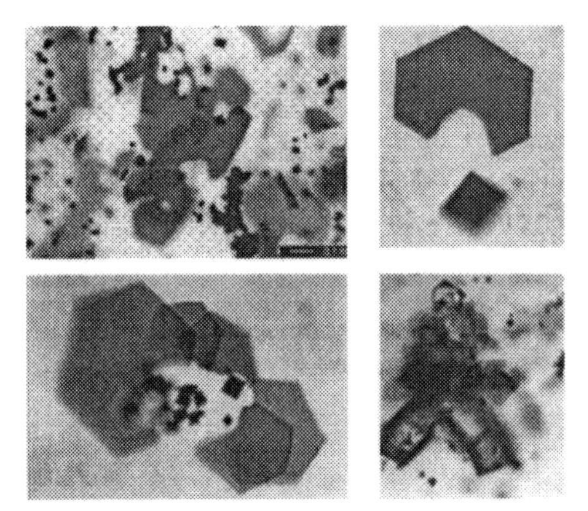

Figure 8. Hematite and spinel interacting in a simulated HLW glass (MS-7, 950°C, 17 h); the bottom right micrograph shows a rare combination of zircon, hematite and spinel in TRU-Al-6 glass (940°C, 24 h).

Another interesting phenomenon, occasionally observed in glasses with high Fe concentration, is the precipitation of hematite that precedes the crystallization of spinel. Hematite often forms large hexagonal platelets (>100 μm) that are very thin and can be bent by convection in the melt. These crystals subsequently dissolve, giving way to spinel (Figure 8; a similar image is shown in [36]). The interaction of spinel with RuO_2 is another curiosity (Figure 9). It was observed in a glass subjected to slow cooling [37].

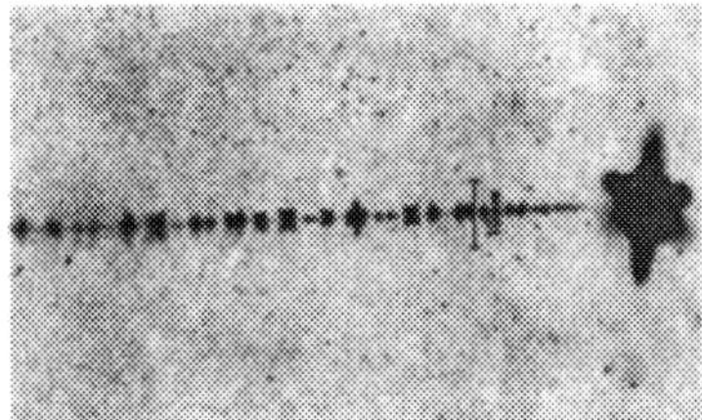

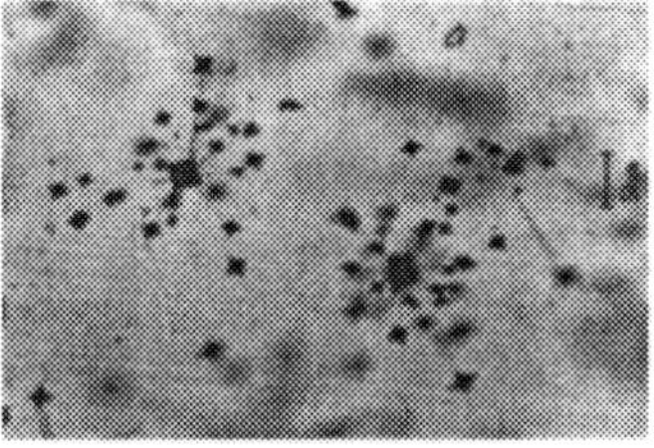

Figure 9. Interaction of RuO_2 needles with spinel crystals in a HLW glass cooled at 33°C/h [37]

CRYSTAL NUCLEATION, GROWTH, AND DISSOLUTION IN HLW GLASS

Early kinetic studies of spinel precipitation from HLW glasses have been conducted by Turcotte et al. [38], Bickford et al. [39], Marra et al. [40], and

Simpson et al. [41]. These authors, who draw time-temperature-transformation (TTT) diagrams, described the sequence in which different phases occurred as the temperature decreased. More detailed studies that include crystal size distribution were recently performed by Orlhac et al. [6] and Stahlavsky [25].

An analytical description of the kinetics of crystal growth was provided by Reynolds and Hrma [23] for spinel, Menkhaus and Hrma [14] for nepheline, and Vienna et al. [42] for acmite. These authors fitted the Kolmogorov-Johnson-Mehl-Avrami (KJMA) equation to data and confirmed that crystal growth was diffusion controlled. The Avrami exponent was 1.5 for spinel and nepheline and 2.5 for clinopyroxene. Spinel and nepheline nucleated virtually instantaneously, whereas acmite crystals nucleated after an incubation time. Li et al. [43] showed, using Raman spectroscopy, that structural units of nepheline pre-exist in glass at temperatures above T_L. The same is likely for spinel; clusters of Cr were observed by Darab et al. [44].

The KJMA equation allows a mathematical construction of the TTT diagram [14,23] (Figure 10). Subsequent studies [45] showed that while crystallization kinetics at high glass viscosity ($\eta > 10^6$ Pa·s) is best described by the KMJA equation, at lower viscosities, crystal settling is unavoidable and the KJMA equation is no longer valid.

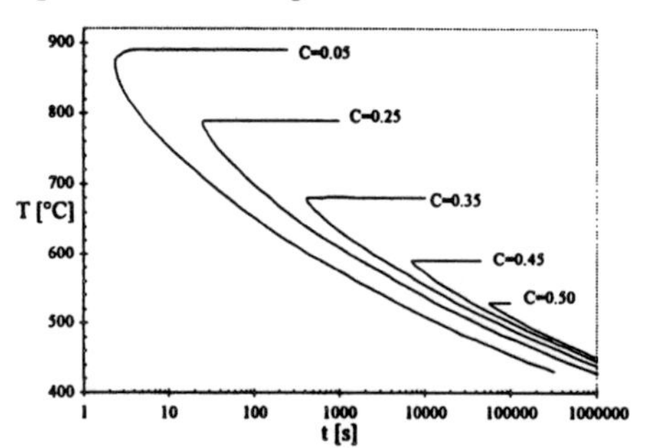

Figure 10. TTT Diagram for Nepheline Crystallization in a HLW Glass [14]

Surface nucleation dominated spinel crystallization in crushed glass (Figure 11). The corresponding Avrami exponent was 0.56, close to the theoretical value of 0.5 [46]. Inoue et al. [47] observed crystal formation on the glass surface, where nucleation is easier and evaporation and oxidation-reduction reactions alter the composition.

For crystallization kinetics under nonisothermal conditions, the KJMA equation must be used in its differential form:

$$\frac{dC}{dt} = nk_A(C_0 - C)\left[-\ln\left(\frac{C_0 - C}{C_0 - C_i}\right)\right]^{(n-1)/n} \tag{2}$$

where C is the fraction of spinel crystals in glass, C_0 is the fraction of spinel crystals in glass at equilibrium, C_i is the initial mass fraction of spinel crystals in glass, n is the Avrami exponent, k_A is the Avrami rate coefficient, and t is time. For nonisothermal calculations, it is important to express in Equation (2) both k_A and C_0 as functions of temperature. If the temperature dependence of C_0 is ignored, which is unfortunately often the case in the literature, the KJMA model is useless for temperatures close to liquidus, where $C_0 \rightarrow 0$ as $T \rightarrow T_L$.

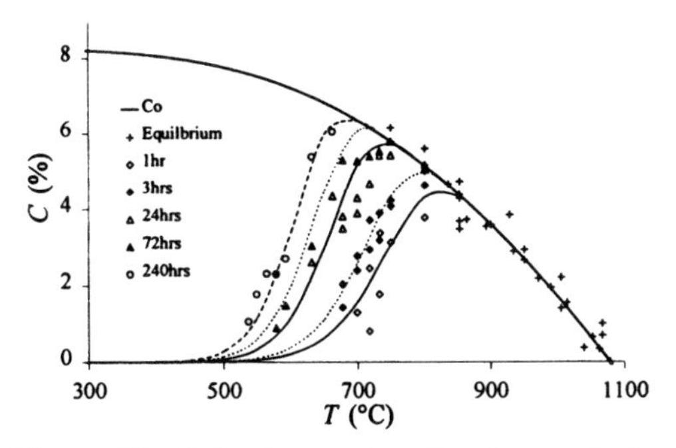

Figure 11. Spinel mass fraction in crushed glass as a function of time and temperature (lines were fitted using Equation (1) with temperature-dependent k_A and C_0) [45]

Using Equation (2), Casler and Hrma [37] calculated the distribution of crystals in a HLW glass canister in which crystallization occurs within a non-uniform and time-dependent temperature field. They also constructed a time-temperature transformation (TTT) diagram for the constant rate of cooling, i.e., at conditions that better represent HLW glass processing history than the isothermal TTT diagrams.

As mentioned above, nucleation is virtually instantaneous if the crystalline structure pre-exists in the form of clusters at $T > T_L$. A study of spinel nucleation [48] showed that platinoids are excellent nucleation agents. This is demonstrated in Figure 12.

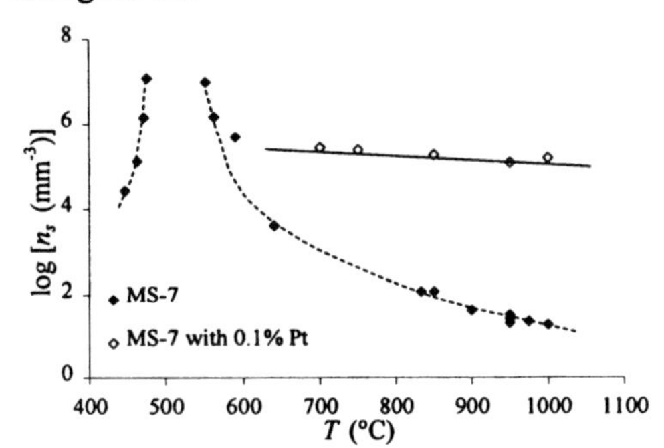

Figure 12. Crystal number density (bulk nucleation) as a function of temperature for MS-7 glass with and without a nucleation agent [51]

Without nucleation agents, the spinel crystal density was 10 to 100 mm^{-3} within the processing temperature interval (for MS-7 glass with T_L = 1078°C) and sharply increased with decreasing temperature. With 0.1 mass% Pt, the number density increased by 3 to 4 orders of magnitude at processing temperatures and remained nearly constant with temperature; by adding Pt, the crystal size decreased from 20 μm to <2 μm. Other platinoids (Ru, Rh, Pd) had a similar effect as Pt.

In HLW glass melters, crystals grow or dissolve while settling. The rate of growth and dissolution under these conditions is described by the Hixson-Crowell equation [48,49]:

$$\frac{da}{dt} = 2k_H(C_0 - C) \quad (3)$$

where a is the crystal size and $k_H = (\rho_g/\rho_s)(D/\delta)$ is the mass transfer coefficient; here D is the diffusion coefficient, δ is the concentration boundary-layer thickness, and ρ_g and ρ_s are glass and spinel densities.

The measured k_H is an Arrhenius function of temperature, the same for growth and dissolution. The measured values of the rates of growth and dissolution for spinel in MS-7 glass are plotted in Figure 13. The concentration distribution of Fe around a dissolving crystal was measured using SEM. It agrees surprisingly well with the boundary-layer theory developed by Levich [50,51,52] (Figure 14).

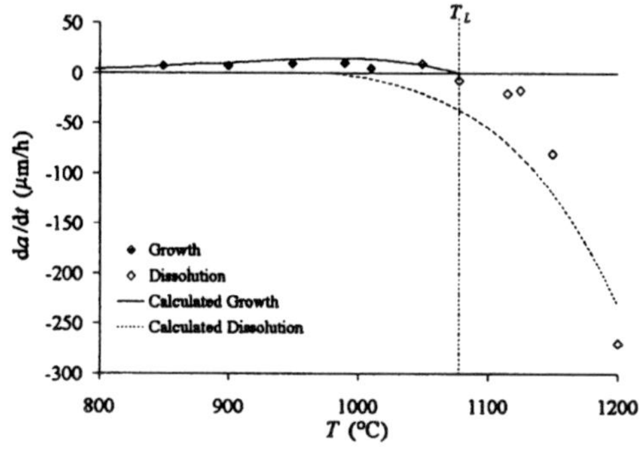

Figure 13. Measured and calculated initial steady-state growth/dissolution rates [50]

The most recent data [53] obtained for high-Cr HLW glass with low concentrations of other transition metals showed a transition from an alkali chromate separated phase at temperatures below 1050°C and eskolaite (Cr_2O_3) at temperatures above 1150°C, leaving a temperature interval in which high-Cr HLW glass can potentially be processed without precipitating any low-solubility chromium phase.

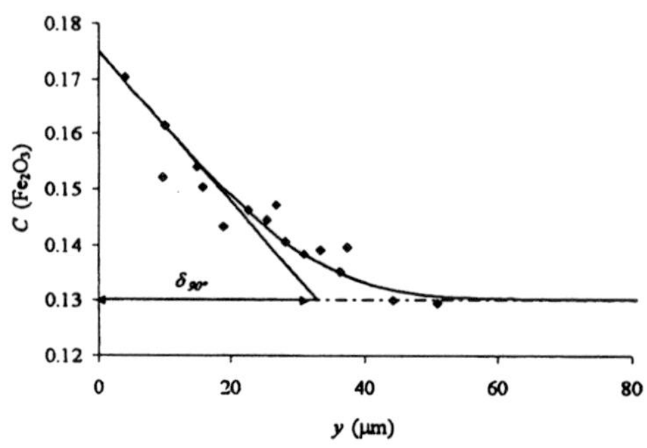

Figure 14. Horizontal concentration profile of Fe_2O_3 at a dissolving spinel crystal [50]

IMPACT OF CRYSTALS ON MELTERS

Only spinel, eskolaite, and Zr-containing phases are likely to precipitate from most U.S. HLW borosilicate melts at $T > 1050$°C and thus interfere with glass processing in continuous melters; other crystals generally precipitate at <1050°C. At low viscosity ($\eta < 10^6$ Pa·s), crystals settle. Their rate of settling is described by the Stokes law for hindered settling [54].

Unless crystals are very small (<10 μm), the settling rate is high enough to produce a sludge layer that can aversely affect the operation of a continuous melter. The development of a sludge layer was analyzed using mathematical modeling [49,55]. To this end, existing glass-melter models were modified by adding the concentration field of the crystalline phase [49].

Crystals grow or dissolve according to their position within the temperature field. The concentration of the crystalline phase affects melt density and viscosity. The advanced model showed that in the melter under analysis (a West Valley type melter) the rate of crystal growth and dissolution was slow as compared to the rate of temperature change to which the crystals were exposed within the circulation flow of the melt. Spinel crystals from the cold region under the cold cap were rapidly distributed within the whole melter volume. Hence, the melter behaved as a nearly ideal mixer with respect to the distribution of spinel crystals [55]. Unfortunately, the circulation flow was not strong enough to prevent settling.

Surprisingly, the T_L has little effect on the sludge layer growth rate in melters with strong circulation flow. Only spinel crystal size has a strong effect: the sludge layer growth rate increases roughly with the square of crystal size. While the sludge layer growth rate is below 1 mm per year for crystals <5 μm, it increases to several cm per year for crystals of ~20 μm in size. As shown above, small crystals form when nucleation agents are present; these crystals settle slowly and leave the melter with the glass.

The absence of evidence that the T_L constraint used for glass formulation impacts the sludge-layer development is rather unsettling. This constraint, which would cost billions of US dollars in the processing and disposal of an excessive amount of HLW glass, does not prevent continuous melters from being vulnerable to failure if nucleation agents happen to be in an insufficient concentration in the waste. Modeling the behavior of crystalline materials in the melter can assist melter design and glass formulation to reduce the impact of crystals on the vitrification process. However, waste loading will dramatically increase only when durable HLW glass can be produced with a crystalline phase, that is, in melters that are not subjected to crystallinity constraints [1].

IMPACT OF CRYSTALS ON GLASS QUALITY

Only those crystalline phases that remove proportionally more glass formers than glass modifiers from the glass have a detrimental effect on the chemical durability of the glass. For example, the Na_2O and SiO_2 contents in acmite, ($NaFeSi_2O_6$) are close to those in a typical HLW glass, and thus the crystallization of acmite has little impact on glass corrosion behavior. On the other hand, the crystallization of nepheline ($NaAlSiO_4$) typically removes little Na_2O from the glass (it contains nearly the same fraction of Na as a typical HLW glass), but removes, if its precipitation is massive, the major glass-forming elements of Al and Si. Other crystals that potentially strongly affect glass durability, such as zircon, do not tend to precipitate during canister cooling because of slow nucleation and growth or precipitate in such small quantities that they do not impact glass properties. The effect of crystallization on the chemical durability of glass is discussed elsewhere in these proceedings [56].

CONCLUSIONS

HLW glasses are more likely to form crystals as their waste loading increases. Fortunately, crystallization is often benign, but it can be detrimental to melters and glass acceptability for the repository. To prevent these undesirable impacts, the glass-composition region must be mapped to determine which crystalline phases are likely to precipitate during glass processing in the melter, during glass cooling in the canister, and during glass storage. For each phase, we need to know its composition, the temperature range at which the phase occurs, and the equilibrium concentration of crystalline phases as functions of temperature and glass composition.

In addition to these basic thermodynamic data, the kinetics of crystal growth and dissolution in non-uniform time-dependent temperature fields should be measured and modeled. In solid or high-viscosity glass, the KJMA equation appears most appropriate for generating TTT diagrams, which are constructed preferably for the constant rate of cooling. The KJMA equation must be used in its differential form, and both the rate constant and equilibrium concentration must be expressed as functions of temperature. The Hixson-Crowell equation is adequate for crystal growth and dissolution in a convective medium (molten glass in the melter) where crystals undergo a free fall (settling).

Mathematical models show that at least some HLW melters operate as a fast mixer that distributes crystals nearly uniformly within the melter. For such melters, constraining T_L may not affect their potential for developing a sludge layer on the bottom. The sludge-layer growth rate increases in proportion to the square of the crystal size. Fortunately, nucleation agents naturally occur in most high-level wastes and keep crystal size small. However, noble metals are not always used in experimental melter tests used in defining property constraints.

The detrimental effects of crystallization on glass quality (corrosion resistance) can be mitigated by careful glass formulation to avoid crystallization that diminishes glass quality. Detrimental effects of crystallization on glass processing can be avoided by using advanced melters that are less sensitive to crystallization or developing sufficient knowledge to control crystal settling with the allowance of small crystal fraction in the melt. The current method of decreasing waste loading is costly.

ACKNOWLEDGMENTS

The author thanks Jesse Alton for taking photographs shown in Figures 1, 5, 6 and 8. Dong Kim and John Vienna kindly reviewed the manuscript and Wayne Cosby provided editorial support. The work described in this paper was funded by the U.S. Department of Energy through the Environmental Management Science and Tanks Focus Area Programs. Pacific Northwest National Laboratory is operated for the U.S. Department of Energy by Battelle under Contract DE-AC06-76RL01830.

REFERENCES

1. D.S. Kim and J.D. Vienna, "Influence of Glass property Restrictions on Hanfrod HLW Glass Volume," *Ceram. Trans.* (this volume).
2. N. Jacquet-Francillon, F. Pacaud, and P. Qeille, "An Attempt to Assess the Long-Term Crystallization Rate of Nuclear Waste Glasses," *Mat. Res. Soc. Symp. Proc.* 11, 249-259 (1982).
3. R.H. Feld and M. Stammler, "Quantitative Determination of Crystalline Phases in Nuclear Waste Glass," *Mat. Res. Soc. Symp. Proc.* 11, 261-271 (1982).
4. A.D. Stalios and R. De Batist, "Crystallization Behavior of a Ferri-Silicate α-Waste Glass," *Mat. Res. Soc. Symp. Proc.* 50, 255-262 (1985).
5. A.C. Buechele, X. Feng, H. Gu, and I.L. Pegg, "Alteration of Microstructure of West Valley Glass by Heat Treatment," *Mat. Res. Soc. Symp. Proc.* 176, 393-402 (1990).
6. X. Orlhac, C. Fillet, and J. Phalippou, "Study of Crystallization Mechanisms in the French Nuclear Waste Glass," *Mat. Res. Soc. Symp. Proc.* 556, 263-270 (1999).
7. D.S. Kim, P. Hrma, D.E. Smith, and M.J. Schweiger, "Crystallization in Simulated Glasses from Hanford High-Level Nuclear Waste Composition Range," *Ceram. Trans.* 39, 179-189 (1994).
8. E. Wang, H. Kuang, K.S. Matlack, A.C. Buchele, and S.S. Fu, "Effect of Fluoride on Crystallization in High Calcium and Magnesium Glasses," *Mat. Res. Soc. Symp. Proc.* 333, 473-479 (1994).
9. B.A. Scholes, D.K. Peeler, and J.D. Vienna, *The Preparation and Characterization of INTEC Phase 3 Composition Variation Study Glasses*, INEEL/EXT-2000-01566, Idaho Engineering and Environmental Laboratory, Idaho Falls, Idaho 2000.
10. D.K. Peeler, T.B. Edwards, I.A. Reamer, J.D. Vienna, D.E. Smith, M.J. Schweiger, B.J. Riley, and J.V. Crum, "Composition / Property Relationships for the Phase 1 Am/Cm Glass Variability Study," *Ceram. Trans. 107*, 427-439 (2000).
11. J.B. Riley, J.D. Vienna, M.J. Schweiger, D.K. Peeler, and I.A. Reamer, "Liquidus Temperature of Rare Earth-Alumino-Borosilicate Glasses for Treatment of Americium and Curium," *Mat. Res. Soc. Symp. Proc.* 608, 677-682 (2000).
12. J.D. Vienna, P. Hrma, A. Jiricka, D.E. Smith, T.H. Lorier, I.A. Reamer, and R.L. Schulz, *Hanford Immobilized LAW Product Acceptance Testing: Tanks Focus Area Results*, PNNL report (in press), Pacific Northwest National Laboratory, Richland, Washington (2001).
13. T.J. Plaisted, F. Mo, B.K. Wilson, C. Young, and P. Hrma, "Surface Crystallization and Composition of Spinel and Acmite in High-Level Waste Glass," *Ceram. Trans.* 119, 317-325 (2001).

14. T.J. Menkhaus, P. Hrma, and H. Li, "Kinetics of Nepheline Crystallization from High-level Waste Glass," *Ceram. Trans.* 107, 461-468 (2000).
15. L.A. Chick and G.F. Piepel, "Statistically Designed Optimization of a Glass Composition," *J. Amer. Ceram. Soc.* 67, 763-768 (1984).
16. J.V. Crum, M.J. Schweiger, P. Hrma, and J.D. Vienna, "Liquidus Temperature Model for Hanford High-Level Waste Glasses with High Concentrations of Zirconia," *Mat. Res. Soc. Symp. Proc.* 465, 79-85 (1997).
17. M. Mika, M.J. Schweiger, and P. Hrma, "Liquidus Temperature of Spinel Precipitating High-Level Waste Glass," *Mat. Res. Soc. Symp. Proc.* 465, 71-78, (1997).
18. Q. Rao, G.F. Piepel, P. Hrma, and J.V. Crum, "Liquidus Temperatures of HLW Glasses with Zirconium-Containing Primary Crystalline Phases," *J. Non-Cryst. Solids* 220 [1] 17-29 (1997).
19. H. Li, B. Jones, P. Hrma, and J.D. Vienna, "Compositional Effects on Liquidus Temperature of Hanford Simulated High-Level Waste Glasses Precipitating Nepheline ($NaAlSiO_4$)," *Ceram. Trans.* 87, 279-288 (1998).
20. P. Hrma, J.D. Vienna, J.V. Crum, G.F. Piepel, and M. Mika, "Liquidus Temperature of High-Level Waste Borosilicate Glasses with Spinel Primary Phase," *Mat. Res. Soc. Proc.* 608, 671-676 (2000).
21. J. D. Vienna, P. Hrma, J. V. Crum, and M. Mika, "Liquidus Temperature-Composition Model for Multicomponent Glasses in the Fe, Cr, Ni, and Mn Spinel Primary Phase Field," *J. Non-Cryst. Solids* 289 (2001).
22. P. Hrma, G.F. Piepel, J.D. Vienna, S.K. Cooley, D.S. Kim, and R L. Russell, *Database and Interim Glass Property Models for Hanford HLW Glasses*, Pacific Northwest National Laboratory, Richland, Washington (2001).
23. J.G. Reynolds and P. Hrma, "The Kinetics of Spinel Crystallization from a High-Level Waste Glass," *Mat. Res. Soc. Symp. Proc.* 465, 261-268 (1997).
24. B.K. Wilson, T.J. Plaisted, J. Alton, and P. Hrma, "The Effect of Composition on Spinel Equilibrium and Crystal Size in High-Level Waste Glass," in preparation.
25. O. Šťáhlavský, *Crystallization of Compounds with Spinel Structure in Borosilicate Glasses for the Disposal of Radioactive Wastes*, Thesis, Institute of Chemical, Department of Glass and Ceramics, Prague 2001.
26. A.C. Buechele, C.-W. Kim, I.S. Muller, H. Gan, I.L. Pegg, and P.B. Macedo, "Solubilities and Redox Effects of Transition Metals in an Alumino-Silicate Waste Glass," *Ceram. Trans* 87 613-620 (1998).
27. P. Izak, P. Hrma, B.K. Wilson, and J.D. Vienna, "Effect of Oxygen Partial Pressure on Liquidus Temperature of a High-Level Waste Glass with Spinel Primary Phase," *Ceram. Trans.* 119, 309-316 (2001).
28. P. Hrma, P. Izak, J.D. Vienna, G.M. Irwin and M-L. Thomas, "Partial Molar Liquidus Temperatures of Multivalent Elements in Multicomponent Borosilicate Glass," submitted to *Phys. Chem. Glasses.*

29. V. Jain and S.M. Barnes, "Effect of Redox on the Crystallization Behavior in the Canistered Product at the West Valley Demonstration Project," *Ceram. Trans.* 23, 251-257 (1991).
30. T.J. Plaisted, P. Hrma, J.D. Vienna, and A. Jiricka, "Liquidus Temperature and Primary Crystallization Phases in High-Zirconia High-Level Waste Borosilicate Glasses," *Mat. Res. Soc. Proc.* 608, 709-714 (2000).
31. J.D. Vienna, D.K. Peeler, T.J. Plaisted, R.L. Plaisted, I.A. Riemer, and J.V. Crum, "Glass Formulation for Idaho National Engineering and Environmental Laboratory Zirconia High-Activity Waste," *Ceram. Trans.* 107, 451-459 (2000).
32. H. Li, J.D. Vienna, P. Hrma, D.E. Smith, and M.J. Schweiger, "Compositional Effect on Precipitation of Nepheline in Glass and its Impact on Resistance of High-Level Waste Glasses Against Corrosion by Water," *Mat. Res. Soc. Symp. Proc.* 465, 261-268 (1997).
33. T.J. Plaisted, J. Alton, B.K. Wilson, and P. Hrma, "Effect of Minor Component Addition on Spinel Crystallization in Simulated High-Level Waste Glass," *Ceram. Trans.* 119, 317-325 (2001).
34. M. Mika, M. Patek, J. Maixner, S. Randakova, and P. Hrma, "The Effect of Temperature and Composition on Spinel Concentration and Crystal Size in High-Level Waste Glass," *The 8th International Conference Proceedings (ICEM'01)*, Bruges, Belgium (2001).
35. M.J. Schweiger, M.W. Stachnik, and P. Hrma, "West Valley High-Level Waste Glass Crystallization in Canisters," *Ceram. Trans.* 87, 335-341 (1998).
36. V. Jain and S.M. Barnes, "Effect of Glass Pour Cycle on the Crystallization Behavior in the Canistered Product at the West Valley," *Ceram. Trans.* 23, 239-249 (1991).
37. D.G. Casler and P. Hrma, "Nonisothermal Kinetics of Spinel Crystallization in a HLW Glass," *Mat. Res. Soc. Proc.* 556, 255-262 (1999).
38. R.P. Turcotte, J.W. Wald, and R.P. May, "Devitrification of Nuclear Waste Glass," *Scientific Basis for Nuclear Waste Management*, Vol. 2, edited by C.J.M. Northrup (Plenum Press, New York. 1980) p. 141-146.
39. D.F. Bickford and C.M. Jantzen, "Devitrification Behavior of SRL Defence Waste Glass," *Mat. Res. Soc. Symp. Proc.* 26, 557-566 (1984).
40. S.L. Marra, M.K. Andrews, and C.A. Cicero, "Time-Temperature-Transformation Diagrams for DWPF Projected Glass Compositions," *Ceram. Trans.* 39, 283-292 (1994).
41. J.C. Simpson, D. Oksoy, T.C. Cleveland, L.D. Pye and V. Jain, "The Statistics of the Time-Temperature-Transformation Diagram for Oxidized and Reduced West Valley Reference 6 Glass," *Ceram. Trans.* 45, 377-387 (1994).
42. J.D. Vienna, P. Hrma, and D.E. Smith, "Isothermal Crystallization Kinetics in Simulated High-Level Nuclear Waste Glass," *Mat. Res. Soc. Symp. Proc.* 465, 17-24 (1997).

43. H. Li, Y. Su, J.D. Vienna, and P. Hrma, "Raman Spectroscopic Study—Effects of B_2O_3, Na_2O, and SiO_2 on Nepheline ($NaAlSiO_4$) Crystallization in Simulated High-Level Waste Glass," *Ceram. Trans.* 107, 469-477 (2000).
44. J.G. Darab, H. Li, D.W. Matson, P.A. Smith, and R.K. MacCrone, "Chemical and Structural Elucidation of Minor Components in Simulated Hanford Low-Level Waste Glasses," *Applications of Synchrotron Radiation in Chemistry and Related Fields*, American Chemical Society 1995.
45. J. Alton, T.J. Plaisted, and P. Hrma. "Modeling the Growth and Dissolution of Spinel Crystals in a Borosilicate Glass Using the KJMA and First Order Equations," in preparation.
46. C.R. MacFrarlane and M. Fragoulis, "Theory of devitrification in multicomponent glass forming systems under diffusion control," *Phys. Chem. Glasses* 27, 228-234 (1986).
47. T. Inoue, H. Yokoyama, T. Onchi, ande H. Koyama, "Surface Layer Crystallization of Simulated Waste Glass at Elevated Temperatures," *Mat. Res. Soc. Symp. Proc.* 26, 535-542 (1984).
48. P. Hrma, J. Alton, J. Klouzek, J. Matyas, M. Mika, L. Nemec, T.J. Plaisted, P. Schill, and M. Trochta, "Increasing High-Level Waste Loading in Glass without Changing the Baseline Melter Technology," *Waste Management '01*, University of Arizona, Tucson, Arizona (2001).
49. P. Schill, M. Trochta, J. Matyas, L. Nemec, and P. Hrma, "Mathematical Model of Spinel Settling in a Real Waste Glass Melter," *Waste Management '01*, University of Arizona, Tucson, Arizona (2001).
50. J. Alton, T.J. Plaisted, and P. Hrma, "Spinel Nucleation and Growth of Spinel Crystals in a Borosilicate Glass," in preparation.
51. P. Hrma and J. Alton, "Dissolution and Growth of Spinel Crystals in a High-Level Waste Glass," *The 8th International Conference Proceedings (ICEM'01)*, Bruges, Belgium (2001).
52. V.G. Levich, *Physicochemical Hydrodynamics*, p. 80-87, Prentice-Hall, Englewood Cliffs, New York, 1962.
53. B.K. Wilson and T.J. Plaisted, private communication.
54. J. Klouzek, J. Alton, T.J. Plaisted, and P. Hrma, "Crucible Study of Spinel Settling in High-Level Waste Glass," *Ceram. Trans.* 119, 301-308 (2001).
55. J. Matyáš, J. Kloužek, L. Němec, and M. Trochta. 2001. "Spinel settling in HLW melters," *The 8th International Conference Proceedings (ICEM'01)*, Bruges, Belgium (2001).
56. B.J. Riley, P. Hrma, J.A. Rosario, and J.D. Vienna, "Effect of Crystallization on High-Level Waste Glass Corrosion," *Ceram. Trans.* (this volume).

EFFECT OF CRYSTALLIZATION ON HIGH-LEVEL WASTE GLASS CORROSION

BJ Riley, P Hrma, J Rosario, and JD Vienna
Pacific Northwest National Laboratory
Richland, WA 99352

ABSTRACT

Crystallization of high-level waste (HLW) glass affects its resistance against corrosion. A large database has accumulated in the literature for HLW glasses subjected to canister-centerline cooling, where the cooling rate is slowest. This database consists of mass fractions of crystalline phase, identified with x-ray diffraction (XRD), and results of the product consistency test (PCT) that measures glass corrosion at 90°C. Using XRD data for heat-treated glasses and PCT data for quenched glasses, PCT B and Na releases from residual glasses were estimated. The estimated values are in reasonable agreement with the measured values from heat-treated glasses. Outliers were attributed to 1) the composition of residual glass being outside the region for which PCT models were established, 2) the low accuracy of semiquantitative XRD, and 3) secondary effects, such as internal stresses or interfacial diffusion.

INTRODUCTION

High-level waste (HLW) contains components that tend to precipitate from glass during cooling. As the glass is poured into a canister, crystallization takes place within the temperature interval between the liquidus temperature (T_L) and glass-transition temperature (T_g). A portion of the glass cast into canisters is quenched on the canister walls, and another portion of glass, near the canister centerline, cools slowly. Thus, the temperature history of the canister centerline cooled (CCC) glass is most favorable for crystalline phases to form.

Above T_g, the average CCC rate is approximately 0.05 K/s.[1-3] This cooling rate allows sufficient time for fast-nucleating and fast-growing phases, such as

To the extent authorized under the laws of the United States of America, all copyright interests in this publication are the property of The American Ceramic Society. Any duplication, reproduction, or republication of this publication or any part thereof, without the express written consent of The American Ceramic Society or fee paid to the Copyright Clearance Center, is prohibited.

spinel and nepheline, to precipitate. Slow-growing phases or those that do not easily nucleate, such as acmite or zircon, are less likely to form crystals in the canister.[4] As a new phase precipitates, it affects the glass matrix, in which it is embedded, both chemically and mechanically. These changes may impact the rate of glass dissolution in water and thus change its chemical durability.[5,6] Consequently, the product consistency test (PCT) responses of glasses quenched on steel plates and those heat-treated according to a simulated CCC schedule are expected to bracket the range of responses expected of all canistered glass.

The acceptability criterion for the HLW glass repository is based on the PCT:[7] the normalized releases of B, Na, and Li (r_B, r_{Na}, and r_{Li}) from the glass must be lower than the r_B, r_{Na}, and r_{Li} from Defense Waste Processing Facility-Environmental Assessment (DWPF-EA) glass.[8] Both quenched and canister centerline glass must meet this constraint. To formulate glasses that satisfy the acceptability criterion, constitutive equations have been developed that relate r_B, r_{Na}, and r_{Li} to glass composition. For the quenched glass, it is sufficient to use first-order equations[4] with coefficients that are estimated from measured data. Several sets of such coefficients have been reported for different HLW glass-composition regions.[4,9]

The effect of CCC on the PCT of HLW glasses was determined for more than 100 glass compositions.[4,5,9] As Kim et al.[5] showed, the r_B values from CCC and quenched glasses were close for most of the compositions regardless of the fraction of crystallinity that formed during heat treatment. For only 18 of these glasses, CCC treatment significantly increased the r_B as compared to quenched glasses. Six compositions exhibited a moderate increase in r_B from CCC glass, and twelve CCC-treated compositions showed a strong increase in r_B. Those glasses with a moderate increase in r_B contained zircon, clinopyroxene, and/or hematite crystals while those with a strong increase in r_B contained alkali aluminosilicates (nepheline and eucryptite) and cristobalite. Cristobalite could be produced by the crystallization of a SiO_2-rich immiscible liquid during cooling.

First- and second-order polynomial functions described satisfactorily r_B and r_{Na} as functions of glass composition for large composition regions of quenched HLW glasses, but failed for CCC-treated glasses. However, when the fractions and compositions of the crystalline phases were known, reasonable agreement was found between measured r_B values for CCC-treated glasses and r_B values estimated for the composition of the residual glass.[5,10-12] A study of high waste-loaded glasses[11] indicated that the residual glass composition was the major factor that controlled the PCT response of glasses with durable crystalline phases. Other chemical or mechanical factors, such as concentration gradients and mechanical stresses, played a secondary role.

This paper is based on the report[13] that evaluated a database of 266 heat-treated (mostly CCC) glasses, for which both the crystalline phases, as determined with semiquantitative XRD, and PCT releases are documented.[4,6,14,15] Glasses were grouped according to the major crystalline phases that precipitated upon heat treatment; three of these groups are described in this paper: 1) Hanford glasses that precipitate nepheline; 2) Hanford glasses that precipitate lithium silicate, cristobalite, hematite, and zircon; and 3) Idaho National Engineering and Environmental Laboratory (INEEL) glasses.

THEORY

For the quenched glass, the following relationship has been established between r_j ($j \equiv$ B, Na, Li) and glass composition[4,9]

$$\ln r_j = \sum_{i=1}^{N} b_{ij} g_i \tag{1}$$

where, g_i is the i-th glass component mass fraction, b_{ij} is the i-th component coefficient for j-th element, and N is the number of components.

The b_{ij} values from quenched glasses, developed for Hanford and INEEL composition regions, are listed for r_B and r_{Na} in Table 1.

Table 1. Component Coefficients for r_B and r_{Na} (ln values for r_i in g/m^2)

	Hanford glasses				INEEL glasses			
	PCT coefficient		Validity range		PCT coefficient		Validity range	
	b_{iB}	b_{iNa}	max	min	b_{iB}	b_{iNa}	max	min
Al_2O_3	-25.4	-25.4	0.00	0.17	-25.4	-25.6	0.13	0.07
B_2O_3	12.0	9.4	0.05	0.20	11.8	9.4	0.15	0.05
CaO	-8.7	-2.0	0.00	0.10	-9.1	-2.0	0.15	0.10
F					0.2	-0.8	0.07	0.04
Fe_2O_3	-3.2	-4.1	0.01	0.15	-3.1	-4.1	0.05	0.00
Li_2O	22.6	19.1	0.01	0.07	23.0	19.8	0.07	0.05
MgO	10.9	11.8	0.00	0.08				
Na_2O	17.6	19.4	0.05	0.20	17.8	19.7	0.13	0.08
P_2O_5					0.2	-0.8	0.03	0.00
SiO_2	-4.3	-4.4	0.42	0.57	-4.3	-4.5	0.47	0.36
ZrO_2	-10.6	-11.4	0.00	0.13	-10.6	-11.6	0.08	0.04
Others	0.2	-0.7	0.01	0.10	0.2	-0.8	0.07	0.01

Crystalline phases identified in Hanford and INEEL glasses by XRD are shown in Figure 1 that displays the expression

$$\ln r_{cjk} = \sum_{i=1}^{K} b_{ij} c_{ik} \tag{2}$$

where r_{cjk} is the equivalent j-th element PCT normalized release from the k-th mineral (the extent to which k-th mineral precipitation impacts the r_j from heat-treated glass), c_{ik} is the i-th oxide mass fraction in the k-th mineral, and K is the number of oxides in the mineral. Roughly, if r_{cjk} value is larger than 2 g/m^2 ($\ln r_{cjk} = 0.69$), the precipitation of the k-th mineral will decrease r_i of the glass. As long as the crystals are fully embedded in glass, their durability does not significantly affect the PCT response of the heat-treated glass.

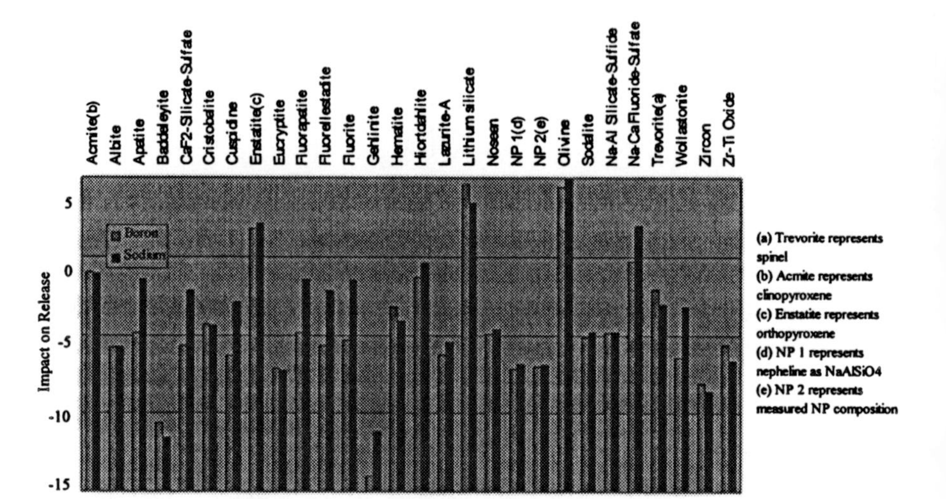

Figure 1. Minerals and their impact on PCT

The residual glass composition is given by the formula[5,10]

$$g_{ri} = \frac{g_i - \sum_{k=1}^{M} c_k c_{ik}}{1 - \sum_{k=1}^{M} c_k} \tag{3}$$

where g_{ri} is the i-th component fraction in residual glass, c_k is the k-th crystalline phase fraction in glass, and M is the amount of crystal phases present in the glass. Rewriting Equation (1) for residual glass,

$$\ln r_{rj} = \sum_{i=1}^{N} b_{ij} g_{ri} \tag{4}$$

where r_{rj} is j-th element normalized PCT release from residual glass; then using Equation (2), we obtain

$$\ln r_{rj} = \frac{\ln r_j - \sum_{k=1}^{M} c_k \ln r_{cjk}}{1 - \sum_{k=1}^{M} c_k} \quad (5)$$

According to Equation (5), minerals with large negative values of $\ln r_{cjk}$ are likely to significantly impact the r_{rj}. Fortunately, only a few minerals reach a high enough fraction in glass to affect PCT.

RESULTS

In this section, we introduce the following notation: r_{Na} and r_B will be used exclusively for quenched glass, whereas subscript h (i.e., r_{hNa} and r_{hB}) will be used to denote measured release values from heat-treated (CCC) glass.

Figure 2 shows that Hanford glasses precipitating nepheline are strongly affected by CCC heat treatment. Figures 3 and 4 compare r_{Na} and r_B with r_{hNa} and r_{hB}. Figure 3 shows that r_{Na} tends to be somewhat higher than r_B from glasses with $r_B < 1$ g/m^2 because of Na leaching at the beginning of the test. Figure 4 shows that r_{hNa} tends to be somewhat lower than r_{hB} from glasses with $r_{hB} > 10$ g/m^2 because part of the Na is bonded in nepheline. Figure 5 demonstrates reasonable agreement between r_{rNa} and r_{hNa}. Only three glasses have substantially overpredicted Na releases ($r_{rNa} << r_{hNa}$).

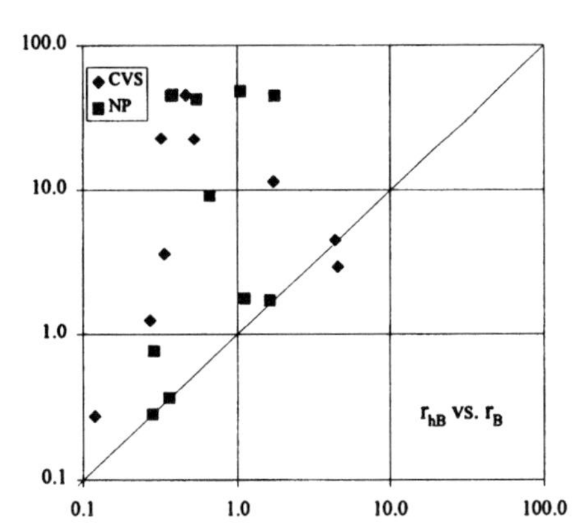

Figure 2. The effect of heat treatment on r_B from nepheline-precipitating glasses

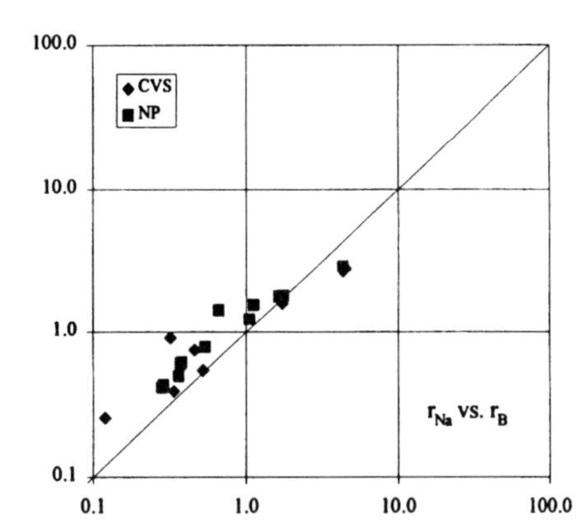

Figure 3. Comparison of r_{Na} with r_B for quenched nepheline-precipitating glasses

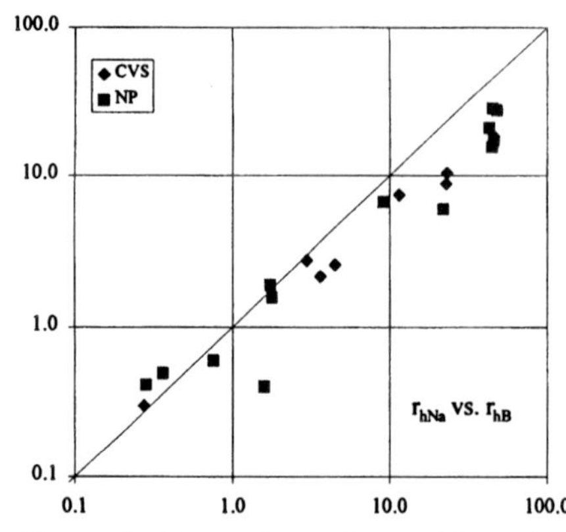

Figure 4. Comparison of r_{hNa} with r_{hB} for nepheline-precipitating glasses

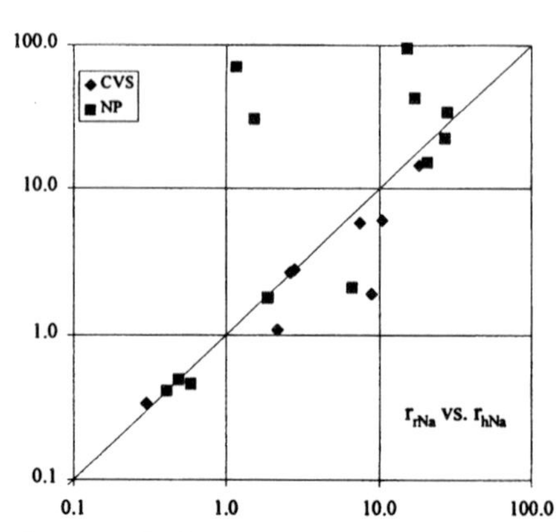

Figure 5. Estimated r_{rNa} vs. measured r_{hNa} for nepheline-precipitating glasses

Similar plots for Hanford glasses precipitating lithium silicate, cristobalite, hematite, and zircon are displayed in Figures 6 to 9. Only for five of these glasses (three glasses precipitating cristobalite and two glasses precipitating lithium silicate), r_{hB} is more than an order of magnitude higher than r_B. Figure 7 shows that r_{Na} tends to be somewhat lower than r_B possibly because of Na absorption on secondary mineral phases that precipitate from the solution exceeded Na leaching at the beginning of the test (a more complex behavior is seen in Figure 8). In Figure 9, sodium releases are underpredicted for three heat-treated glasses with cristobalite (perhaps associated with phase separation).

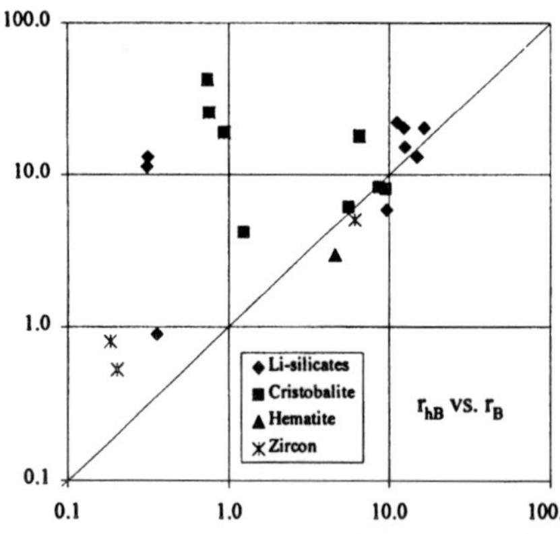

Figure 6. The effect of heat treatment on r_B from glasses precipitating lithium silicate, cristobalite, hematite, and zircon

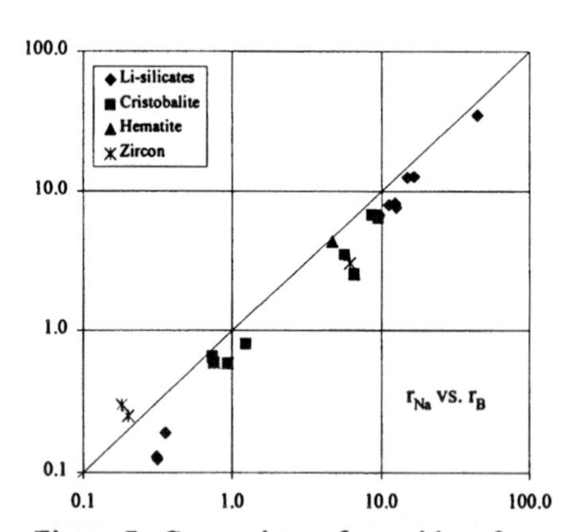

Figure 7. Comparison of r_{Na} with r_B for quenched glasses precipitating lithium silicate, cristobalite, hematite, and zircon

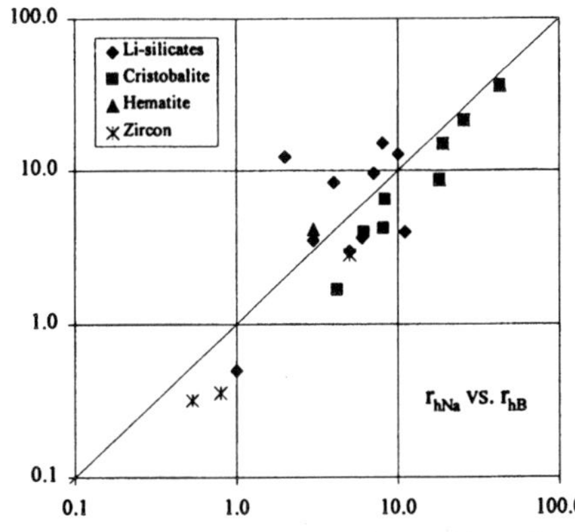

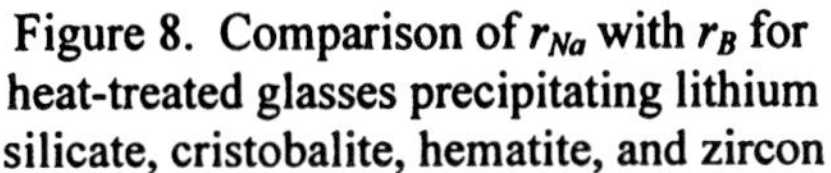

Figure 8. Comparison of r_{Na} with r_B for heat-treated glasses precipitating lithium silicate, cristobalite, hematite, and zircon

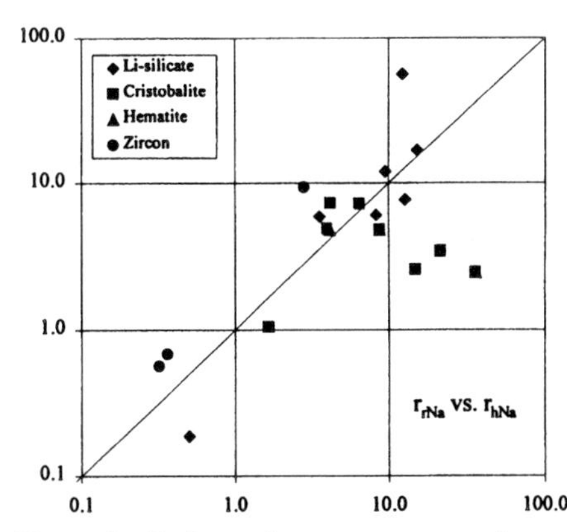

Figure 9. Estimated r_{rNa} vs. measured r_{hNa} for glasses precipitating lithium silicate, cristobalite, hematite, and zircon

Results for INEEL glasses are displayed in Figures 10 to 13. The five glasses with $r_{hB} >> r_B$ precipitated aluminosilicates, baddeleyite, and calcium fluoride silicate sulfate. Measurement error is probably the cause of $r_{hB} << r_B$ for two glasses. With one exception (probably a measurement error), good agreement exists between r_B and r_{Na} (Figure 11). Slightly higher r_{Na} from most glasses are a consequence of Na ion exchange at the beginning of dissolution. As seen in Figure 12, crystallization results in an increased difference between r_{hB} and r_{hNa}. Reasonable agreement exists between calculated r_{rNa} and measured r_{hNa} (compare Figures 10 and 13).

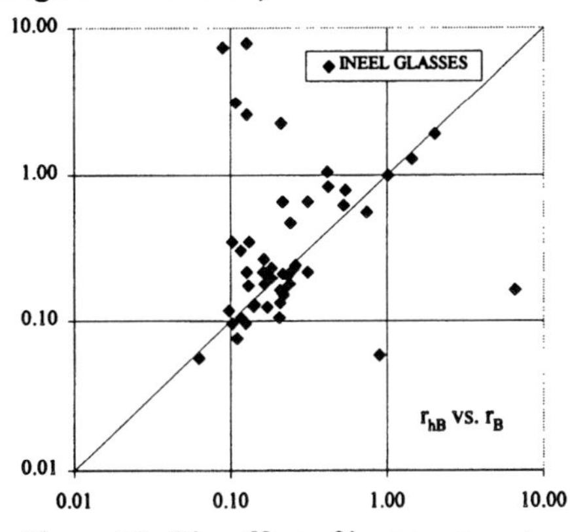

Figure 10. The effect of heat treatment on r_B from INEEL glasses

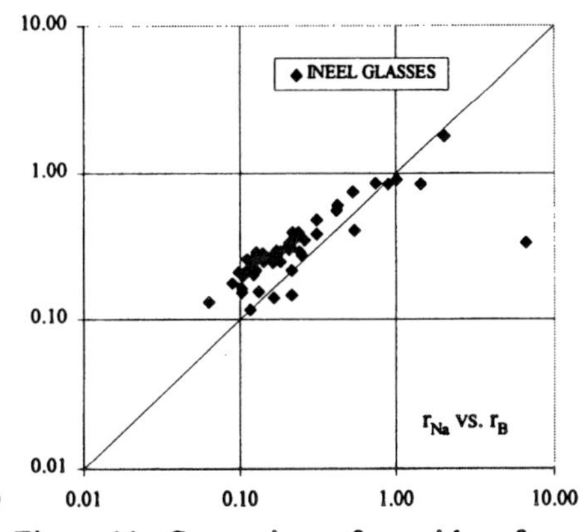

Figure 11. Comparison of r_{Na} with r_B for quenched INEEL glasses

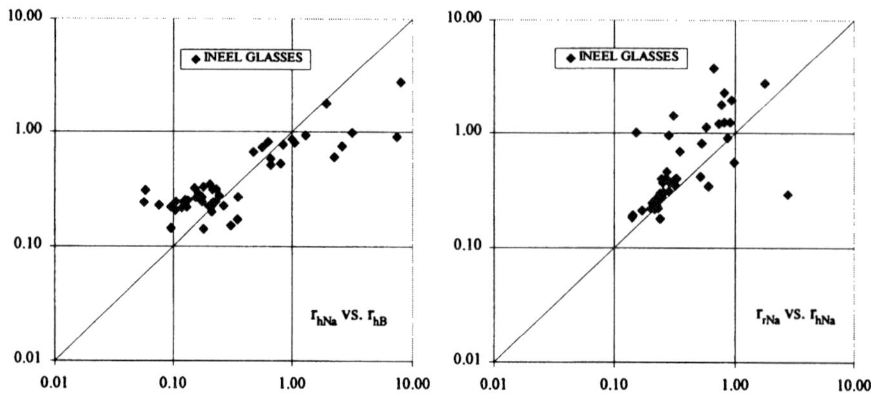

Figure 12. Comparison of r_{Na} with r_B for heat-treated INEEL glasses

Figure 13. Estimated r_{rNa} vs. measured r_{hNa} for INEEL glasses

CONCLUSIONS

Precipitation of cristobalite and certain aluminosilicates could increase r_{hj} by several orders of magnitude, whereas severe crystallization of other phases has little or even opposite effect. This study demonstrates that reasonable agreement exists between estimated release values based on estimated composition of residual glass (r_{rj}) and measured values from heat-treated glass (r_{hj}). This indicates that the effect of heat-treatment on glass dissolution can be attributed to changes of the residual glass composition associated with crystallization as the primary cause. Though other chemical or mechanical factors (internal stresses or interfacial diffusion) are not ruled out as secondary effects, outliers are likely associated with the composition of residual glass being outside the region for which the models were established, the lack of composition data for mineral phases, and the low accuracy of semiquantitative XRD.

To develop models usable for glass formulation, the following three steps are proposed: 1) use quantitative XRD and SEM/EDS to obtain more accurate values for the crystallinity and the composition of crystalline phases; 2) perform PCT on the composition region of residual glasses to establish valid b_{ij} values; and 3) develop a method for predicting compositions and concentrations of mineral phases that precipitate during canister-centerline cooling.

REFERENCES

1. Edwards, R.E. 1987. *SGM Run 8 - Canister and Glass Temperature During Filling and Cooldown*, DPST-87-801, Savannah River Laboratory, Aiken, South Carolina.

2. Lee, L. 1989. *Thermal Analysis of DWPF Canister During Pouring and Cooldown*, DPST-89-269-T_L, Savannah River Laboratory, Aiken, South Carolina.

3. Casler, D. G., and P. Hrma. 1999. "Nonisothermal Kinetics of Spinel Crystallization in a HLW Glass," *Mat. Res. Soc. Proc.* 556, 255-262.
4. Hrma, P., G. F. Piepel, M. J. Schweiger, D. E. Smith, D-S. Kim, P. E. Redgate, J. D. Vienna, C. A. LoPresti, D. B. Simpson, D. K. Peeler, and M. H. Langowski. 1994. *Property/Composition Relationships for Hanford High-Level Waste Glasses Melting at 1150°C*, PNL-10359, Vol. 1 and 2, Pacific Northwest Laboratory, Richland, Washington.
5. Kim, D-S., D. K. Peeler, and P. Hrma. 1995. "Effects of Crystallization on the Chemical Durability of Nuclear Waste Glasses," *Ceram. Trans.* 61, 177-185.
6. Cicero, C.A., S. L. Mara, and M. K. Andrews. 1993. *Phase Stability Determinations of DWPF Waste Glasses (U)*, WSRC-TR-93-227, Westinghouse Savannah River Company, Aiken, South Carolina.
7. American Society for Testing and Materials (ASTM). 1998. "Standard Test Methods for Determining Chemical Durability of Nuclear, Hazardous, and Mixed Waste Glasses: The Product Consistency Test (PCT)," C 1285-97 in *1998 Annual Book of ASTM Standards* Vol. 12.01, ASTM, West Conshohocken, Pennsylvania.
8. Jantzen, C. M., N. E. Bibler, D. C. Beam, C. L. Crawford, and M. A. Pickett. 1993. *Characterization of the Defense Waste Processing Facility (DWPF) Environmental Assessment (EA) Glass Standard Reference Material*, WSRC-TR-92-346, Rev. 1, Westinghouse Savannah River Company, Aiken, South Carolina.
9. Hrma, P., G. F. Piepel, J. D. Vienna, P. E. Redgate, M. J. Schweiger, and D. E. Smith. 1995. "Prediction of Nuclear Waste Glass Dissolution as a Function of Composition," *Ceram. Trans.* 61, 497-504.
10. Hrma, P., and A. W. Bailey. 1995. "High Level Waste at Hanford: Potential for Waste Loading Maximization," *Proc. 1995 Int. Conf. Nucl. Waste Manag. and Environ. Remediation (ICEM'95)*, Vol. 1, pp. 447-451.
11. Bailey, A. W., and P. Hrma. 1995. "Waste Loading Maximization for Vitrified Hanford HLW Blend," *Ceram. Trans.* 61, 549-556.
12. Li, H., J. D. Vienna, P. Hrma, D. E. Smith, and M. J. Schweiger. 1997. "Nepheline Precipitation in High-Level Waste Glasses - Compositional Effects and Impact on the Waste Form Acceptability," *Mat. Res. Soc. Proc.* 465, 261-268.
13. Riley, B. J., J. A. Rosario, and P. Hrma. 2001. *Impact of HLW Glass Crystallinity on the PCT Response*, Pacific Northwest National Laboratory, Richland, Washington.
14. Vienna, J.D., P. Hrma, M. J. Schweiger, M. H. Langowski, P. E. Redgate, D-S. Kim, G. F. Piepel, D. E. Smith, C. Y. Chang, D. E. Rinehart, S. E. Palmer, and H. Li. 1996. *Effect of Composition and Temperature on the Properties of High-Level Waste (HLW) Glass Melting above 1200°C*, PNNL-10987, UC-810, Pacific Northwest Laboratory, Richland, Washington.
15. Marra, S. L., and C. M. Jantzen. 1993. *Characterization of projected DWPF glasses heat treated to simulate canister centerline cooling*, WSRC-TR-92-142, Rev. 1, Westinghouse Savannah River Company, Aiken, South Carolina.

THE EFFECT OF GLASS COMPOSITION ON CRYSTALLINITY AND DURABILITY FOR INEEL RUN 78 CALCINE WASTE SIMULANT

J. V. Crum and J. D. Vienna
Pacific Northwest National Laboratory
PO Box 999, K6-24
Richland, WA 99352

D. K. Peeler and I. A. Reamer
Westinghouse Savannah River Company
Savannah River Technology Center
Aiken, SC 29808

D. J. Pittman
University of South Carolina at Aiken
Aiken, SC 29808

ABSTRACT

Past glass-formulation efforts for Idaho National Engineering and Environmental Laboratory Run 78 calcine waste have shown that waste loading is limited by the formation of crystalline F-containing phases.[a] Crystallization generally occurs rapidly during cooling. Crystallization in glass upon cooling can result in a non-durable waste form, depending upon the type and extent of crystallization. Thus, crystallinity in waste glass is usually constrained to a small mass% of the waste form. Understanding the effect of crystallinity on the durability of the resulting waste form is necessary to determine if increases in waste loading are possible for Run 78 calcine and similar waste types.

This study measured the mass% crystallinity of glass heat treated according to the canister centerline cooling profile as a function of glass composition and the resulting effect of crystallinity on durability. Crystallinity during CCC heat treatment was strongly influenced by changes in concentration

(a) Musick, C.A., B.A. Scholes, R.D. Tillotson, D.M. Bennert, J.D. Vienna, J. V. Crum, D.K. Peeler, I.A. Reamer, D. F. Bickford, J. C. Marra, N.L. Waldo. 2000. *Technical Status Report: Vitrification Technology Development Using INEEL Run 78 Pilot Plant Calcine*, INEEL\EXT-2000-00110, Idaho National Engineering and Environmental Laboratory, Idaho Falls, Idaho.

To the extent authorized under the laws of the United States of America, all copyright interests in this publication are the property of The American Ceramic Society. Any duplication, reproduction, or republication of this publication or any part thereof, without the express written consent of The American Ceramic Society or fee paid to the Copyright Clearance Center, is prohibited.

of Al_2O_3, ZrO_2, SO_3, F, Li_2O, Na_2O, and SiO_2. For the majority of glasses tested, crystallization had only a minor effect on normalized release rates. However, a limited number of glasses that crystallized nepheline, hiortdahlite, and/or sodalite had dramatically increased normalized releases.

INTRODUCTION

Crystallization upon cooling is a primary area of interest that may limit the loading of INEEL calcine waste in glass. Prior formulation efforts for Run 78 calcine have self-imposed a ≤ 2 mass% crystallinity constraint after canister centerline cooling (CCC) heat treatment. This reduces the risk of a glass having poor durability due to excessive crystallization during slow cooling. Crystallization has been shown to result in a non-durable residual glass, depending upon the type and the extent of crystallinity [(1)(2)]. Past studies on Run 78 calcine waste simulant have shown that as waste loading is increased, the likelihood of crystallization upon cooling increases [(3)]. The purpose of this study was to examine the effects of glass composition on crystallization during CCC heat treatment and the resulting effect of crystallization on glass durability.

TEST MATIX DESIGN

Two test matrixes were designed to study the effects of components expected to have the most influence on crystallization during CCC. The first test matrix (using the "DZr-CV" nomenclature) was designed around a baseline glass, DZr-CV-BL2. DZr-CV-1 through -24 glasses were designed to vary Al_2O_3, B_2O_3, CaO, F, Fe_2O_3, La_2O_3, Li_2O, Na_2O, P_2O_5, SO_3, SiO_2, and ZrO_2 one-component-at-a-time while the remaining components were kept in constant proportion to the baseline glass. All of the components that were not independently varied were grouped into one component called *others* that was not varied.

A second test matrix (using the "DP" nomenclature) was designed around a centroid glass, DP-Cent. Using the same components selected to vary in the first test matrix, the second test matrix was statistically designed to vary many-components-at-a-time in relation to the centroid glass composition. Unlike the first test matrix, these glasses did not contain an *others* component. In addition, DZR-CV-BL1 and -BL2 were added to the second test matrix (named DP-BL1 and -BL2, respectively).

RESULTS

A complete listing of the glass compositions, crystal fractions, and Product Consistency Test (PCT) responses of quenched and heat-treated samples were reported by Riley et al.[(4)]

Crystallinity of Quenched Glasses

The quenched glasses were examined after the second melt to determine the homogeneity. Several of the quenched glasses visually showed signs of crystallization. XRD was used to determine the type and concentrations of the crystalline phases present in the quenched glasses. Total crystallinity ranged from 0-22 mass%. Table 1 shows the crystalline phases, their range of concentration, and the number of glasses in which each was identified.

Table 1 Semi-quantitative X-ray diffraction results for quenched glasses

Phase	Concentration, mass%	# of glasses
Fluorite, CaF_2	0 – 16	14
Fluorellestadite, $Ca_{10}(SiO_4)_3(SO_4)_3F_2$	0 – 15	6
Baddeleyite, ZrO_2	0 – 5	5
Zircon, $ZrSiO_4$	0 – 4	3
$LiNa(SO_4)$	0 – 1	1
NaF	0 – 0.5	2

Crystallinity of CCC glasses

Results of the CCC heat treatment[(5)] showed glasses ranging from amorphous to glasses that were highly crystallized. Table 2 shows the crystalline phases identified in the CCC samples by XRD, their concentrations, and the number of different specimens that contained these phases.

Table 2 Semi-quantitative X-ray diffraction results for CCC heat treated glasses

Phase	Concentration, mass%	# of glasses
Nosean, $Na_8(AlSiO_4)_6SO_4$	0 – 33	4
Nepheline, $NaAlSiO_4$	0 – 21	4
Lazurite, $Na_6Ca_2(AlSiO_4)_6SO_4$	0 – 10	2
Sodalite, $Na_8(AlSiO_4)_6(ClO_4)_2$	0 – 9	2
Hiortdahlite, $(Na,Ca)_3Zr_{1-x}(Si_2O_7)(F,O)_2$	0 – 30	4
Fluorellestadite, $Ca_{10}(SiO_4)_3(SO_4)_3F_2$	0 – 16	2
Zircon, $ZrSiO_4$	0 – 16	5
Fluorite, CaF_2	0 – 11	14
Baddeleyite, ZrO_2	0 – 9	7
Fluorapatite, $Ca_5(PO_4)_3F$	0 – 8	15
Cuspidine, $Ca_4F_2Si_2O_7$	0 – 6	7
Apatite, $Ca_{10}(PO_4)F_2$	0 – 6	1
$LiSiO_3$	0 – 5	2

Figure 1 is a plot of total crystallinity vs. change in glass composition for CCC DZr-CV glasses. Also shown are the major crystalline phases identified in the CCC samples. Hiortdahlite and sodalite group phases were present in all DZr-CV-glasses that were more than 10 mass% crystallized. DP-glasses that formed high crystal fractions (>20 mass%) during CCC contained nepheline, hiortdahlite, and/or sodalite group phases.

The one-component-at-a-time variation glasses (DZr-CV) indicate that SO_3, Al_2O_3, ZrO_2, P_2O_5, and F promoted crystallization when added to the solution. Adding Li_2O, SiO_2, and Na_2O resulted in decreased crystallization. When the concentration of ZrO_2, P_2O_5, Fe_2O_3, and to some extent SiO_2 were added or removed, it resulted in more crystallization than the baseline glass.

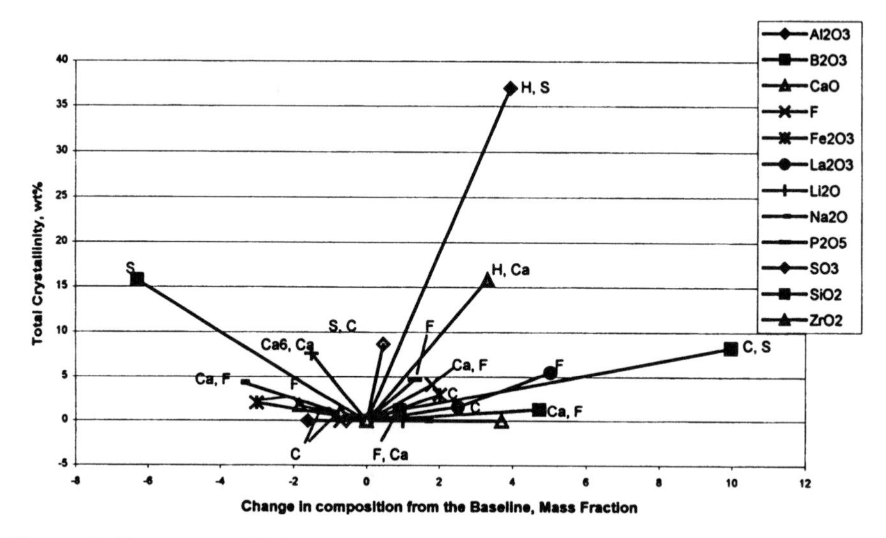

Figure 1. Total crystallinity as a function of change in composition of CCC samples for DZr-CV-glasses. Major phases are given for each sample. C = Cuspidine, Ca = Fluorite, $Ca_6 = Ca_6Na_4(PO_3F)_6O_2$, F = Fluorapatite, and S = Sodalite type phases. In locations where two phases are shown, the first is the highest concentration of the two.

PCT Results

The minimum, maximum, and average normalized releases of B, Li, and Na, as tested by the 7 day PCT[6], are given in Table 3 for the quenched and CCC samples. The majority of the quenched and CCC glasses had release rates well below the limits set by WAPS[7] with values shown in Table 3. The average normalized releases for the quenched and CCC glass samples were below 1 g/m^2.

The glasses with normalized releases >1g/m^2 were DP-08, -09, -11, and DZr-CV-09 quenched glasses and DP-1, -2, -6, -8, -9, -11, -13, -22, -23, DZr-CV-2, and -5 CCC heat-treated glasses.

Table 3 Normalized release rates of B, Li, and Na in g/m^2

	r_B(Q,T)	r_{Li}(Q,T)	r_{Na}(Q,T)	r_B(CCC,R)	r_{Li}(CCC,R)	r_{Na}(CCC,R)
Minimum	0.09	0.00	0.12	0.05	0.00	0.13
Average	0.30	0.38	0.37	0.65	0.78	0.45
Maximum	2.03	1.99	1.79	6.69	5.42	2.42
EA glass	8.35	4.78	6.67	—	—	—

Q = quenched, CCC = Heat treated according to canister centerline cooling profile,
T = normalized to target composition, R = normalized to residual composition.

The Effect of Crystallinity on Durability

Figure 2 shows logarithm r_B of quenched versus CCC glasses along with the major crystalline phase identified in the CCC heat-treated samples. Based upon the plot, the effect of crystallinity depends on the composition and concentration of the crystalline phase formed during CCC. Nepheline, Hiortdahlite, Nosean, and Fluorapatite were present in the glasses that had a significant increase in r_B after CCC.

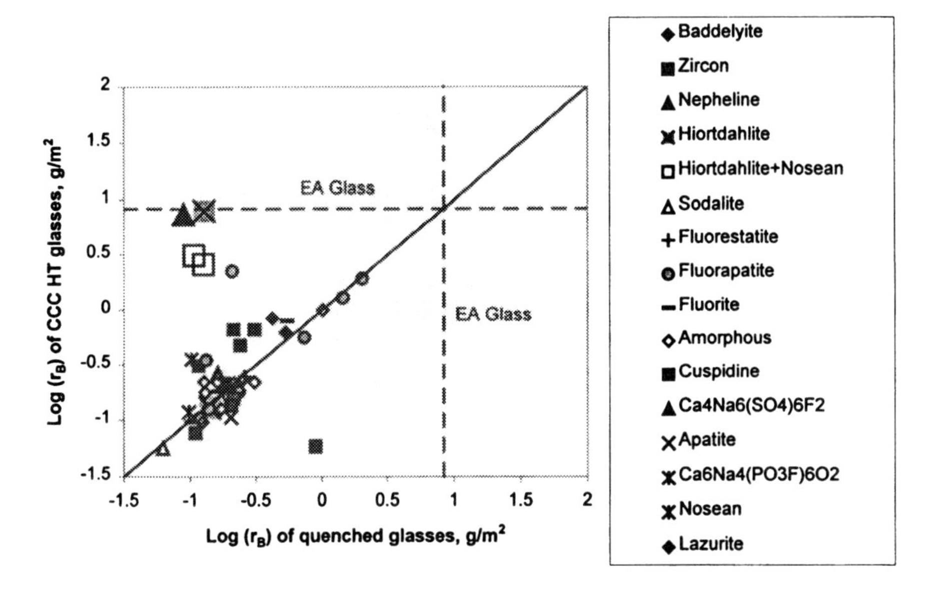

Figure 2 Log (r_B) of quenched versus CCC glasses plotted with major crystalline phases in CCC glasses

To examine the effect of crystallinity on PCT response, delta total crystallinity was plotted versus delta Log (r_B), where delta is the difference between quenched and CCC samples in Figure 3.

The plot shows that total crystallinity moderately predicts the change in r_B for most of the data with some exceptions. Four glasses (which precipitated hiortdahlite, nepheline, nosean or combinations of the three) plot well above the rest of the data, while one glass (which precipitated cuspidine) plotted well below the rest of the data.

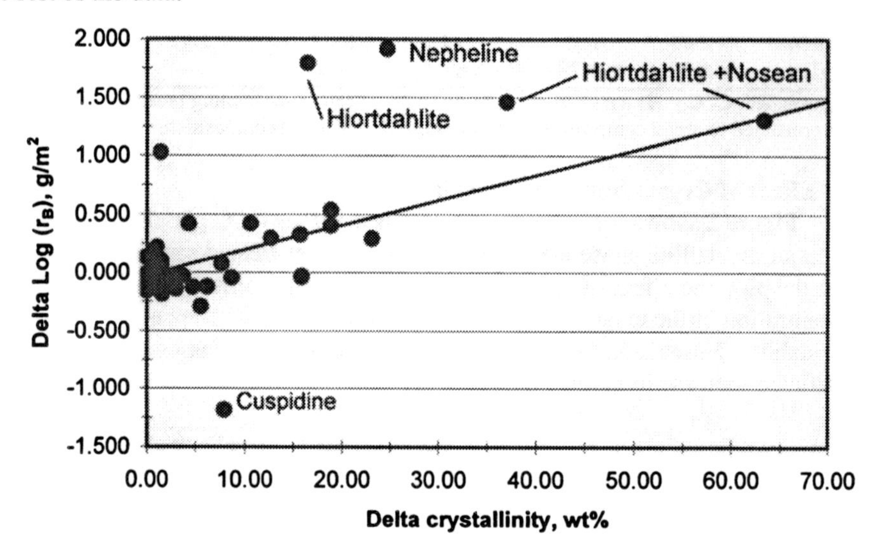

Figure 3 Delta crystallinity versus delta Log (r_B) where delta is the difference between quenched and CCC glasses.

Surface-contour plots were created to show the variation in glass durability with composition. Durability (ln(r_{Na}) from PCT results) was plotted as the vertical surface with glass composition plotted in a pseudo-ternary phase space of Al_2O_3-Na_2O+Li_2O-SiO_2 as the horizontal plane. The results for quenched and CCC glasses are shown in Fig. 4 and 5, respectively. From these plots, it is evident that for quenched glasses, the region high in Al_2O_3 and SiO_2 concentration is the most durable glass-forming region over the composition region tested. However, comparing this to the CCC glasses, the high Al_2O_3 region of the plot becomes the least durable region. This is because the high Al_2O_3 region also contains all of the glasses that highly crystallized nepheline, hiortdahlite, and/or sodalite phases during CCC heat treatment.

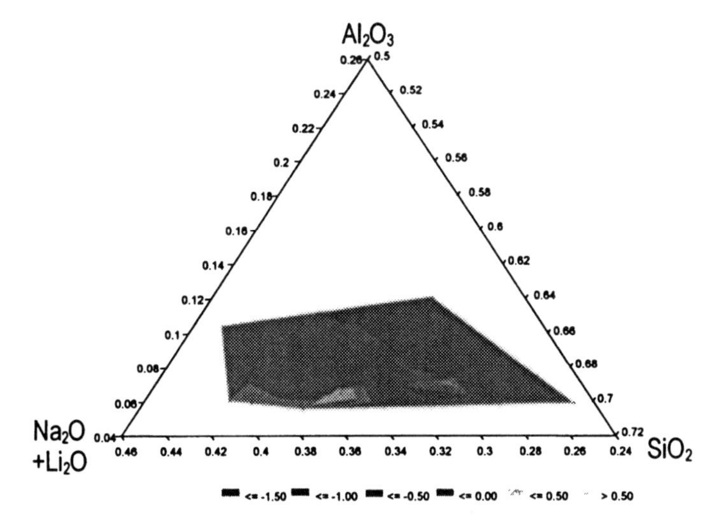

Figure 4. Ln (r_{Na}) of quenched glasses plotted in Al_2O_3-Li_2O+Na_2O-SiO_2 sub-mixture space

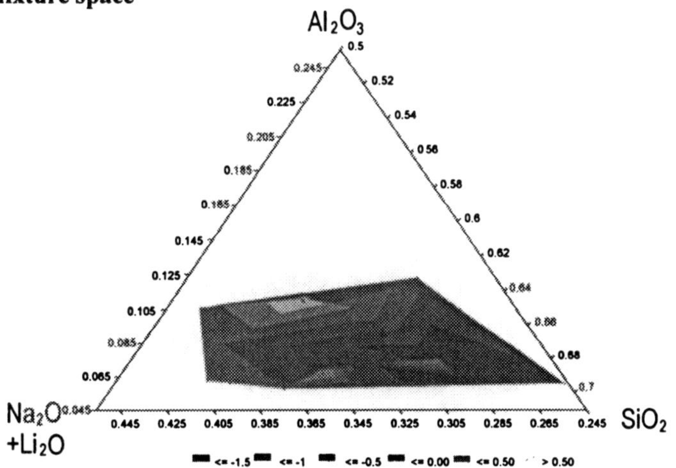

Figure 5. Ln (r_{Na}) of CCC glasses plotted in Al_2O_3-Li_2O+Na_2O-SiO_2 sub-mixture space

CONCLUSIONS

A more through evaluation of the data is needed, but based upon the initial examination, some areas of concern were identified for glass-formulation efforts. Glasses formulated low in SiO_2 and high in alkalis allow higher waste loading of Run 78 calcine[3]. However, if frit compositions low in SiO_2 and high in alkalis are combined with high Al_2O_3 concentration (contributed as a waste component), it may result in a glass composition prone to crystallize nepheline, hiortdahlite, or sodalite group phases. High concentrations of these crystalline phases in CCC glasses were observed to result in large increases in normalized PCT releases, which is an indication of low durability. Consequently, future formulation efforts to increase waste loading of INEEL calcine waste or similar wastes should avoid glass-forming regions that favor these silicate phases.

ACKNOWLEDGEMENTS

The authors would like to acknowledge the following people: Brian Riley and Jose Rosario for help in completing the study and E. William Holtzscheiter for management and guidance.

This study was funded by the Department of Energy's (DOE's) Office of Science and Technology, through the Tanks Focus Area, and office of Waste management through the Idaho National Environmental and Engineering Laboratory high-level waste program. This study was performed as a collaborative effort by Pacific Northwest National Laboratory (operated for DOE by Battelle under Contract DE-AC06-76RL01830), Savannah River Technology Center (operated for DOE by Westinghouse Savannah River Company under Contract DE-AC09-96SR18500), and the Idaho National Engineering and Environmental Laboratory (operated for DOE by Bechtel, Babcox and Wilcox Incorporated under Contract DE-AC07-941D 13223).

REFERENCES

[1]Li, H., J. D. Vienna, P. Hrma, D. E. Smith, and M. J. Schweiger. 1997. "Nepheline Precipitation in High-Level Waste Glasses - Compositional Effects and Impact on the Waste Form Acceptability," *Scientific Basis for Nuclear Waste Management* (Editors W. J. Gray and I. R. Triay), Vol. 465, pp. 261-268, Material Research Society, Pittsburgh, Pennsylvania.

[2]Kim, D-S, D. K. Peeler, and P. Hrma. 1995. "Effects of Crystallization on the Chemical Durability of Nuclear Waste Glasses," *Ceram. Trans.* 61, pp. 177-185, American Ceramic Society, Westerville, Ohio.

[3]Musick, C. A., B. A. Scholes, R. D. Tillotson, D. M. Bennert, J. D. Vienna, J. V. Crum, D. K. Peeler, I. A. Reamer, D. F. Bickford, J. C. Marra, N. L. Waldo. 2000. *Technical Status Report: Vitrification Technology Development Using INEEL Run 78 Pilot Plant Calcine,* INEEL\EXT-2000-00110, Idaho National Engineering and Environmental Laboratory, Idaho Falls, Idaho.

[4]Riley, B. J., J. A. Rosario, and P. Hrma. 2001. *Impact of HLW Glass Crystallinity on the PCT Response*, PNNL-13491, Pacific Northwest National Laboratory, Richland, Washington.

[5]Marra, S. L. and C. M. Jantzen. 1993. *Characterization of Projected DWPF Glasses Heat Treated to Simulate Canister Centerline Cooling (U),*

WSRC-TR-92-142, Rev. 1, Westinghouse Savannah River Company, Aiken, South Carolina.

[6]American Society for Testing and Materials (ASTM). 1998. "Standard Test Method for Determining Chemical Durability of Nuclear Waste Glasses: The Product Consistency Test (PCT)," ASTM-C-1285-97, in *Annual Book of ASTM Standards*, Vol. 12.01, ASTM, WEST Conshohocken, Pennsylvania.

[7]U.S. Department of Energy (DOE). 1993. *Waste Acceptance Product Specifications for Vitrified High-Level Waste Forms,* Office of Environmental Restoration and Waste Management, Washington D.C.

Chemical Durability and Characterization

LONG-TERM CORROSION TESTS WITH HANFORD GLASSES

W. L. Ebert, M. A. Lewis, and N. L. Dietz
Argonne National Laboratory
9700 South Cass, Ave.
Argonne, IL 60439

ABSTRACT

Five glasses are being subjected to long-term product consistency tests (PCTs) and vapor hydration tests (VHTs) to (1) study the effect of glass composition on corrosion rates, (2) compare the relative responses in short-term and long-term tests, and (3) provide a data base to support performance assessment calculations for the Hanford low-activity waste disposal system. These glasses were selected for detailed study from a suite of 56 glasses that were formulated to span the composition range of possible low-activity waste glasses for Hanford tank wastes and are being subjected to short-term PCTs and VHTs. The results of PCTs conducted through 140 days are discussed in this paper.

INTRODUCTION

This report summarizes progress to date in laboratory tests being conducted to evaluate the corrosion behavior of low-activity waste glasses made with Hanford tank wastes. The DOE Office of River Protection has contracted with Bechtel National, Inc., to design, construct, and commission the Hanford Tank Waste Treatment and Immobilization Plant.[1] Although the contractor is responsible for formulating and producing waste forms to immobilize the high-level and low-activity fractions of Hanford tank wastes, DOE must determine that the waste forms are acceptable for transfer, handling, and, in the case of the immobilized low-activity waste (ILAW), disposal in the Hanford disposal system. Acceptable ILAW must meet several contractual requirements that address chemical, physical, and radiological properties. Physical and radiological properties are measured directly. The chemical durability of the ILAW glasses is evaluated by three test methods specified in Section C of the contract:

2.2.2.17.1 Leachability Index: The waste form shall have a sodium leachability index greater than 6.0 when tested for 90 days in deionized water using the ANSI/ANS-16.1 procedure.

2.2.2.17.2 Product Consistency Test: The normalized mass loss of sodium, silicon, and boron shall be measured using a seven day product consistency test run at 90°C as defined in ASTM C1285-98.[2] The test shall be conducted with a glass to water ratio of 1 gram of glass (-100 +200 mesh) per 10 milliliters of water. The normalized mass loss shall be less than 2.0 grams/m^2.

To the extent authorized under the laws of the United States of America, all copyright interests in this publication are the property of The American Ceramic Society. Any duplication, reproduction, or republication of this publication or any part thereof, without the express written consent of The American Ceramic Society or fee paid to the Copyright Clearance Center, is prohibited.

2.2.2.17.3 *Vapor Hydration Test: The glass corrosion rate shall be measured using a seven day vapor hydration test run at 200°C as defined in the DOE concurred upon Product and Secondary Waste Plan. The measured glass alteration rate shall be less than 50 grams/(m^2 day).*

These tests address regulatory requirements (the leachability index), the chemical durability of the waste glass (the product consistency test), and the propensity for rate-increasing alteration phases to form (the vapor hydration test). Although these screening tests indicate which waste forms are expected to perform acceptably in the disposal system, additional testing is required to provide the information needed to calculate the performance of the waste forms and the disposal system.[3] Key information needs include the corrosion behavior and the release rates of radionuclides from the waste forms.

A Task was initiated under the auspices of the DOE Tanks Focus Area/Immobilization to determine if the ILAW composition and the short-term PCT or VHT required for product acceptance can be used to identify ILAWs that are likely to be acceptable for disposal.[4] Many glasses were formulated to study the effects of composition on the PCT and VHT responses. Four glass compositions from that suite of glasses plus a reference glass formulated by BNFL, Inc., were selected for more extensive testing to support performance assessment calculations. These glasses are expected to bound the concentrations of key glass components in the waste glasses, based on the waste stream composition and the likely processing flow sheet. The results of long-term PCTs completed through 140 days are summarized in this paper.

EXPERIMENTAL

The nominal compositions of the five glasses selected for long-term testing are given in Table I. (HLP-51 and HLP-56 are also referred to elsewhere as LAWBP1 and LAWA44, respectively.) Glasses were made by fusing component oxides, carbonates, and salts at about 1200°C for about one hour then cooled at about 21°C/h to simulate the cooling profile of the waste forms. Glasses were examined with scanning electron microscopy (SEM) to verify the absence of inclusions and glass/glass phase separation.

Product consistency tests (PCTs) are conducted with crushed, sieved glass (–100 +200 mesh) at glass/water mass ratios of 1/10 and 5/5. The crushed glass was carefully washed with demineralized water to remove fine particles. The particle size and absence of fines were verified with SEM examination. The glass/water mass ratios can be expressed in terms of the glass surface area/solution volume (S/V) ratios based on the densities of the glasses, and are about 2000 and 20,000 m^{-1}, respectively. All tests are being performed with demineralized water in Type 304L stainless steel vessels sealed with Teflon gaskets. At the end of the test, test solutions were passed through 450-nm-pore filters to remove any suspended glass particles then analyzed with inductively couple plasma-atomic emission spectroscopy (ICP-AES). The extent of reaction in the PCTs was determined by normalizing the solution concentration of a glass component to the S/V ratio of the test and the concentration of that component in the glass as

$$NL(i) = \{C(i) - C^{o}(i)\} / \{(S/V) \bullet f(i)\} \quad (1)$$

where NL(i) is the normalized mass loss based on element i, C(i) is the concentration of element i measured to be in the test solution, C°(i) is the background concentration of element i (which was measured in a series of blank tests), S/V is the glass surface area/solution volume ratio, and f(i) is the mass fraction of element i in the glass. The values of C°(i) were negligible relative to the values of C(i). The surface area (S) of glass in a test was calculated using the size fraction and the specific surface area, which was about 0.020 m^2/g for all glasses, and the mass of crushed glass in the test. The volume of test solution (V) was determined based on the mass of water added to the vessel during test assembly. The values of f(i) were calculated from the nominal glass compositions.

The values of NL(B), NL(Na), and NL(Si) are plotted in Fig. 1 to compare the relative releases of these elements. (Uncertainty bars are drawn at 15% of each value to represent analytical uncertainty.) Sodium is released preferentially to boron in most tests at 20 and 40°C, but boron is released preferentially to sodium in tests at 70 and 90°C. The relative normalized releases of B, Na, and Si indicate the corrosion mechanism. At low test temperatures, Na is released by ion-exchange reactions faster than B is released by network hydrolysis reactions. The greater release of B than Na in tests at higher temperatures may indicate that ion exchange has slowed to a greater extent than the hydrolysis reaction or that sodium is being sequestered in alteration phases. Samples of reacted glass are being examined to search for alteration phases, and a few small crystallites of precipitated phases have been detected with SEM. These will be compared with phases detected on samples reacted in VHTs.

Table I. Nominal Glass Compositions, in mass %

	HLP-09	HLP-12	HLP-33	HLP-51	HLP-56
Al_2O_3	6.84	6.74	4.00	10.00	6.20
B_2O_3	12.0	9.63	6.00	9.25	8.90
CaO	0.01	0.01	0.01	--[a]	1.99
NaCl	0.45	0.27	0.53	0.58	0.65
Cr_2O_3	0.07	0.07	0.09	0.02	0.02
NaF	0.01	0.01	0.02	0.04	0.01
Fe_2O_3	5.38	9.00	5.90	2.50	6.98
K_2O	0.4	0.4	0.47	2.20	0.50
La_2O_3	--	--	--	2.00	--
MgO	1.47	1.44	1.61	1.00	1.99
MoO_3	--	--	--	--	0.01
Na_2O	19.3	19.3	22.7	20.0	20.0
P_2O_5	0.05	0.05	0.06	0.08	0.03
ReO_2	0.01	0.01	0.01	--	0.10
SO_3	0.07	0.07	0.08	0.10	0.10
SiO_2	48.0	47.3	52.0	41.9	44.6
TiO_2	2.93	2.89	3.21	2.49	1.99
ZnO	1.47	1.44	1.61	2.60	2.96
ZrO_2	1.47	1.44	1.61	5.25	2.99
Density[b]					

[a]Component not included in glass formulation.
[b]Measured by water displacement, in g/cm^3.

The values of NL(i) in the 7-day test at 90°C and 2000 m^{-1} for soluble components are summarized in Table II. The responses of HLP-09, HLP-12, HLP-51, and HLP-56 are the same, within testing uncertainty, and all are significantly lower than the response for HLP-33. The responses for all glasses are less than 2.0 g/m^2, however, and all glasses meet the PCT requirement for ILAW glass. The values of NL(B) for tests at 2000 and 20,000 m^{-1} are plotted against the test duration in Fig. 2 to compare the relative responses of the five glasses. The average dissolution rates (based on the release of boron) for the five glasses in 7- and 140-day tests at 2000 m^{-1} in 14- and 98-day tests at 20,000 m^{-1} are summarized in Table III. The values in the table are the normalized release rates, which are calculated as NR(i) = NL(i)/t, where t is the test duration. This gives the average dissolution rate over the test duration and includes the effects of the evolving pH.

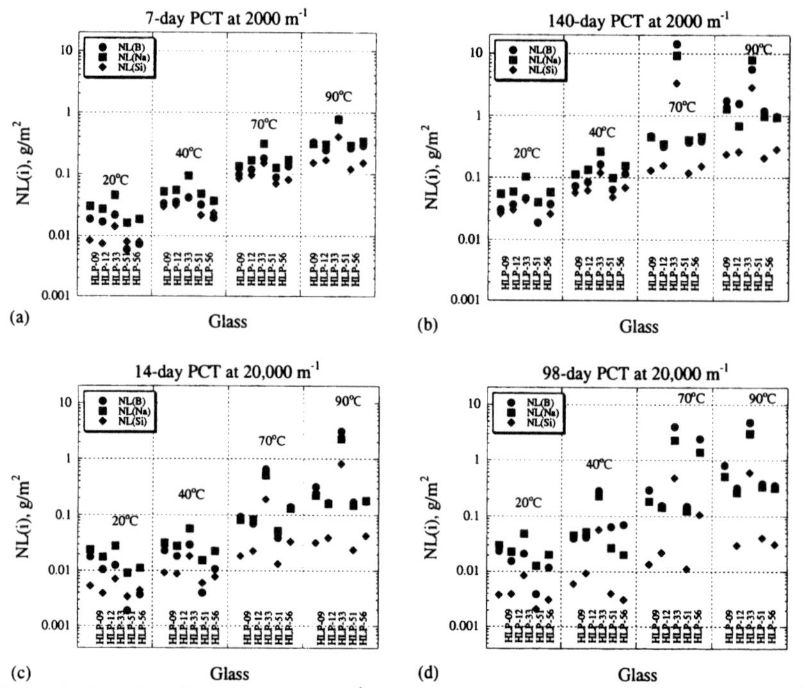

Figure 1. Results of PCTs at 2000 m^{-1} for (a) 7 and (b) 140 days and at 20,000 m^{-1} for (c) 14 and (d) 98 days.

Table II. Results of 7-day PCT at 90°C and 2000 m^{-1}, in g/m^2

Glass:	HLP-09	HLP-12	HLP-33	HLP-51	HLP-56
NL(Al)	0.14	0.16	0.071	0.12	0.11
NL(B)	0.33	0.25	0.81	0.27	0.29
NL(K)	<0.8	<0.8	<0.7	0.17	<0.6
NL(Na)	0.31	0.31	0.78	0.29	0.35
NL(Si)	0.15	0.17	0.41	0.12	0.15

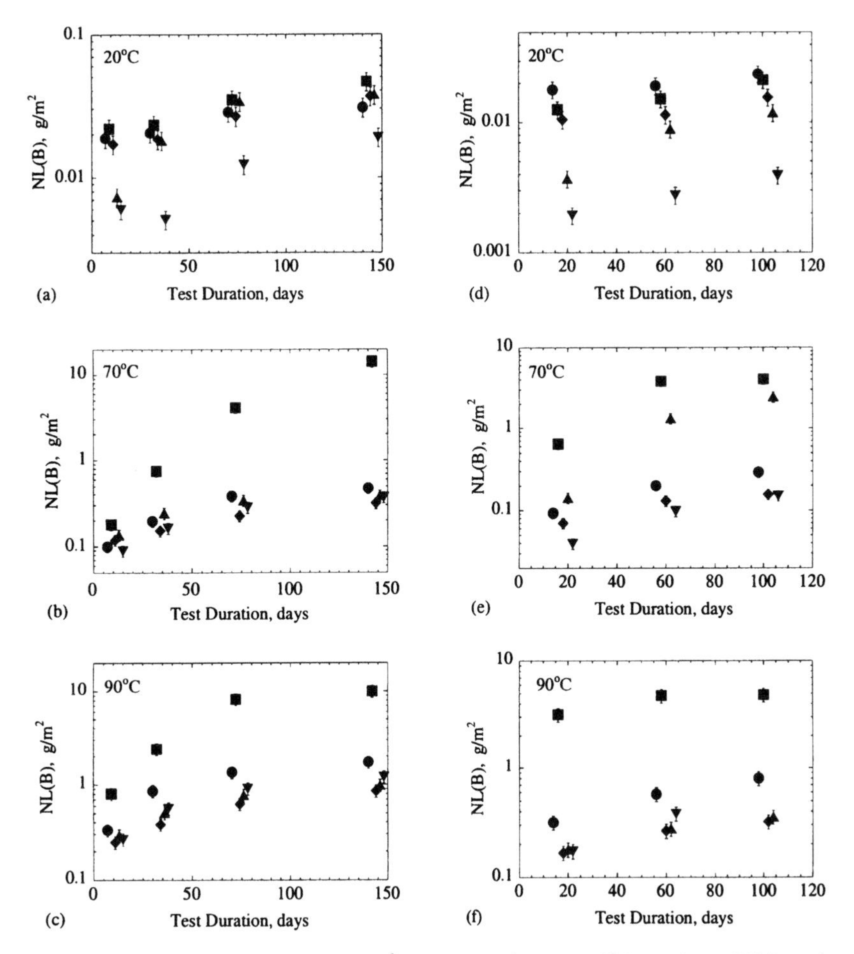

Figure 2. Results of PCTs at 2000 m^{-1} and (a) 20°C, (b) 70°C, and (c) 90°C and at 20,000 m^{-1} and (d) 20°C, (e) 70°C, and (f) 90°C: (●) HLP-09, (◆) HLP-12, (■) HLP-33, (▼) HLP-51, and (▲) HLP-56.

Although the value of NR(i) calculated in this way cannot be related to a particular process or long-term behavior, it does provide a useful way to compare the relative extents of corrosion of the different glasses. Glass HLP-33 dissolves at the highest rate at all temperatures and at both S/V ratios except at 20°C and 20,000 m^{-1}, but the relative dissolution rates of the other glasses change with the test conditions. Note that HLP-56 has a greater extent of dissolution in the test at 70°C than in the test at 90°C, based on the B, K, Na, and Si concentrations. Glass HLP-33 shows the same inverse temperature dependence in tests conducted for 70 and 140 days at 70 and 90°C at 2000 m^{-1}. This behavior is not currently understood.

Table III. Average Dissolution Rate as NR(B)[a], mg/(m^2•d)

Glass:	HLP-09	HLP-12	HLP-33	HLP-51	HLP-56
7-d and 2000 m^{-1}					
20°C	2.7	2.4	3.1	<0.9	1.1
40°C	4.8	5.0	5.9	2.8	4.6
70°C	14	17	26	13	19
90°C	48	35	120	38	42
140-d and 2000 m^{-1}					
20°C	0.22	0.26	0.31	0.14	0.27
40°C	0.52	0.61	1.2	0.46	0.83
70°C	3.4	2.3	102	2.6	2.8
90°C	12	6.1	71	8.6	7.0
14-d and 20,000 m^{-1}					
20°C	1.3	0.75	0.89	0.14	0.26
40°C	1.7	1.3	2.1	0.29	0.77
70°C	6.7	5.0	46	2.8	10
90°C	23	12	230	12	13
98-d and 20,000 m^{-1}					
20°C	0.24	0.16	0.22	0.040	0.12
40°C	0.40	0.42	2.9	0.66	0.71
70°C	3.0	1.6	41	1.5	25
90°C	8.2	3.3	49	4.5	3.6
score	54	36.5	78	24.5	47

[a] NR(B) = NL(B)/test duration.

To compare the relative durabilities of the five glasses, the rates for each test condition in Table III were used to rank the glasses: a value of 1 was assigned to the glass with the lowest rate and a value of 5 for the glass with the highest rate. The rankings were totaled for the 16 test conditions, and the scores are given in Table III. The order of increasing durability (i.e., lowest to highest scores) is HLP-33 < HLP-09 < HLP-56 < HLP-12 < HLP-51. The order from the 7-day test at 90°C and 2000 m^{-1}, which are the test conditions specified in the contract, is HLP-33 < HLP-09 ≈ HLP-56 ≈ HLP-51 ≈ HLP-12. These results indicate that the response in the 7-day PCT successfully identified the least durable of the five glasses prior to the formation of rate-increasing alteration phases. The propensity of these five glasses to generate alteration phases is being evaluated with VHTs; those results will be reported elsewhere.

CONCLUSIONS

Product consistency tests (PCTs) are being conducted with five glasses within the composition space being studied for low-activity waste glasses made with Hanford tank wastes. Those glasses are referred to as HLP-09, HLP-12, HLP-33, HLP-51 (LAWABP1), and HLP-56 (LAWA44). This report summarizes PCTs completed through 140 days for tests at 2000 m^{-1} and through 98 days for tests at 20,000 m^{-1}. The available PCT results indicate that HLP-33 is the least durable glass under all test conditions; the other glasses have higher but similar durabilities. However, all five glasses meet the contractual requirement that response in a 7-day PCT at 90°C and 2000 m^{-1} as NL(i) be less than 2.0 g/m^2.

ACKNOWLEDGMENTS

Glasses HLP-51 and HLP-56 were provided by Pacific Northwest National Laboratory staff. Glasses HLP-09, HLP-12, and HLP-33 were made at ANL. Test samples were prepared by J. Falkenberg, L. Hafenrichter, R. Miles, and V. Zyryanov. Solutions were analyzed by S. Lopykinski and D. Huff. Solids were analyzed by P. Johnson (XRD) and V. Zyryanov (SEM). This work is being conducted in support of the DOE Office of River Protection under DOE Interoffice Work Order M0CH20E10R. Work at Argonne National Laboratory is supported by the U.S. Department of Energy Office of Environmental Management under contract W31–109-ENG-38.

REFERENCES

[1]Office of River Protection contract for the Design, Construction, and Commissioning of the Hanford Tank Waste Treatment and Immobilization Plant, contract DE-AC27-01RV14136, January 2001.

[2]Standard Test Methods for Determining Chemical Durability of Nuclear Waste Glasses: The Product Consistency Test (PCT) Standard C1285, American Society for Testing and Materials, West Conshohocken, PA (1998).

[3]B.P. McGrail, W. L. Ebert, D.H. Bacon, and D.M. Strachan, A Strategy to Conduct an Analysis of the Long-Term Performance of Low-Activity Waste Glass in a Shallow Subsurface Disposal System at Hanford, Pacific Northwest National Laboratory Report PNNL-11834 (February 1998).

[4]J. D. Vienna, A. Jiricka, B. P. McGrail, B. M. Jorgensen, D. E. Smith, B. R. Allen, J. C. Marra, D. K. Peeler, K. G. Brown, I. A. Reamer, and W. L. Ebert, Hanford Immobilized LAW Product Acceptance: Initial Tanks Focus Area Testing Data Package, Pacific Northwest National Laboratory Report PNNL-13101 (1999).

DISSOLUTION KINETICS OF HIGH-LEVEL WASTE GLASSES AND PERFORMANCE OF GLASS IN A REPOSITORY ENVIRONMENT

Y.-M. Pan, V. Jain, and O. Pensado
Center for Nuclear Waste Regulatory Analyses
Southwest Research Institute
6220 Culebra Road
San Antonio, TX 78238

ABSTRACT

The chemistry of the water contacting the waste form after corroding the waste package may have a significant influence on waste form degradation. The dissolution kinetics of two simulated high-level waste (HLW) glasses (WVDP Ref. 6 and DWPF Blend 1), produced at the West Valley Demonstration Project and the Defense Waste Processing Facility, respectively, were measured in aqueous solutions of $FeCl_2$ and $FeCl_3$ at temperatures of 40, 70, and 90 °C using a modified product consistency test (PCT) method. These species simulate the presence of steel corrosion products. An empirical expression accounting for the intrinsic dissolution rate, pH dependence coefficient, and activation energy was obtained for modeling waste glass dissolution. Enhanced glass dissolution was observed in the presence of corrosion products. This expression was used to evaluate the impact of glass waste form on radionuclide release within the context of a performance assessment (PA) for the proposed HLW geologic repository at Yucca Mountain (YM), Nevada. Comparisons between the HLW glass cases and the spent fuel Nominal Case indicate that the enhanced glass dissolution by corrosion products results in a maximum dose contribution of about 30 percent of the Nominal Case dose at 52,000 yr, but the magnitude of the predicted mean dose rates is small in all cases.

INTRODUCTION

The long-term dissolution behavior of waste glass immersed in aqueous solutions can be expressed by a general rate equation shown in Eq. 1 that was originally developed for silicate hydrolysis reactions.[1,2] A similar form of the rate expression has been adopted by the Department of Energy (DOE) to model waste glass degradation.

To the extent authorized under the laws of the United States of America, all copyright interests in this publication are the property of The American Ceramic Society. Any duplication, reproduction, or republication of this publication or any part thereof, without the express written consent of The American Ceramic Society or fee paid to the Copyright Clearance Center, is prohibited.

$$\text{Rate} = S\left\{k \cdot \left[1 - \frac{Q}{K}\right]\right\} \qquad (1)$$

where

S = surface area of glass immersed in solution
k = forward dissolution rate
Q = concentration of dissolved silica in the solution
K = a quasi-thermodynamic fitting parameter equal to the apparent silica saturation value for the glass

In a closed-system leaching test, glass dissolution follows a nonlinear behavior as a function of time. The glass components are initially released at a fast rate, but the rate decreases with time due to a combined effect of increased silica concentration in the contact solution and increased diffusion path due to development of a hydrated surface layer on the surface of the waste glass. The silica concentration effect can be quantified by the chemical affinity term (1-Q/K). The forward dissolution rate, which represents the high initial rate in the absence of the build-up of dissolved silica and provides an upper bound to the dissolution rate, can be measured in experiments by maintaining the affinity term close to unity. The forward rate depends on glass composition, solution pH, and temperature, and can be expressed as

$$k = k_o \cdot 10^{\eta \cdot pH} \cdot \exp(-E_a / RT) \qquad (2)$$

where

k_o = intrinsic dissolution rate, in unit of g/m^2-day
η = pH dependence coefficient
E_a = activation energy, in unit of kJ/mol
R = gas constant, which is 8.314 kJ/mol-K
T = temperature in Kelvin

The model parameter values (k_o, η, E_a) in the rate expression in Eq. 2 have been determined by the DOE by regression of experimental data from single-pass flow-through and MCC-1 static leach tests for various waste glass compositions in buffer solutions prepared with deionized (DI) water.[3] However, the DOE model abstraction ignores the presence of corrosion products from the dissolution of waste package internal components such as FeOOH, $FeCl_2$, and $FeCl_3$ that could influence glass degradation. As previously reported in Pan et al.[4], the presence of aqueous solutions of $FeCl_2$ and $FeCl_3$ significantly enhances dissolution of HLW glasses. Substantially high initial boron and alkali release rates, approximately 50 to 70 times greater than those in DI water, were measured in 0.25 M $FeCl_3$ solution at 90 °C.

This work describes experimental investigations into dissolution kinetics of two simulated waste glass samples from WVDP and DWPF. The glasses were subjected to leaching tests in the presence of $FeCl_2$ and $FeCl_3$ at temperatures of 40, 70, and 90 °C. While the test environment does not represent YM repository J-13 groundwater environment, the literature shows that dissolution behavior of glass observed in DI water is more aggressive than that in J-13 water.[5] The relatively high leach rates observed in DI water are attributed to the solution composition effects, as elemental concentrations increase in the leachate, elemental release rates decrease. Therefore, the DI water bounds the effect of J-13 environment. From the glass leaching results, model parameters, including the intrinsic dissolution rate, pH dependence coefficient, and activation energy, were determined. An empirical rate expression accounting for the effect of iron compounds on the glass dissolution behavior was used for PA calculations to evaluate the impact of glass waste form on radionuclide release for the proposed YM repository.

EXPERIMENTAL PROCEDURES

Two simulated HLW glasses, WVDP Ref. 6 glass frit and DWPF Blend 1 glass, were used for dissolution study. These glasses are referred hereafter as WVDP glass and DWPF glass, respectively, and their compositions are given in Table I. To simulate internal WP environment, DI water enriched with either ferrous or ferric chlorides of concentrations 0.0025 M and 0.25 M was used. All the tests were conducted using a modified product consistency test (PCT) method in accordance with ASTM Standard Test Method C1285-97[6] except that the solutions were replaced at regular intervals. In these tests, 60-cm^3 perfluoroalkoxy (PFA) Teflon® vessels were used. Approximately 3 g of crushed glass specimen with a particle size distribution between -100 to +200 mesh was placed in each vessel. A 30 cm^3 test solution was added to the vessel. This gives a glass surface area to solution volume ratio (S/V) of 2,000 m^{-1}, as calculated in the ASTM standard. The test specimens were placed in ovens held at temperatures of 40, 70, and 90 °C. The solution was replaced entirely with an identical volume of fresh solution twice every week, at an interval of alternate three-day and four-day cycles, for a total test time of four weeks. At the end of each test period, the vessels were removed from the oven and allowed to cool. A small portion of leachate was used to measure pH. The leachate was then filtered with a 0.45 μm syringe filter for cation analysis using the inductively coupled plasma (ICP) atomic emission spectrometry technique.

The normalized concentration for element i, NC_i, and the normalized leach rate for element i, NLR_i, at the nth solution replacement can be calculated by the following equations:

Table I. Chemical compositions of test glasses (in weight percent)

Oxide	WVDP Ref. 6[a]	DWPF Blend 1[b]	Oxide	WVDP Ref. 6[a]	DWPF Blend 1[b]
Al_2O_3	6.67	4.16	MnO	0.51	—
B_2O_3	11.48	8.05	MnO_2	—	2.05
BaO	—	0.18	MoO_3	—	0.15
CaO	0.66	1.03	Na_2O	11.94	9.13
Cr_2O_3	—	0.13	Nd_2O_3	—	0.22
Cs_2O	—	0.08	NiO	—	0.89
CuO	—	0.44	P_2O_5	2.01	—
FeO	—	—	RuO_2	—	0.03
Fe_2O_3	11.95	10.91	SO_3	0.25	—
K_2O	5.15	3.68	SiO_2	42.28	51.9
La_2O_3	—	—	TiO_2	1.04	0.89
Li_2O	4.84	4.44	ZnO	—	—
MgO	0.18	1.41	ZrO_2	1.28	0.14
—	—	—	Total	100.24	99.91

[a]Composition provided by West Valley Nuclear Services Co., Inc.
[b]Composition provided by Westinghouse Savannah River Company.

$$NC_i = \frac{C_i}{F_i} \quad (3)$$

$$NLR_i = \frac{(NC_i)_n - (NC_i)_{n-1}}{(S/V)(t_n - t_{n-1})} \quad (4)$$

where NC_i is in g/m^3, C_i is the concentration of element i in solution in g/m^3, F_i is the mass fraction of element i in glass, NLR_i is in g/m^2-day, t_n-t_{n-1} is the time in days between the n-1th and nth solution replacements, and S/V is the surface-to-volume ratio in m^{-1}.

RESULTS AND DISCUSSION

Glass Leaching Results

As generally observed in closed-system leaching tests, the release rates for glass components decreased continuously with time for all test conditions used in this study, even though the decrease in release rate varied with the test conditions. Chemical analysis on glass surface indicated that the high initial leaching results in the formation of an alkali-depleted surface, and consequently leaching rate decreases due to an extended diffusion path.[4] Leach rates calculated after the first solution replacement were used as initial leach rates to provide a conservative account of the effect of glass dissolution and are plotted as a function of final leachate pH in Figure 1. The initial leach rates for all major components were pH dependent. While the leach rates for B and alkali decreased with an increase in pH, the Si release rate remained relatively constant. Al leach rate showed a minimum for both the 70 and 90 °C tests. The pH value at which the minimum rate occurred varied with both the glass composition and test temperature. For the 40 °C tests, however, the leach rates for all elements decreased continuously with increasing leachate pH. Figure 2 shows the log NLR_B as a function of leachate pH at each temperature for both the WVDP and DWPF glasses, based on the initial release of boron. A linear regression between log NLR_B and leachate pH was performed, and the regression equations are also given in Figure 2. A good correlation between leach rate and pH is found for all test temperatures, as indicated by the high correlation coefficients (R^2).

In contrast to a linear pH dependence with a negative slope observed for B release in this study, Knauss et al.[7] and Abraitis et al.[8] conducted experiments in controlled pH environments and reported V-shaped dissolution rate versus pH curves with minima at near-neutral pH for both B and Si release. McGrail et al.[9] also studied glass dissolution kinetics at pH values between 6 and 12 and showed that glass dissolution increases with increasing pH. These observations suggest that the effect of pH on glass dissolution depends on the glass compositions and test conditions. While all the experiments in the cited references were conducted in controlled pH environments in the absence of the influence of the affinity term, solution pH was initially set by hydrolysis of $FeCl_2$ or $FeCl_3$ in this study and the final pH values were generally higher than the initial ones. In the case of glass leaching in DI water, the leachate pH drifted from acidic to the basic range. The observed discrepancy in the effect of pH on dissolution rate in the basic range could be the result of variations in leachate pH. It is also noted that the initial leach rates for B and most alkali elements are much higher in comparison with the Si release rate in the low pH ranges. HLW glass dissolution rates could be significantly underestimated if based on measured Si release rates. In addition, the influence of the affinity term has not been evaluated under the current test conditions. Nevertheless, the use of B release rates provides a conservative upper bound to the release rate of radionuclides.

Determination of Model Parameters

From Figure 2 the slopes of the linear regression equations provide the pH dependence coefficients (η). It is apparent that η does not significantly change with the test temperature and glass composition. Activation energy (E_a) can be regressed from the experimental data in a plot of ln NLR_B versus 1/T based on the rate equation in Eq. (2), and the intrinsic dissolution rate (k_o) can then be determined using the values of η and E_a. The model parameters are summarized in Table II. The E_a values

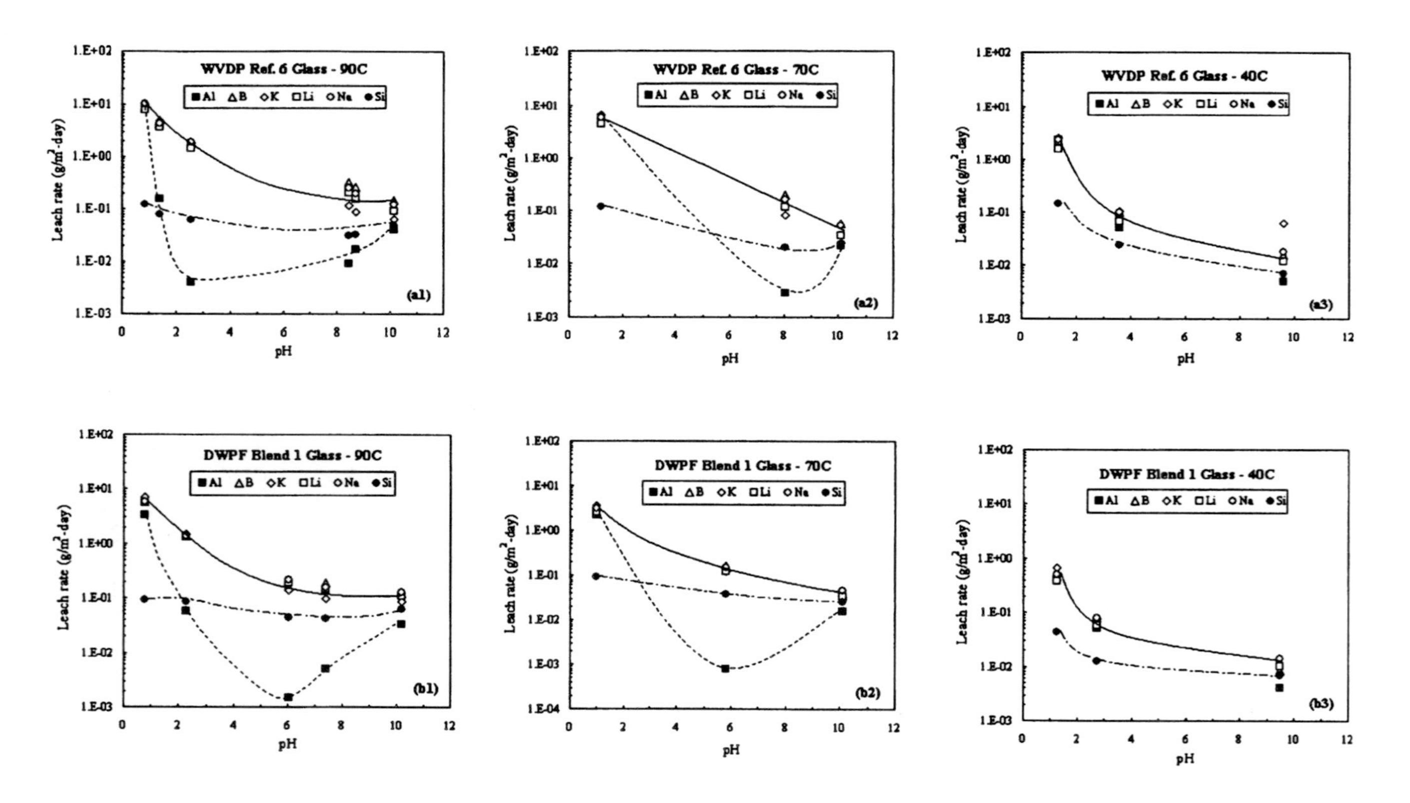

Figure 1. Normalized leach rate for various elements versus leachate pH after first solution replacement for (a) WVDP and (b) DWPF glasses at various temperatures

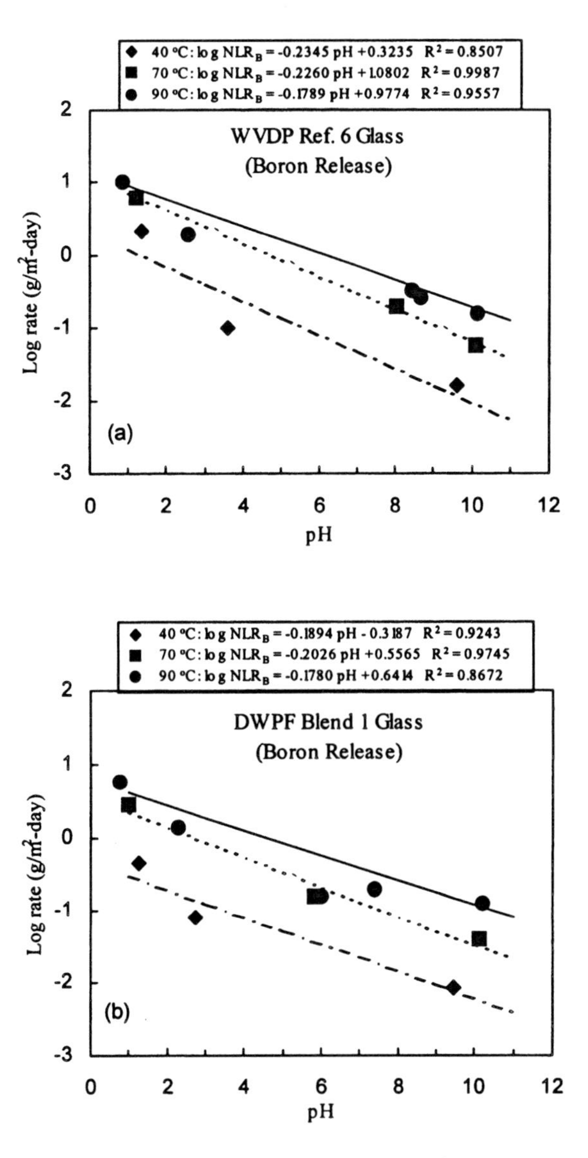

Figure 2. Linear regression of normalized leach rate for boron versus leachate pH after first solution replacement for (a) WVDP and (b) DWPF glasses at various temperatures

for both waste glasses are consistent with the values reported in the literature for borosilicate waste glasses.[7-10] The mean and standard deviation values calculated for a combined case are listed in Table III and compared to other parameter values published elsewhere.[3] Note that the rate expression used by the DOE for the basic leg has combined the affinity term with the intrinsic dissolution rate constant. A significantly higher k_o value, $\log_{10} k_o = 7.91 \pm 0.16$ g/m²-day, was measured from the forward rates.[3] The model parameters in Table III were used in PA calculations and discussed later in this paper.

Table II. Glass dissolution model parameters summary

Galss	Temperature (°C)	$\log_{10} k_o$ (g/m²-day)	η	E_a (kJ/mol)
WVDP Ref. 6	40	8.53	-0.2345	49.2±7.4
WVDP Ref. 6	70	8.57	-0.2260	49.2±7.4
WVDP Ref. 6	90	8.22	-0.1789	49.2±7.4
DWPF Blend 1	40	8.49	-0.1894	52.8±2.5
DWPF Blend 1	70	8.60	-0.2026	52.8±2.5
DWPF Blend 1	90	8.37	-0.1780	52.8±2.5

Table III. Comparison of glass dissolution model parameters

Parameter	Case A (CNWRA Rate Expression)	Case B (DOE Rate Expression)[3]
k_o (g/m²-day)	$10^{8.46 \pm 0.14}$	$10^{14.0 \pm 0.5}$ if pH < 7.1
		$10^{6.9 \pm 0.5}$ if pH ≥ 7.1
η	-0.20 ± 0.02	-0.6 ± 0.1 if pH < 7.1
		0.4 ± 0.1 if pH ≥ 7.1
E_a (kJ/mol)	51.0 ± 5.6	80 ± 10 if pH < 7.1
		80 ± 10 if pH ≥ 7.1

The rate expression for dissolution of waste glasses has been evaluated by the DOE,[3] primarily based on the experimental results by Knauss et al.[7] Because the glass dissolution rates were found to have a V-shaped pH dependence, with minima at near-neutral pH, separate rate expressions were obtained for dissolution under acidic or basic conditions, as reflected in Table III. In addition, the release of B occurred faster than that of Si under some test conditions. The model parameter values based on B release rates were determined to bound the range of HLW glass compositions and environmental conditions. Figure 3 compares the calculated dissolution rates on the basis of Equation (2) and various model parameters in Table III. It is seen that the model parameters supported by experimental data in this paper are associated with higher dissolution rates than those reported elsewhere, in the near-neutral pH range. The higher dissolution rates are consistent with the observations of enhanced glass dissolution in the presence of iron-containing corrosion products.[4]

Performance Assessment Analyses

The Nuclear Regulatory Commission (NRC) and Center for Nuclear Waste Regulatory Analyses (CNWRA) have been developing a tool, the Total-system Performance Assessment (TPA) code,[11] intended to support review activities for a potential license application, by the DOE, for construction of a HLW repository at YM. Based on a Monte Carlo scheme, the TPA code is used to compute the expected annual dose to the average member of the critical group in the event of failure of the waste packages to isolate radionuclides. A Nominal Case (defined as a particular set of models and model parameters describing likely behaviors of the YM repository system) is generally selected to perform sensitivity and uncertainty analyses and to study the performance of the system, as simulated by the TPA code. In the Nominal Case model or description, it is assumed that 70,040 metric tons of spent nuclear fuel is packaged in 7,176 waste packages and emplaced in the proposed repository. Each waste package contains, on average, 9.8 metric tons of spent nuclear fuel. The Nominal Case description does not account for the presence of HLW in glass form. Equation (2) was incorporated into the TPA code to account for glass leaching in the event glass is contacted by water. Computations presented in this section are aimed at evaluating the relative importance of glass dissolution with respect to spent nuclear fuel dissolution, in units of dose.

In the computer simulations, the model parameters obtained from this work (Case A) and from the DOE(Case B) in Table III were considered. The current TPA model version 4.1e does not allow for the simultaneous inclusion of two waste forms in the estimation of the dose. To circumvent that problem, it was assumed that glass was the only kind of waste form in the system and that a total of 4,667 metric tons of HLW glass[12] contained in 3,910 waste packages[13] was emplaced in the proposed repository. The initial radionuclide inventory was selected from available data.[13] In order to obtain appropriate comparison of the dose deriving from glass dissolution to the Nominal Case dose, a temperature versus time curve for the latter case was

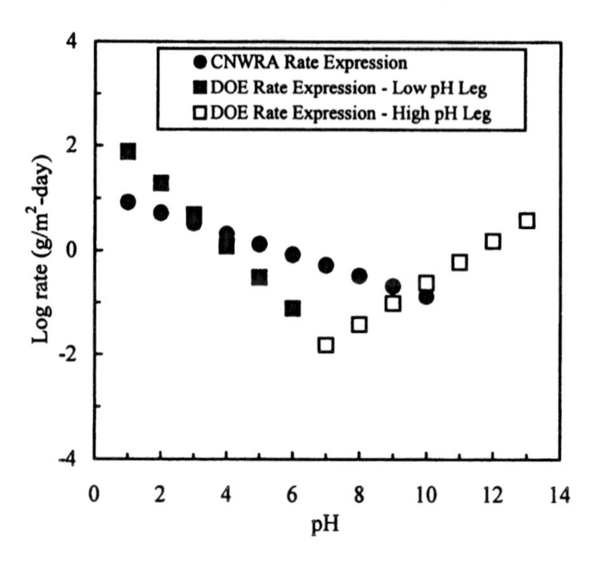

Figure 3. Comparison of calculated glass dissolution rates using different rate expressions

employed. The justification is that in the mixed system, spent fuel—glass form, the temperature of the repository is dictated by the thermal activity of the spent fuel. The total surface area of the exposed glass form per waste package was set equal to 99 m^2, as suggested by other studies.[14] In the Monte Carlo analysis, the pH was uniformly sampled in the range 4.8 to 10, consistent with predictions of the chemistry inside the waste package.[15] Results of the computations are summarized in Figure 4.

The mean dose rates for these runs were computed by averaging the total effective dose equivalent for the nominal scenario at an instant of time from 200 Monte Carlo realizations and are reported in Figure 4 for each case. As shown in Figure 4, the predicted mean dose rates for Case A are higher than those computed for Case B (Cases A and B refer to the notation in Table III), which are a direct consequence of the higher dissolution rates in the near-neutral pH associated to Case A that accounts for the effect of corrosion products (see Figure 3). The impact of glass dissolution on the predicted mean dose rate in Case A could be up to 30 percent of the Nominal Case at times of the order of 52,000 yrs. Within the first 10,000 yr, the dose deriving from glass dissolution is of the same order of magnitude as the Nominal Case dose. Nevertheless, the magnitude of these predicted mean dose rates is less than 10^{-3} rem/yr at times $t < 60,000$ yr in Figure 4. The dose at earlier times are a consequence of the assumption that initial defects could be present in waste containers. At times greater than 40,000 yr, a significant number of waste packages

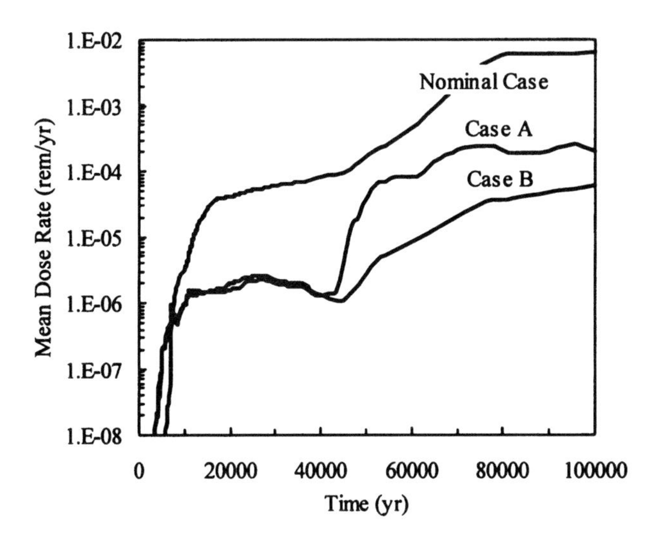

Figure 4. Predicted mean dose rates for three cases from 200 realizations. The Cases A and B that account for 4,667 metric tons of HLW glass refer to the parameters in Table III. The Nominal Case of the TPA code models 70,040 metric tons of spent nuclear fuel in the repository system.

fail due to general corrosion; thus, additional sources of radionuclide release are available in the system, causing the increase in the dose. The most important radionuclides contributing to the mean dose rate are ^{237}Np at all times, and ^{99}Tc and ^{129}I at early times (< 15,000 yr).

CONCLUSIONS

Rate expressions for glass dissolution were developed taking account of the effect of corrosion products. The model parameters including the intrinsic dissolution rate, pH dependence coefficient, and activation energy, were determined based on the initial release of boron from leaching of two simulated waste glass compositions in aqueous solution of $FeCl_2$ and $FeCl_3$. The observed discrepancy in the effect of pH on dissolution rate in the basic range could be the result of variations in leachate pH in comparison with the experiments in the cited references under controlled pH environments. Nevertheless, the present model parameters provide a conservative account of glass dissolution. Dissolution rates calculated on the basis of Equation (2) and model parameters derived in this paper (Case A in Table III) are higher, in the near-neutral pH range, than those computed with parameter values suggested by other studies. Performance assessment analyses of the proposed HLW repository at YM, indicate that the predicted mean dose rate deriving from glass dissolution could be as high as 30 percent of the dose rate caused by leaching of the spent nuclear fuel. The presence of corrosion products enhances glass dissolution and subsequent release of radionuclides to the environment.

ACKNOWLEDGMENTS

This paper was prepared to document the work performed by the CNWRA for the NRC under contract No. NRC–02–97–009. This paper is an independent product of the CNWRA and does not necessarily reflect the views or the regulatory position of the NRC.

REFERENCES

[1]P. Aagaard and H.C. Helgeson, "Thermodynamic and Kinetic Constrains on Reaction Rates Among Minerals and Aqueous Solutions I. Theoretical Considerations," *American Journal of Science*, **282**, 237–85 (1982).

[2]W.L. Bourcier, *Critical Review of Glass Performance Modeling*. ANL–94/17, Argonne National Laboratory, Argonne, IL, 1994.

[3]U.S. Department of Energy, *Defense High Level Waste Glass Degradation*. ANL–EBS–MD–000016 Revision 00 ICN 01, Civilian Radioactive Waste Management System, Management & Operating Contractor, Las Vegas, NV, 2000.

[4]Y.M. Pan, V. Jain, M. Bogart, and P. Deshpande, "Effect of Iron Chlorides on the Dissolution Behavior of Simulated High-Level Waste Glasses," *Ceramic Transactions of the Environmental Issues and Waste Management Technologies in the Ceramic and Nuclear Industries*, **Vol. 119** (2001).

[5]G.L. McVay and C.Q. Buckwaler, "Effect of Iron on waste-Glass Leaching," *Journal of American Ceramic Society*, **66** [3] 170–74 (1983).

[6]ASTM C 1285-97, "Determining Chemical Durability of Nuclear, Hazardous, and Mixed Waste Glasses: the Product Consistency Test (PCT)," *Annual Book of ASTM Standards*, **Vol. 12.01**, 774–91, American Society for Testing and Materials, West Conshohocken, PA, 1999.

[7]K.G. Knauss, W.L. Bourcier, K.D. McKeegan, C.I. Merzbacher, S.N. Nguyen, F.J. Ryerson, D.K. Smith, H.C. Weed, and L. Newton, "Dissolution Kinetics of a Simple Analogue Nuclear Waste Glass as a Function of pH, Time and Temperature," *Materials Research Society Symposium Proceeding*, **Vol. 176**, 371–81 (1990).

[8]P.K. Abraitis, D.J. Vaughan, F.R. Livens, L. Monteith, D.P. Trivedi, and J.S. Small, "Dissolution of a Complex Borosilicate Glass at 60 °C: The Influence of pH and Proton Adsorption on the Congruence of Short-Term Leaching," *Materials Research Society Symposium Proceeding*, **Vol. 506**, 47–54 (1998).

[9]B.P. McGrail, W.L. Ebert, A.J. Bakel, and D.K. Peeler, "Measurement of Kinetic Rate Law Parameters on a Na-Ca-Al Borosilicate Glass for Low-Activity Waste," *Journal of Nuclear Materials*, **249**, 175–89 (1997).

[10]T. Advocat, J.L. Crovisier, E. Vernaz, G. Ehret, and H. Charpentier, "Hydrolysis of R717 Nuclear Waste Glass in Dilute Media: Mechanisms and Rate a Function of pH," *Materials Research Society Symposium Proceeding*, **Vol. 212**, 57–64 (1991).

[11]S. Mohanty and T.J. McCartin (Coordinators), *Total-system Performance Assessment (TPA) Version 4.0 Code: Module Description and User's Guide*. Center for Nuclear Waste Regulatory Analyses, San Antonio, TX, 2000.

[12]U.S. Department of Energy, *Waste Form Degradation Process Model Report*. TDR–WIS–MD–000001 REV 00 ICN 01, Civilian Radioactive Waste Management System, Management & Operating Contractor, Las Vegas, NV, 2000.

[13]U.S. Department of Energy, *Inventory Abstraction Analysis Model Report*. ANL–WIS–MD–000006 REV 00, Civilian Radioactive Waste Management System, Management & Operating Contractor, Las Vegas, NV, 2000.

[14]U.S. Department of Energy, *Total System Performance Assessment-Viability Assessment (TSPA-VA) Analyses Technical Basis Document. Waste Form Degradation, Radionuclide Mobilization, and Transport through the Engineered Barrier System*. B00000000–01717–4301–00006 Rev 00, Civilian Radioactive Waste Management System, Management & Operating Contractor, Las Vegas, NV, 1998.

[15]U.S. Department of Energy, *Summary of In-Package Chemistry for Waste Forms*. ANL–EBS–MD–000050 REV 00, Civilian Radioactive Waste Management System, Management & Operating Contractor, Las Vegas, NV, 2000.

ANALYSIS OF LAYER STRUCTURES FORMED DURING VAPOR HYDRATION TESTING OF HIGH- SODIUM WASTE GLASSES

Andrew C. Buechele, Frantisek Lofaj, Cavin Mooers, and Ian L. Pegg
Vitreous State Laboratory, The Catholic University of America
Washington, DC 20064

ABSTRACT

The Vapor Hydration Test (VHT) is presently specified as one of the durability tests that low-activity waste (LAW) glasses for the Hanford site must meet. This study reports on the effects of glass compositional variations on VHT durability and follows the kinetics and microstructure of layer development in selected glass formulations. A variety of layer structures and thickness have been observed by scanning electron microscopy (SEM) in specimens of different compositions subjected to the VHT at 200°C for 24 days. X-ray Diffraction (XRD) patterns taken from reacted surfaces show a predominance of analcime and zeolite-type phases. Additionally, coupons of a high-sodium formulation were subjected to the VHT in a series spanning 1 to 24 days to follow the kinetics of the reaction; two plateaus were seen in rate of layer development. Transmission electron microscopy (TEM) cross-sectional specimens of the corroded layers prepared using a precision ion polishing system (PIPS) and ultramicrotomy techniques reveal details of the microstructure of the interface. TEM observations were correlated with those of SEM and energy dispersive spectroscopy (EDS.)

INTRODUCTION

Tank wastes stored at the Hanford site will be separated into low-activity waste (LAW) and high-level waste (HLW) fractions, each of which will be vitrified separately. The resulting LAW glass products are required to meet a variety of durability requirements[1-3], one of which is the Vapor Hydration Test (VHT). The VHT is particularly effective in accelerating the reaction progress to the point at which the role of secondary phases becomes important; such phases are expected to be influential in determining the long-term alteration behavior of the glass. The VHT consists essentially of contacting a glass coupon with saturated water vapor at elevated temperature and pressure under conditions that

To the extent authorized under the laws of the United States of America, all copyright interests in this publication are the property of The American Ceramic Society. Any duplication, reproduction, or republication of this publication or any part thereof, without the express written consent of The American Ceramic Society or fee paid to the Copyright Clearance Center, is prohibited.

minimize the possibility of reflux. In this work, corroded coupons of different glasses subjected to the VHT for the same time-temperature schedule were examined by scanning electron microscopy (SEM) and optical microscopy (OM) to determine the thickness and microstructure of the reacted layers. The thickness of the reacted layer is used to determine the rate of alteration of the glass for comparison of the durabilities of different glass formulations. The microstructures of the reacted layers vary widely among different formulations and provide useful information with respect to the underlying reaction mechanisms.

A wide range of glass formulations developed at the Vitreous State Laboratory (VSL) for Hanford LAW streams have been subjected to the VHT[3]. Tests have been run both on glasses produced in small-scale crucible melts and in pilot-scale melter runs. The present study reviews some of the more interesting compositional effects on the alteration rate and layer microstructure.

EXPERIMENTAL PROCEDURE

Material Characterization

The results from nine LAW glass formulations are discussed in this paper; the compositions of these glasses, all of which were prepared in crucible melts, are shown in Table I. Also shown in Table I are the measured VHT alteration rates,

Table I. Formulated compositions in weight %.

	LAWA-						LAWC-	
	23R	33	44	53	57	105	21S	25
Al_2O_3	9.70	11.97	6.20	6.09	6.09	7.03	6.13	5.79
B_2O_3	4.32	8.85	8.9	6.11	6.11	8.28	10.10	9.54
CaO	4.46	0.00	1.99	7.77	2.84	1.85	6.41	6.06
Fe_2O_3	7.43	5.77	6.98	7.40	7.40	6.49	6.48	6.12
K_2O	2.31	3.10	0.50	0.49	0.49	0.60	0.15	8.09
Li_2O	2.08	0.00	0.00	0.00	0.00	0.00	2.74	0.00
MgO	2.08	1.99	1.99	1.46	1.46	1.85	1.51	1.43
Na_2O	20.00	20.00	20.00	19.72	19.72	24.00	11.88	11.22
SiO_2	40.52	38.21	44.55	41.66	41.66	41.42	46.76	44.18
SnO_2	0.00	0.00	0.00	0.00	4.93	0.00	0.00	0.00
TiO_2	0.00	2.49	1.99	1.09	1.09	1.85	1.12	1.06
ZnO	3.34	4.27	2.96	2.95	2.95	2.76	3.02	2.85
ZrO_2	3.05	2.49	2.99	2.95	2.95	2.78	3.02	2.86
SO_3	0.04	0.10	0.095	1.48*	1.48*	0.11	2.50*	0.41
Other	0.67	0.76	0.855	0.83	0.83	0.98	0.68**	0.39
Alteration Rate, $g \cdot m^{-2} d^{-1}$	>113	61.1	1.0	0.8	<0.1	39.7	2.9	<0.1

*Present in the raw formulation at this level to produce saturation of SO_3 in melt.

**Formulation was normalized to 102.5%

which are discussed below. Two of the glasses, LAWA23R and LAWA33, are early formulations that pre-date the VHT requirement. While these glasses performed substantially better on the 7-day Product Consistency Test (PCT) requirement, they are towards the poorer end of the range of VHT performance for current LAW formulations[2,3]. The glass LAWA44 is representative of current LAW formulations and exhibits a VHT response that is about a factor of 50 below the present acceptance criterion[1,3] of 50 $g{\cdot}m^{-2}d^{-1}$. The remaining 5 glasses can be regarded as variations on LAWA44.

Vapor Hydration Testing

Test coupons were cut with approximate dimensions of 10 mm x 10 mm x 2 mm with one of the large surfaces cut roughly parallel to a fractured surface. A 1.3 mm hole was drilled through the coupon with a diamond drill bit. Samples were subsequently ultrasonicated in high-purity acetone, allowed to air dry, and weighed. Coupons typically weighed about 0.2 g. The actual dimensions of each test coupon were measured with a micrometer to determine its surface area since this enters into the calculation of the amount of water to be added to the test vessel.

VHT experiments were conducted in Parr series 4700 stainless steel pressure bombs with a volume of 22 cm^3. A Teflon™ gasket was utilized to seal the pressure vessels. The coupons were suspended in the center of the vessels on stainless steel wires to avoid contact with the water or vessel walls. The vessels were flushed with argon prior to sealing. The volume of deionized water added to the pressure vessel was calculated to be sufficient to saturate the volume of the vessel at the test temperature, plus an additional amount to provide for the formation of a non-dripping layer on the surface of the coupon, as suggested by Ebert and Bates[4]. The total volume was typically around 0.2 ml of deionized water.

The vessels were prepared, sealed, weighed and placed in an oven at 200°C. At the conclusion of the test, the vessels were removed from the oven, cooled, weighed and then opened. The coupons were then removed, weighed, subjected to XRD of the corroded surface, and next prepared for SEM, OM and/or TEM observation.

Three series of experiments were performed for this study. In the first series different glass formulations were tested at 200°C for 24 days. In the second series, eight coupons of the same formulation were tested for different durations ranging from 1 to 24 days. The second series was performed in order to investigate the kinetics of layer formation and the dissolution rate behavior of a particular glass formulation. The third series involved four coupons of another formulation tested for durations ranging from 24 to 206 days to follow kinetics of layer formation one of the most durable glasses.

Microstructure Observation

Optical and Scanning Electron Microscopy. The reacted coupons were air dried, again weighed, and examined by X-ray diffraction (XRD). The coupons were then cut in half using a low-speed diamond saw. One half of the coupon was mounted vertically in epoxy with the glass-corrosion layer in cross section. Standard metallographic techniques were employed for grinding and polishing, with a 0.05 μm diamond paste utilized during the final polish. These samples were first observed optically using polarized light at 50x to 500x, and then carbon coated for SEM and EDS evaluation.

SEM assessments were conducted at 25 kV with a working distance of 39 mm, primarily using BSE imaging to characterize general differences in chemical composition and layer morphology. More precise elemental distribution was determined with EDS semi-quantitative analysis, mapping, and line-scans.

Transmission Electron Microscopy. Following SEM and EDS analysis, the mounted samples were re-polished and etched for 5 to 10 seconds in 5% HCl. Two-stage extraction replicas were taken for more precise EDS analysis and possible diffraction work on microcrystalline phases.

The unmounted portion of the VHT coupons were subsequently halved again and adhered, surface-layer to surface-layer, with G1 epoxy. The adhered material was sectioned into approximately 2 mm x 2 mm x 0.5 mm pieces, mounted to slotted nickel grids with M-Bond 610™, and polished to about 100 μm. Dimpling was then performed to reduce the thickness in the sandwiched leached layers region to ~15 μm. Thin foils were pre-milled on a Gatan DuoMill 600 at an angle of 15°, followed by milling on a precision ion polishing system (PIPS) at angles of 5° and 3° and accelerating voltages of 5 kV and 3 kV, respectively, until a small perforation was produced.

A small amount of VHT material was reserved for ultramicrotomy. This material was ground into 1 to 20 μm particles and embedded in BEEM capsules using a hard Spurr's resin. After trimming and facing of the blocks, gold-to-gray sections were obtained using a 45° diamond knife on a Leica Supernova ultramicrotome employing cutting speeds of 0.5 mm/s to 10 mm/s.

The thin foils, two stage replicas, and microtomed sections were investigated using a Phillips EM430 TEM with EDS. Various accelerating voltages were used, but it was generally found that 150 kV was optimal for VHT specimens. Diffraction and EDS methods were employed to characterize the various phases observed.

RESULTS

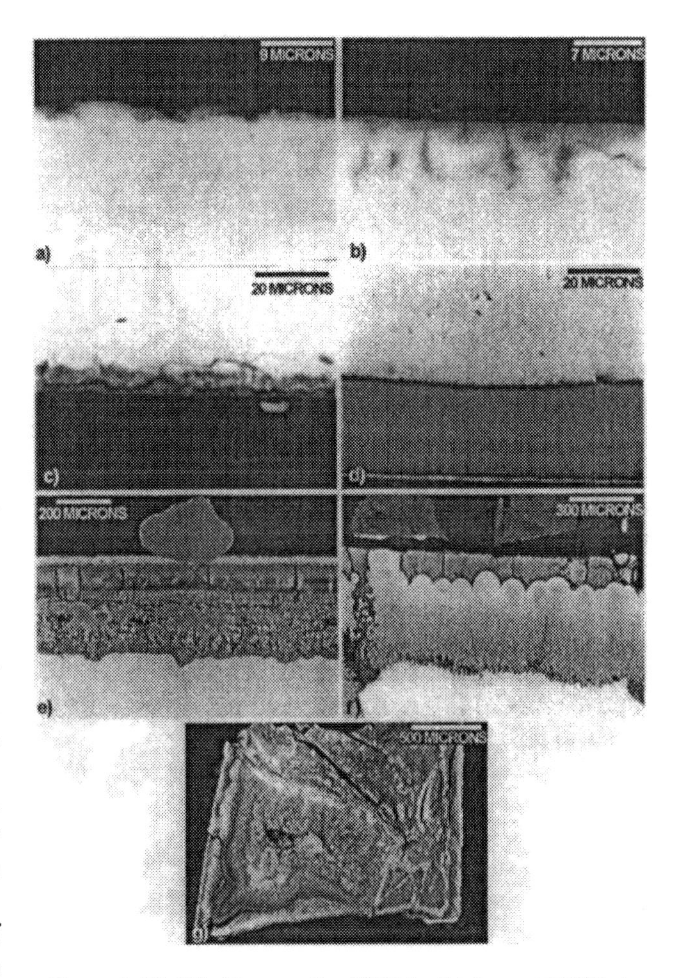

Figure 1. Modified layers after VHT for 24 days at 200°C a) LAWA57, b) LAWC25, c) LAWA53, d) LAWC21S, e) LAWA105, f) LAWA33, g) LAWA23R.

Figure 1 displays SEM micrographs showing the modified layers of seven of the 24-day VHT coupons. The variety in thickness and microstructure of the modified layers is immediately evident. The most durable glasses on the VHT are LAWA57 and LAWC25 (Figures 1a and 1b respectively). The addition of 5 wt% SnO_2 to LAWA53 (Figure 1c) as a partial substitution for CaO produced LAWA57, which shows a substantial improvement in durability over LAWA53. It is worth noting that LAWA53 with a SO_3 content of 0.6 wt% exhibits a nine-fold smaller dissolution rate than a lower-sulfur version of the same glass (LAWA52, 0.1 wt% SO_3, not shown) under the same test conditions. In general, it was found that, other things equal, greater sulfur content in these glasses glass usually improved the VHT durability, probably due to pH buffering in the surface layer. Figure 1d shows LAWC21S, which is similar to LAWC25 in composition except that it contains 2.7 wt% Li_2O in place of 8 wt% K_2O (approximately equal on a molar basis) with the rest of the composition renormalized. The Li_2O clearly produces a less-durable glass, as measured by

VHT. The glass LAWA105 (Figure 1e) is a high-sodium (24 wt% Na_2O) glass formulated to explore the high waste-loading extreme for certain LAW streams. A highly-structured layer about 360 μm thick develops on LAWA105; however, this glass still meets the VHT acceptance criterion (Table I). The glasses LAWA33 and LAWA23R (Figure 1f and 1g) are at the low-durability end of the range of VHT performance spanned by this group of glasses. As mentioned above, these glasses were formulated before the VHT requirement was instated and were designed instead to meet the 7-day room temperature PCT requirement[2, 3]. A layer in excess of 500 μm developed on LAWA33, while LAWA23R reacted all the way through the coupon.

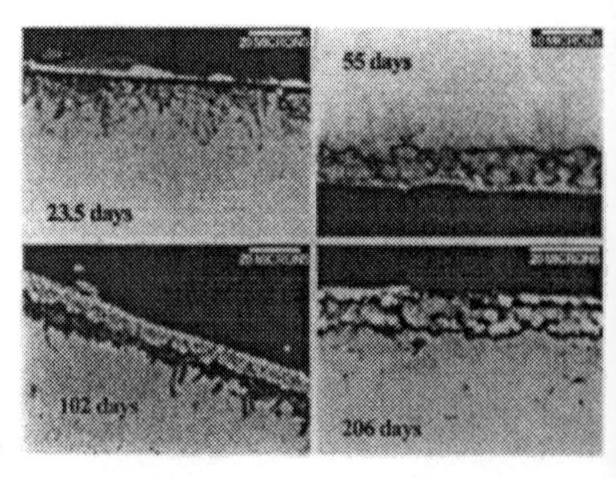

Figure 2. Modified layers on LAWA44 after VHT at 200°C for a) 23.5 days, b) 55 days, c) 102 days, d) 206 days.

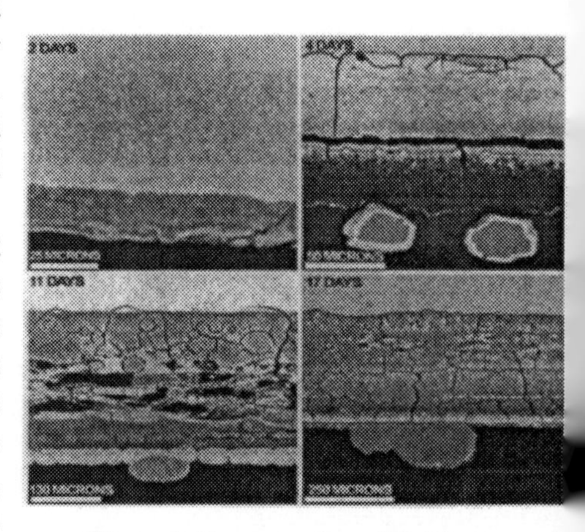

Figure 3. Progress of modified layer formation on LAWA105 during VHT at 200°C.

LAWA44 is typical of current formulations for Hanford LAW streams and has a somewhat lower sodium content (20 wt% Na_2O) than LAWA105. Figure 2 shows specimens of LAWA44 that were subjected to 200°C VHT for durations of 23.5, 55, 102, and 206 days. Although the modified layers show an evolution of microstructure, it is clear that the thickness changes little as the test progresses. This is in contrast to the case for LAWA105, as shown in the sequence of micrographs comprising Figure 3. The kinetics of layer formation are clearly quite different for the two glasses. It is common to see an incubation period for in the time dependence of the VHT layer growth[2,5,6]; initially, there is a period during which there is very little layer development, which is followed by a sharp increase in rate. The plot of the kinetics in Figure 4 indicates that LAWA105 has a very short incubation period followed by a gradual increase in layer growth rate until 3.5 to 4 days, at which point a brief, sudden acceleration occurs followed by a leveling off of the growth rate. A second rapid acceleration occurs at about 11 days followed again by a leveling off. Such rate excursions have been seen in data from a variety of test methods and are thought to be related to the onset of the formation of new secondary phases[2,7].

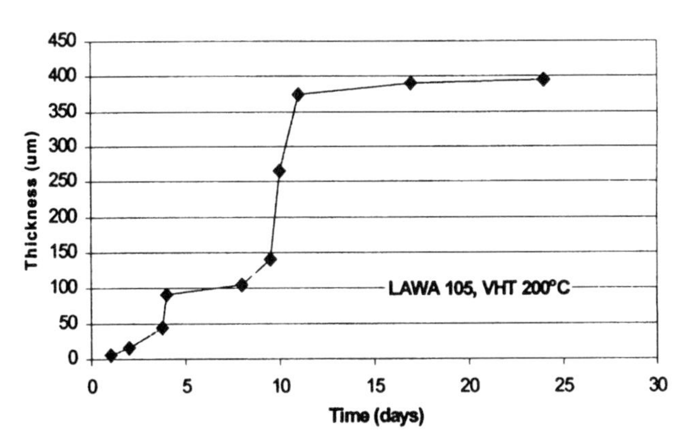

Figure 4. Kinetics of modified layer formation LAWA105 during VHT at 200°C.

The XRD patterns taken from the LAWA105 samples show analcime, forming first, with pectolite and zirconolite developing soon thereafter. These phases are also clearly evident in EDS x-ray mapping.

Figure 5 shows a representative TEM micrograph of a thin-foil specimen of the early stages of layer development in the VHT of LAWA23. Note the 200-300 nm thick gel layer at the reaction interface. Beneath this gel layer are "wormhole"-type tubules that penetrate into the glass for another 500-1000 nm. Similar tubules were found in TEM specimens of other VHT glasses. Figure 6 shows a thin-foil specimen of LAWC25, also showing tubules extending into the glass from the base of the gel layer. The mechanism of formation of such tubules, or whether there development is unique to the VHT, is not immediately apparent.

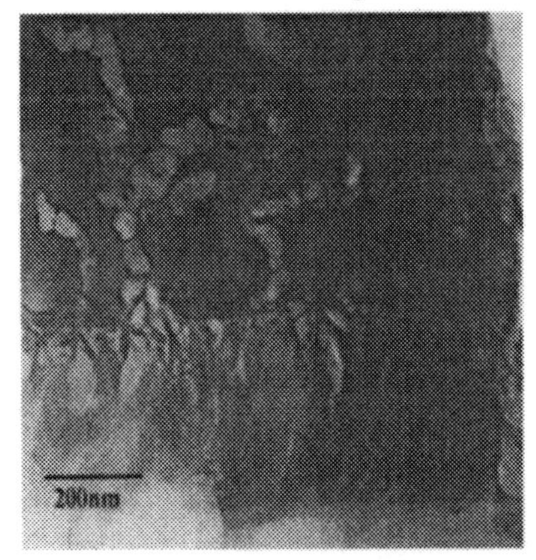

Figure 5. TEM micrograph of modified layer on LAWA23R after VHT for 10 days at 150°C.

Figure 6. TEM micrograph of modified layer on LAWC25 after VHT for 24 days at 200°C.

It would be useful to look for similar tubules in glass samples from other durability tests (e.g., long-term PCT powders) as their presence indicates localized extensions of the reaction interface deep into the glass beyond the majority of the reaction front. These features further highlight the complexity of the overall reaction mechanism.

Figure 7. TEM micrograph of modified layer on LAWA57 after VHT for 24 days at 200°C.

Figure 7 shows the reaction interface in a thin-foil specimen of LAWA57 (tin-containing glass). However, the TEM results do not reveal any indications as to the origin of the protective action of the tin addition in terms of diminished VHT alteration rate.

TEM specimens were also produced by ultramicrotomy and two-stage replication but these samples added no new insights on layer structure. Some secondary crystalline phases were found in several of these samples, but the phases either could not be positively identified, or were ones already identified by XRD.

DISCUSSION AND CONCLUSIONS

Observations of layer structures that develop in various high-sodium glasses under VHT conditions show strong dependence on the glass composition. The results indicate that addition of SnO_2 or SO_3 tends to reduce the VHT alteration rates and that the substitution of K_2O for Li_2O substantially improves the VHT durability. The kinetics of the VHT response for a glass at the high-sodium end of LAW loadings show two plateaus in alteration rate over the usual 24-day period of the test. This is in sharp contrast to the behavior of a glass formulation that is more representative of current nominal LAW formulations (LAWA44), which exhibits very little layer growth beyond the 24-day level, even when the test is extended to 206 days. In many glasses, TEM observations show tubules extending from the base of the obvious reaction interface deep into the unreacted glass.

The high rate of reaction occurring at the glass surface under VHT conditions makes the test quite sensitive to the effects of glass composition variations, even for comparatively short test times. However, perhaps because of this sensitivity, the VHT response exhibits a relatively complex dependence on glass composition[6], which is still relatively poorly understood. A complete mechanistic model to support prediction of the long-term behavior of such glasses would ideally, not only explain the alteration kinetics for selected glass compositions, but also the effects of glass composition variations. This latter effect, however, remains particularly challenging. The characterization of layer microstructure and of secondary phase formation in response to specific glass composition changes should provide one of the many pieces of information that will be necessary to address that challenge. The mode of formation of the tubules that were identified in this work and their importance, or otherwise, in the overall reaction mechanism, also warrants further study.

ACKNOWLEDGMENTS

The authors would like to thank the staff of the VSL for support with glass preparation and analytical work and M.C. Paul for manuscript preparation. This work was support in part by the Department of Energy, Office of River Protection through Duratek, Inc.

REFERENCES

[1]Bechtel National, Inc. Design, Construction, and Commissioning of the Hanford Tank Waste Treatment and Immobilization Plant, Contract Number: DE-AC27-01RV14136, U.S. Department of Energy, Office of River Protection (12/11/00).

[2]A.C. Buechele, S.T.- Lai, and I.L. Pegg, "Alteration Phases on High-Sodium Waste Glasses After Short- and Long-Term Hydration," Ceramic Transactions, Eds. G.T. Chandler and X. Feng, **107,** 251, American Ceramic Society (2000).

[3]I.S. Muller, A.C. Buechele, and I.L. Pegg, "Glass Formulation and Testing with RPP-WTP LAW Simulants," Final Report, VSL-00R3560-2, February 23, 2001.

[4]W.L. Ebert and J.K. Bates, "The Reaction of Synthetic Nuclear Waste Glass in Steam and Hydrothermal Solution," Mat. Res. Soc. Symp. Proc. **176,** 339-346 (1990).

[5]J.D. Vienna, A. Jiricka, B.P. McGrail, B.M. Jorgensen, D.E. Smith, B.R. Allen, J.C. Marra, D.K. Peeler, K.G. Brown, and I.A. Reamer, "Hanford Immobilized LAW Product Acceptance: Initial Tanks Focus Area Test Data Package," PNNL-13101, February 2000.

[6]X. Lu, F. Perez-Cardenas, H. Gan, A.C. Buechele, and I.L. Pegg, "Kinetics of Alteration in Vapor Phase Hydration Tests on High-Sodium Waste Glasses," paper in these Transactions.

[7]A.C. Buechele, S.T.- Lai, and I.L. Pegg, "Alteration Layers on Glasses After Long-Term Leaching," Ceramic Transactions, Eds. D.K. Peeler and J.C. Marra, **87,** 423, American Ceramic Society (1998).

KINETICS OF ALTERATION IN VAPOR PHASE HYDRATION TESTS ON HIGH SODIUM WASTE GLASS

Xiaodong Lu, Fernando Perez-Cardenas, Hao Gan, Andrew C. Buechele, and Ian L. Pegg
Vitreous State Laboratory, The Catholic University of America
Washington, DC 20064

ABSTRACT

The time dependence of the extent of alteration in a glass vapor hydration process typically displays a characteristic "sigmoidal" form, consisting of incubation, fast-rising, and final stages. Several models have been proposed to describe the individual stages of the overall process, but a consistent representation of the entire transformation kinetics is lacking. In this paper, we propose a *generalized* Avrami model to describe the vapor hydration of low-activity nuclear waste glasses subjected to vapor phase hydration at elevated temperatures. The model is demonstrated on extensive data from 61 glasses from two laboratories tested at a range of times and temperatures and spanning a wide range of compositions. The parameters introduced in the model are interpreted in terms of the underlying mechanisms of the vapor hydration transformation. Of special importance is the introduction of the retardation exponent, m, which indicates the degree of difficulty for the system to complete the phase transformation ("frustration"). Microscopically determined composition profiles of reacted glass samples are generally consistent with this interpretation.

INTRODUCTION

Tank wastes stored at the DOE Hanford site will be separated into high-level waste (HLW) and low-activity waste (LAW) fractions, each of which will be separately vitrified in the River Protection Project Waste Treatment Plant (RPP-WTP)[1]. The glass formulations that will be used in that facility are being developed and characterized at the Vitreous State Laboratory (VSL)[2-5]. The LAW stream is high in sodium and represents the overwhelming majority of the total mass. The LAW glass product, which will be stored in a shallow-land disposal facility on the Hanford site, is subject to a variety of durability requirements[1], one of which is the vapor hydration test (VHT). In the VHT, glass samples are

To the extent authorized under the laws of the United States of America, all copyright interests in this publication are the property of The American Ceramic Society. Any duplication, reproduction, or republication of this publication or any part thereof, without the express written consent of The American Ceramic Society or fee paid to the Copyright Clearance Center, is prohibited.

exposed to water vapor at high temperatures (nominally 200°C) in a pressure vessel. Reaction occurs in a thin film of condensed water on the glass surface that rapidly becomes saturated due to the inherent extremely high effective values of the surface-to-volume ratio (S/V). As a result, the later stages of the overall reaction progress are achieved in a relatively short period of time. Hydrous mineral phases or hydrous amorphous materials form and grow as the reaction progresses towards the interior of the glass sample; such phases are believed to be influential in determining the long-term alteration rate of the glass. Understanding the kinetics of the VHT alteration process is one aspect of the development of the basis for projecting the long-term release from the LAW disposal facility. In addition, however, an understanding of both the kinetics and the dependence of the VHT rate on glass composition is important in the development of quantitative glass property-composition models that are needed to define qualified LAW glass composition regions for operation of the RPP-WTP facility.

Several models have been proposed to describe the kinetics of the alteration of glass samples during vapor hydration process; these models are either power-laws in time[6-7] or simple linear regressions used to estimate rates[8-9]. However, none of these approaches is able to represent the entire VHT alteration process and, instead, each seeks to represent a particular stage of the overall process. Suppose that t represents the time and y the volumetric fraction of the glass sample that has been transformed by reaction in the VHT, a linear model $y \sim t$ often describes fairly well, but separately, the intermediate or the later stage of the kinetics. Conversely, a power-law model, $y \sim t^n$, sometimes fits the data satisfactorily during the early stages[6-7]. However, either type of model deviates significantly from the data during some portion of the reaction progress. Clearly, development of an improved kinetic model for the VHT process is desirable from several perspectives including improved understanding of the nature of the glass-vapor hydration process, as well as the simple pragmatic aspects of data reduction (i.e., extraction of "rates" from VHT vs. time data) and the need to relate VHT rates quantitatively to glass composition.

The approach used in the present work is based on the Avrami equation[10-11], which has previously been shown to describe the VHT kinetics quite well[12]. Furthermore, that approach has previously been used to develop models for the dependence of the VHT response on the composition of LAW glasses[12]. The Avrami equation is a semi-empirical equation that has been widely applied to metallurgical and ceramic systems to represent quantitatively phase transformations governed by nucleation and crystal growth[10-11]. As discussed below, the Avrami equation gives a qualitatively better representation of the entire VHT process than do simple linear or power-law models. In this paper we describe the use of a variant of the Avrami equation, the so-called *generalized* Avrami equation, which provides an even better representation of the VHT data.

The potential physical significance of the parameters involved in the model is discussed in an effort to better understand the mechanisms involved in the phase transformation occurring during vapor hydration.

The VHT data used in the present work are from simulated LAW glasses studied previously in this laboratory[3,13-14] and at Pacific Northwest National Laboratory (PNNL)[8-9]. These data were supplemented by new data on four of the PNNL formulations that were re-melted at VSL, subjected to the VHT procedure, and then examined by using X-Ray Diffraction (XRD) and Scanning Electron Microscopy/Energy Dispersive Spectroscopy (SEM/EDS) techniques.

VAPOR HYDRATION TEST METHOD

The details of the VHT methods that are in use have been presented previously[3,8-9,13-15]. Briefly, a glass coupon with known dimensions is suspended in a sealed stainless steel pressure vessel with sufficient water to saturate the atmosphere at the test temperature and provide for a non-dripping surface layer on the coupon[15]. After the prescribed test duration, the coupon is removed, dried, sectioned, and polished for microscopic characterization. Typically, the hydration rate is calculated on the basis of the depth of the alteration of the glass sample as determined either by optical microscopy (from the thickness of the remaining glass[8-9]) or by SEM (from the layer thickness[3,13-15]). In the present work, to describe the evolution of the hydration process, y is defined as the fraction of the sample volume that has been transformed by reaction, which corresponds to the thickness of the altered layer(s) divided by half the sample thickness (0.7 mm in this study), since both parallel sides undergo hydration. Alteration phase compositions and the elemental concentration profiles perpendicular to the glass surface were analyzed by SEM/EDS. Typically, one or more layers of hydrated phases will form near the glass-vapor interface. The most commonly observed hydrated phase in these glasses is analcime ($Na(AlSi_2O_6)H_2O$).

AVRAMI EQUATION APPLIED TO VHT KINETICS

The kinetics of the vapor hydration process typically exhibits three characteristic stages (Figure 1), referred to herein as *incubation, fast rising, and final*[8-9,13-14]. The hydration rate in the incubation period is slow, as characterized by the gradual appearance of transformed regions on the surface of the glass, signaling the initiation of the phase transformation. The growth of the altered layer during this period is, in most cases, negligible. This is followed by the *fast-rising* stage, during which the transformation process accelerates considerably. The *final* stage is characterized by the gradual slow-down of the hydration rate. Depending on the glass composition, the final slowing-down of the hydration rate varies considerably and, in some case, lasts orders of magnitude longer than the first two stages together. Qualitatively, it is striking that the overall sigmoidal curve shape observed from the VHT measurement resembles a typical time evolution curve of the Avrami equation[10-12].

The Avrami equation, which is commonly represented in the form

$$1 - y = e^{-k^n t^n}, \tag{1}$$

has been extensively used to describe phase transformation process occurring by nucleation, growth, and impingement. Its widespread use can be attributed to the fact that, at least qualitatively, it reproduces rather well the commonly observed behavior shown in Figure 1. This expression is a solution to the following differential equation under the assumption of constant k:

$$\frac{dy}{dt} = (1 - y)k^n t^{n-1}. \tag{2}$$

It is important to point out that the factor *(1-y)* is responsible for the characteristic slowdown of the *final* stage. In the derivation of this equation, the factor *(1-y)* describes the retardation caused by either the spatial impingement (hard impingement as crystals grow into each other) or concentrational depletion (soft impingement as the depleted regions overlap), or both[10-11]. In several instances, where (1) has been used to describe a particular crystallization mechanism (for example, the spherulitic crystalline growth in some solids, among them polymers), it can be assumed that *n* is dictated by the geometry of the transformation process and that *k* reflects the rate constant of the transformation[16].

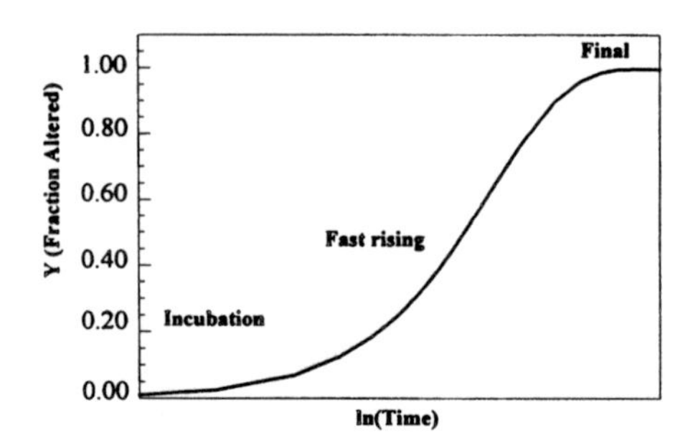

Figure 1. A representative sigmoidal time-evolution curve of Avrami equation.

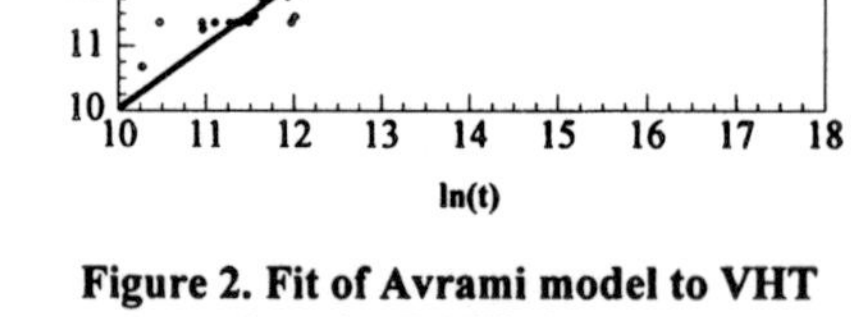

Figure 2. Fit of Avrami model to VHT data for LAW glasses.

Despite its simplicity, in many cases, the Avrami model describes the experimental VHT data satisfactorily over the entire measured time dependence. A total 61 distinct sets of vapor hydration test data (each set consists of a series of vapor hydration tests of different durations for the same glass at a given temperature), were fitted to the linearized (by taking logarithms twice) version of Eq. (1), with *k* and *n* as the regression parameters for each set. Of these, 56 sets of the tests were described reasonably well by the Avrami equation with $R^2>0.7$, as shown in Figure 2, where, according to Eq. (1), the function on the *y*-axis should be linearly dependent on *ln t*. More than 80 % of the 56 sets have R^2 greater than or equal to 0.80. The five tests that were less-well represented ($R^2 \leq 0.7$) are those that are either rather flat in the dependence of *y* on *t* (with R^2 ranging from 0.43, to 0.54 to 0.69), or those with a negative slope[8,9]. Figure 3 shows examples of the individual fits of the Avrami equation to VHT data for the LAW glasses.

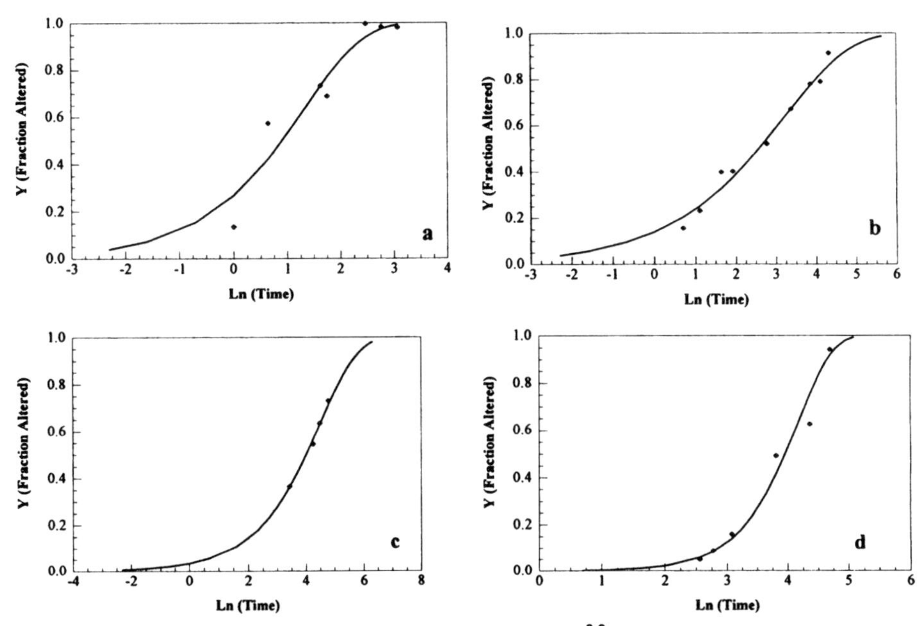

Figure 3. Fit of Avrami equation to 200°C VHT data[8-9] for LAW glasses: a) HLP-27, $k = 3.1 \times 10^{-6}$, $n = 0.8876$; b) HLP-35, $k = 5 \times 10^{-7}$, $n = 0.5958$; c) HLP-11, $k = 1.3 \times 10^{-7}$, $n = 0.7479$; and d) HLP-07, $k=1.8 \times 10^{-7}$, $n = 1.7417$.

Of particular interest in the present work is the roughly 10 to 20 % of the variance in the VHT data that cannot be explained by the Avrami model. Figure 4 shows one instance of the inadequacy of the Avrami equation in this respect. For some glasses, the *final* stage of the vapor hydration process can be significantly different than described by Eq. (1). In some cases, the *final* stage can be several orders of magnitude longer than the initial stages, while in others, it can be shorter than that described by the Avrami equation. These distinctive VHT-time curves of various LAW glasses illustrate the important effect of the glass composition as well as the test conditions. From the above discussion, it appears that the primary limitation of the Avrami equation derives from the linear retardation factor *(1-y)* in Eq. (2), which, in its original form, represents the fraction of the space still available for growth of randomly nucleated crystals. It is not unreasonable to suppose that the nature of the impingement process in the VHT is somewhat different, since the fraction of remaining glass does not scale in a simple way to the available reaction sites or the growth volume for hydration reaction. Accordingly, we have investigated a *generalized Avrami* differential equation, which includes a non-linear retardation factor:

$$\frac{dy}{dt} = (1 - y)^m k^n t^{n-1}. \tag{3}$$

A similar equation with a value of *m* greater than one has been proposed by Line et al.[17] to describe the dehydration kinetics of polycrystalline analcime. The physical meaning of the exponent *m* was, however, not addressed in that study. A superficial inspection of (3) suffices to illustrate the role played by *m*. If $m > 1$, the main effect is to slow down (with respect to the Avrami model) the increase of *y* since the hydration rate is smaller for the same *k* and *n*. Conversely, if $m < 1$, an acceleration, with respect to the Avrami model, occurs. In any case for which $m > 0$ for a given *n* and *k*, the factor $(1 - y)^m$ determines the overall shape of the reaction progress curve in a plot *y(t) vs. t,* especially the later stage as the retardation term goes to zero. Consequently, the extent of the slow-down of the reaction progress curve is controlled by *m* (Figure 5). The retardation exponent *m* is therefore used in this study as a measure of the level of the slow-down or acceleration effect in the VHT kinetics.

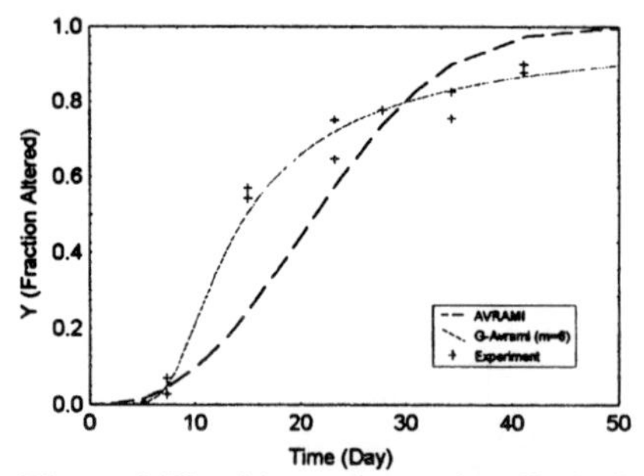

Figure 4. Fit of Avrami equation (dashed curve) and generalized Avrami equation (G-Avrami) (solid curve) to 150°C VHT data[8-9] for HLP-46 glass.

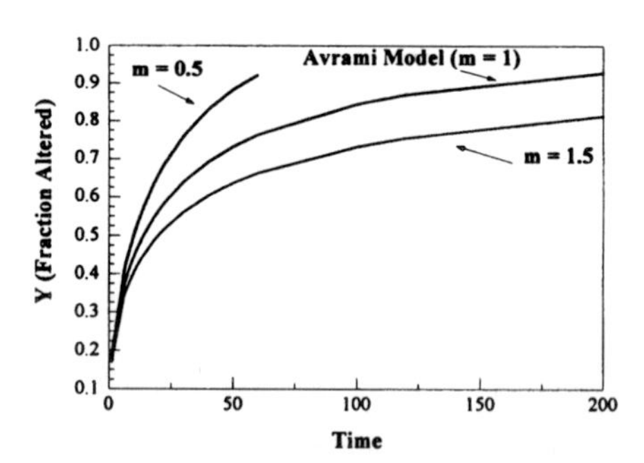

Figure 5. Illustration of the effect of the retardation exponent m on the alteration kinetics for fixed values of n and k.

The generalized Avrami equation gives significantly improved fits to the data, as compared to the Avrami equation. For the same 61 sets of data modeled by the Avrami equation, the population of the $R^2>0.7$ group increased by 15% when the the generalized Avrami equation was used. Moreover, the R^2 values of many data sets increased from the 0.7 to 0.8 range to equal to or greater than 0.95. The improvement afforded by including the exponent *m* is clearly demonstrated in Figure 6. Figure 7 shows the fits to VHT data at 200°C for two other LAW glasses: the transformation presented in Figure 7a has a large *m* exponent and, therefore, a long-lasting *final* stage; whereas that in Figure 7b exhibits an exponent *m* of less than one and, therefore, shows a rapid (as compared with the Avrami model) approach to completion.

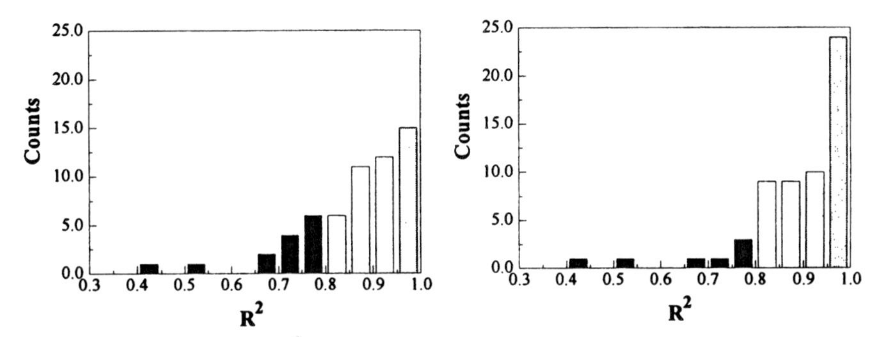

Figure 6. Distribution of R^2 values from regression of VHT data with the Avrami equation (left) and the generalized Avrami equation (right).

The glasses that have low hydration rates are usually those that show signs of slow-down (as compared to the Avrami equation) at a relatively early time of their history and are characterized by larger values of *m* and *n*. While this supports the assertion concerning the role of the exponent *m*, the positive correlation between *n* and *m* blurs the physical meaning usually assigned to the values of *n* for the classic Avrami equation. For example, during a typical vapor hydration test, from a macroscopic perspective, the glass alteration occurs mostly in a one-dimensional fashion. However, *n* obtained from regression analysis using the generalized Avrami equation can be significantly larger than one. The meaning of *n* for the Austin-Rickett equation (a special case of the generalized Avarmi equation with $m=2$) has been reported as representing the actual physical dimensionality of the transformation processes[18]. The relationship between *n* and the dimensionality of crystal growth for a system with $m>2$, however, has not been reported.

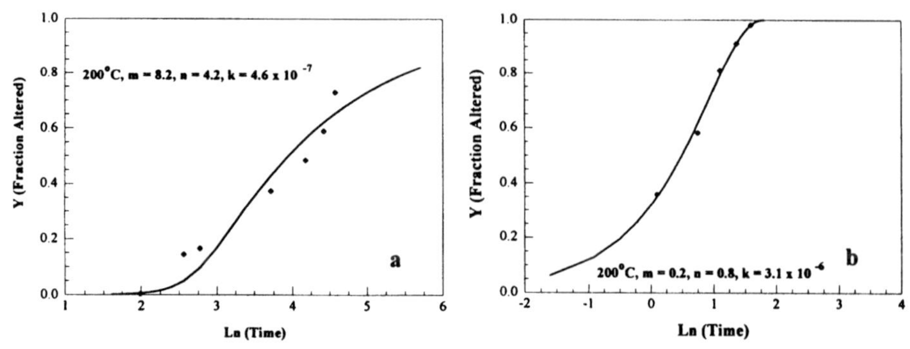

Figure 7. Fit of generalized Avrami equation to the 200°C VHT data of LAW glasses: a) HLP-12 and b) HLP-46.

GLASS COMPOSITION AND THE RETARDATION EXPONENT

The circumstances under which the transformation occurs in the vapor hydration experiments are of a different nature than those corresponding to the crystallization described in the derivation of the Avrami equation[10-11]. In the original Avrami model, the early stages of transformation are dominated by nucleation, followed by a rapid growth of the crystalline domains, and the *final* stage is dominated by an impingement process (as in the crystallization of linear polyethylene[19]), or a depletion of reactants (as in the recrystallization of some dilute salts[11]). As a result, geometrical factors determine the value of n. In VHT experiments, it has been found (at least from a macroscopic viewpoint) that, once the *incubation* stage is completed, the transformation progresses following a frontal planar interface. Therefore, geometrical impingement in the classic Avrami sense is not responsible for the retardation factor $(1-y)^m$ in Eq. (3). Furthermore, computer simulations of the vapor hydration process[20] clearly show that this retardation factor cannot be the result of depletion alone. It is postulated that, instead, the slow-down of the hydration process, especially in the later stages of the transformation, is due to a combination of two factors: a composition mismatch and the consequent reaction shift between the initial glass and the product phases; and a concentration depletion of certain key elements. Microscopic analysis of the hydration test samples was used to investigate the involvement of these two possible slow-down mechanisms.

VHT samples of three glasses of different compositions and hydration rates were selected for study by SEM/EDS. The experimental times were selected to produce an approximately 200-micron-deep hydrated layer for all three glasses at 200°C (Table I). After 5 days (VHLP-48) or 25 days (VHLP-12 and –36) hydration, all three glasses developed a two-layer structure in the reacted part of the samples. On the outer surface is always a polycrystalline analcime phase that is on top of another sodium-aluminosilicate phase of a different composition (Figure 8). Herein, for the purpose of discussion, analcime is labeled the stable phase, while the aforementioned sodium-aluminosilicate phase is labeled the metastable phase. Analysis of the EDS line scans normal to the glass surface leads two important observations: (*i*) The analcime layer developed on the surface and maintained its stoichiometry closely, while the underlying metastable phase shows very large variations in its constituents; this is especially true for Al or Si if the ratio of these elements in the host glass deviated significantly from the typical analcime stoichiometry, and also for elements that are incompatible with the stable analcime phase. The line scans for Ti and Zr (both incompatible with analcime) revealed the buildup of these elements in the metastable phase (Figure 9). (*ii*) There is no significant depletion of the key glass

Table I. Target compositions of three glasses subjected to VHT (wt%).

Oxide	VHLP-48	VHLP-12	VHLP-36
Al_2O_3	11.97	6.74	11.94
B_2O_3	8.85	9.63	12
Cl	0.58	0.27	0.22
CaO	0	0.01	0.01
Cr_2O_3	0.02	0.07	0.06
F	0.04	0.01	0.01
Fe_2O_3	5.77	9	9.87
K_2O	3.1	0.4	0.33
MgO	1.99	1.44	2.69
Na_2O	20	19.26	16
P_2O_5	0.08	0.05	0.04
SO_3	0.1	0.07	0.06
SiO_2	38.25	47.25	36
TiO_2	2.49	2.89	5.38
ZrO_2	2.49	1.44	2.69
ZnO	4.27	1.45	2.69
m	0.2	8.2	3.4
n	0.58	4.2	1.89
k (s^{-1})	4×10^{-7}	4.6×10^{-7}	2.1×10^{-7}
Projected time needed for 90% hydration completion	32 days	800 days	600 days

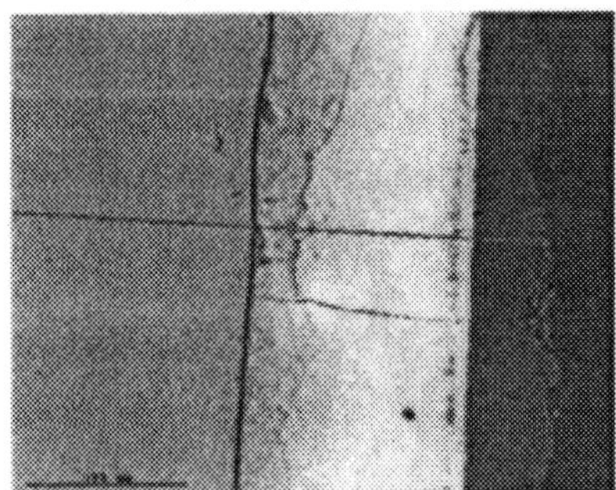

Figure 8. SEM image of VHLP-36 glass after 25-day VHT at 200°C.

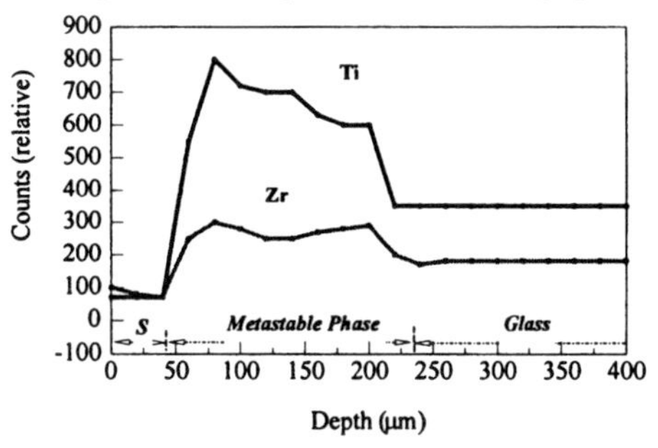

Figure 9. Concentration profiles of Ti and Zr normal to the sample surface for VHLP-36 after 25 day-VHT at 200°C. S = Stable phase.

constituents (e.g., Na_2O) within the glass near the metastable phase. This second observation suggests that during up to 30% of the transformation, soft-impingement as caused by the concentration depletion in the glass is not responsible for the observed slowdown in the vapor hydration test of these three glasses. However, a large concentration gradient is indeed present for many elements in the *hydrated* metastable phase, which will inevitably slow-down the growth of the analcime phase. The relevance of the slow-down of the formation of analcime phase for the vapor hydration measurements is discussed below.

The first observation suggests a number of possibilities. Superficially, the gradual buildup of the mismatched elements and the incompatible elements within the metastable phase (e.g., Fe, Ti, Zr, and Ca, etc.) would hinder the growth of the stable phase. However, it is not clear why the slow-down of the growth of the stable phase would slow-down the growth of the metastable phase, since the hydration rate is always measured on the basis of the interface displacement between the metastable phase and the glass.

Table I lists the compositions of the selected glasses together with the fitting parameters for the generalized Avrami equation. A careful comparison suggests that the *frustration* caused by the composition mismatch between the analcime and the host glass, which resulted in the large disparity of the major analcime-building constituents and the buildup of the incompatible elements in the neighboring metastable phase, correlates closely with the value of the retardation exponent *m*. The higher the level of the *frustration* by mismatch or incompatibility, the larger the value of the *m* exponent, and the slower the overall vapor hydration rate for the glass (Table I). For instance, if we compare glass VHLP-48 and VHLP-36, which have similar Al_2O_3/SiO_2 ratios, it is found that VHLP-36, the sample with higher concentrations of Fe, B, and Ti (incompatible elements for analcime formation), has the larger frustration exponent, *m*. Glasses VHLP-36 and VHLP-12 have similar levels of incompatible elements B and Fe, but the Al_2O_3/SiO_2 ratio in VHLP-12 presents a larger mismatch with respect to analcime. A larger *m* for this latter glass is consistent with this fact (Table I). Although the vapor hydrated depth is not dimensionally determined by the growth of the stable analcime phase, the kinetics of the analcime growth likely still directly or indirectly influence the vapor phase hydration progress. For instance, the composition change of the metastable phase as a result of the growth of the analcime phase would likely affect the Gibbs free energy of the hydration reaction or the diffusivity of water species through the stable and the metastable phases. This work is ongoing to further elucidate these issues.

CONCLUSIONS

The vapor phase hydration of high-sodium LAW glasses at elevated temperature is controlled by complex kinetic processes as characterized by a sigmoidal time evolution curve with a slow incubation period, followed by a fast rising period, and a final period of gradually decreasing alteration rate. For the first time, the kinetics of this complex vapor hydration process for glasses spanning a wide composition range are described satisfactorily by a semi-empirical equation, the generalized Avrami equation. The generalized Avrami equation allows the characterization of vapor phase hydration data obtained at different times and temperatures quantitatively on a common basis. This, in turn, permits a more detailed evaluation of the effects of the glass constituents on the vapor hydration rate[12]; the development of such property-composition models is important in the definition of qualified glass composition regions for operation of the LAW vitrification facility.

The isothermal vapor phase hydration process is described by three semi-empirical parameters, the rate constant k, the dimension exponent n, and the retardation exponent m. All three parameters depend strongly on the glass composition, and each influences the hydration rate in different ways. For many of the glasses studied in this paper, the rate constants are similar (around 10^{-7} s^{-1}), while the hydration rates differ by order of magnitude as a result of the combined effects of n and m. A unique postulate of the present work is that the retardation exponent m in the proposed generalized Avrami equation is associated with the level of the compositional frustration in the transformation of the host glass to hydrous crystalline phases. Both compositional mismatch and compositional incompatibility increase the level of frustration, and hence the value of the m, and consequently, slow down the vapor hydration rate.

ACKNOWLEDGMENTS

The authors would like to thank the staff of the VSL for support with glass preparation and analytical work and M.C. Paul for manuscript preparation. This work was support in part by the Department of Energy, Office of River Protection through Duratek, Inc.

REFERENCES

[1]Bechtel National, Inc. Design, Construction, and Commissioning of the Hanford Tank Waste Treatment and Immobilization Plant, Contract Number: DE-AC27-01RV14136, U.S. Department of Energy, Office of River Protection (12/11/00).

[2]I.S. Muller and I.L. Pegg, "Glass Formulation and Testing with TWRS LAW Simulants," VSL Final Report, January 1998.

[3]I.S. Muller, A.C. Buechele, and I.L. Pegg, "Glass Formulation and Testing with RPP-WTP LAW Simulants," Final Report, VSL-00R3560-2, February 23, 2001.

[4]S.S. Fu and I.L. Pegg, "Glass Formulation and Testing with TWRS HLW Simulants," VSL Final Report, January 1998.

[5]W.K. Kot and I.L. Pegg, "Glass Formulation and Testing with RPP-WTP HLW Simulants," Final Report, VSL-01R2540-2, February 16, 2001.

[6]J.S. Luo, et al, *Natural Analogues of Nuclear Waste Glass Corrosion*, ANL-98/22, Argonne National Laboratory, Argonne, Illinois (1998).

[7]J. J. Mazer, J. K. Bates, C. M. Stevenson and C. R. Bradley, Obsidians and tektites: natural analogues for water diffusion in nuclear waste glasses, *Materials Research Society Symposium Proceedings* Vol. 257, pp. 513-520 (1992).

[8]J.D. Vienna, et al, *Hanford Immobilized LAW Product Acceptance: Initial Tanks Focus Area Testing Data Package*, PNNL-13101, Pacific Northwest National Laboratory, Richland, Washington (2000).

[9]R.L. Schulz, et al, *Hanford Immobilized LAW Product Acceptance: Tanks Focus Area Testing Data Package II,* PNNL-13344, Pacific Northwest National Laboratory, Richland, Washington (2000).

[10]J. Burke, *The Kinetics of Phase Transformations in Metals*, Pergamon Press, Oxford (1965).

[11]J.W. Christian, *The Theory of Transformations in Metals and Alloys*, 2nd edition, Pergamon Press, Oxford (1975).

[12]H. Gan and I.L. Pegg, "Development of Property Composition Models for RPP-WTP LAW Glasses," Final Report, VSL-01R6600-1, August 6, 2001.

[13]A.C. Buechele, S.T.- Lai, and I.L. Pegg, “Alteration Phases on High-Sodium Waste Glasses After Short- and Long-Term Hydration,” Ceramic Transactions, Eds. G.T. Chandler and X. Feng, vol. 107, p. 251, American Ceramic Society (2000).

[14]A.C. Buechele, F. Lofaj, C. Mooers, and I.L. Pegg, “Analysis Of Layer Structures Formed During Vapor Hydration Testing Of High-Sodium Waste Glasses,” paper in these Transactions.

[15]W.L. Ebert and J.K. Bates, “The Reaction of Synthetic Nuclear Waste Glass in Steam and Hydrothermal Solution,” Mat. Res. Soc. Symp. Proc. **176**, (1990) 339-346.

[16]B. Wunderlich, Macromoclecular Physics, Vol. 2: Crystal Nucleation, Growth, Annealing, Academic Press, New York (1976).

[17]C. M. Line, et al, *American Mineralogist*, **80**, pp. 268-279 (1995).

[18]M. J. Starink, “Kinetic Equations for Diffusion-Controlled Precipitation Reactions,” Journal of Materials Sciences, **32** 4061, (1997)

[19]R. H. Doremum, B. W. Roberts, and d. Turnbull (eds), Growth and Perfection of Crystals, John Wiley, New York, (1958).

[20]F.C. Perez-Cardenas, H. Gan, X. Lu, and I.L. Pegg, “Mechanism of Vapor Phase Hydration in High-Sodium Waste Glasses from Computer Simulations,” *Materials Research Society*, Annual Meeting, Abstract and Proceedings, 2001 (*to be published*).

TCLP LEACHING PREDICTION FROM THE "THERMO™" MODEL FOR BOROSILICATE GLASSES

J. B. Pickett, and C. M. Jantzen
Westinghouse Savannah River Company
Aiken, SC 29808

ABSTRACT

The Savannah River Technology Center (SRTC) of the Westinghouse Savannah River Co. (WSRC) has been conducting glass property modeling for many years to support the High Level Waste (HLW) vitrification process in the Defense Waste Processing Facility (DWPF). SRTC also demonstrated the first mixed waste glass for the site's M-Area plating line sludge, which met the EPA's Land Disposal Restriction (LDR) storage and disposal criteria. A mechanistic glass durability model, the Thermodynamic Hydration Energy Reaction MOdel (THERMO™) was developed to predict glass durability from composition and to control the DWPF production process, using the ASTM C1285-97 Product Consistency Test (PCT) as the leaching test. During the treatment of the M-Area sludge, the model (THERMO™) was successfully used to predict the leaching properties of the M-Area glass, using the EPA Toxicity Characteristic Leaching Procedure (TCLP). This was the first successful prediction of glass leaching results using the TCLP procedure. The THERMO™ model can be used to correlate the PCT and TCLP leaching test results for radioactive and mixed waste glasses.

BACKGROUND

The Savannah River Site (SRS) is a 300 square mile complex, which has been dedicated to the production of national defense materials since 1952. The SRS is operated by the Westinghouse Savannah River Co. (WSRC) for the U.S. Department of Energy (DOE). One of the legacies of the production activities was a mixed waste plating line sludge, generated in the Reactor Materials Area (M-Area). This plating line sludge was generated during the wastewater treatment of effluents from the nickel plating of depleted uranium cores, which were subsequently irradiated in the site's reactors to form plutonium-239.

The M-Area plating line sludge was an EPA hazardous waste (F-006 listed waste). Since it contained a significant amount of depleted uranium, it was a "mixed" waste as defined by the Federal Facilities Compliance Act (FFCA) of 1992. In order to comply with the RCRA (Resource Conservation and Recovery Act) laws, the waste had to be treated to meet the Land Disposal Restriction (LDR) regulations [1].

To the extent authorized under the laws of the United States of America, all copyright interests in this publication are the property of The American Ceramic Society. Any duplication, reproduction, or republication of this publication or any part thereof, without the express written consent of The American Ceramic Society or fee paid to the Copyright Clearance Center, is prohibited.

The Savannah River Technology Center (SRTC) conducted numerous treatability studies to determine how the M-Area wastes could be stabilized to meet the regulatory requirements. The initial scouting studies demonstrated that both vitrification or cementitious techniques [2, 3, 4] resulted in stabilized wasteforms which met the hazardous waste disposal requirements.

VITRIFICATION STUDIES

SRTC Crucible Studies

The initial vitrification studies with the mixed M-Area waste were conducted by the Savannah River Technology Center (SRTC) in 1992. The studies indicated that either borosilicate or soda-lime-silica glass (SLS) formulations could be used, with at least a 75% waste (sludge) loading on a calcine oxide basis. The resulting glasses melted at <1150°C and demonstrated an 81% volume reduction from original waste to final glasses. The borosilicate glasses were shown to meet the hazardous waste leaching requirements in place at that time [4]. Additional treatability studies were conducted in 1993 on a wide range of borosilicate and SLS glasses to optimize the glass formulation to be used for the stabilization of the M-Area sludge [5].

In 1993, the M-Area wastewater treatment facility was continuing to treat a supernate stored in the M-Area waste tanks. The resulting wastewater filtercake, a silica rich perlite, was then returned to the waste storage tanks. At the completion of the supernate treatment, it was planned to combine the contents of all of the storage tanks into one homogenous sludge mixture. Since the final volume of the high silica filtercake could not be accurately established, test glasses were prepared which spanned the range of "high" silica, "nominal" silica, and "low" silica concentrations in the eventual sludge. Forty-four different glass formulations were tested, to determine the optimum formulation for melting temperature (1150°C maximum), waste loading, and durability (leachability) [5].

Office of Technology Development (OTD) Pilot Scale Melter Tests

In 1994, SRTC conducted a scale-up treatability test using the 10 kg/day test melter at the Catholic University of America (CUA) Vitreous State Laboratory (VSL). The test was conducted for the DOE's Office of Technology Development (OTD). Actual M-Area sludge samples were used, from Tanks 7, 8, and 10. By 1994, the blending scheme for the M-Area tanks had changed from one "master" blend of all tanks, to two homogenous mixtures. One feed tank would be Tank #7, with ½ of the filtercake in Tank 10, the other would be Tank #8, with the remainder of the filtercake. Seven glass samples were produced during the test run at VSL, using glass formulations provided by SRTC.

Vendor Treatment Facility (VTF) Glass Results

Between 1996 and 1999, all of the M-Area plating line sludge was vitrified by GTS Duratek, in a melter called the Vendor Treatment Facility (VTF) at the SRS. There were 44 different feed batches prepared and treated throughout the run, and each of these was tested to ensure that the final glass met the contractual TCLP leaching requirements for Ni, Cr, Pb, and U. The TCLP leach results from the

VTF glasses were used as "validation" data points for the TCLP vs. ΔG_f (THERMO™) prediction.

EXPERIMENTAL

SRTC Crucible Tests

Samples of the M-Area sludges were collected from two storage tanks, #8 and #10. The sludge in Tank 8 was generated from the nickel-plating operations in M-Area, and contained primarily aluminum hydroxide, depleted uranium (as a sodium uranate), sodium nitrate, silica, a zeolite (aluminosilicate), and aluminum phosphate [6]. The material in Tank 10 was the filtercake from the wastewater treatment of the supernate, and Zeolite-A, which had formed in-situ in the tanks. The volume of all of the stored nickel-plating sludges was estimated to be ~244,000 gallons in 1993. The volume of the perlite in Tank 10, after treatment of the supernate was completed, was assumed to range from 200,000 gallons at 17% solids, to 450,000 gallons at 30% solids. The crucible studies used glass formulations, which spanned the range of the potential feed. The range of glass formulations is summarized in Table I; the individual glass formulations were described previously [5].

Table I. Tank Ratios and Glasses Studied During SRTC Treatability Studies

Formulation	Assumed Volume (Kgals)		% Solids		No. of Glasses Studied
	Tanks 1-8	Tank 10	Tanks 1-8	Tank 10	
Nominal	244	412	30	17	26
Low Silica	244	200	30	17	11
High Silica	244	450	30	17	7

It is interesting to note that the final tank volumes, when measured in 1995, were most similar to the "low silica" formulation, i.e., 330,000 gallons of plating line sludge and 228,000 gallons of Tank 10 filteraid.

The glasses were size reduced to prepare them for the TCLP and PCT leaching tests. A solid sample for TCLP leaching test (the EPA's SW-846 Method 1311) is size reduced to pass a 3/8 inch sieve. In the SRTC tests, the crucible samples were size reduced until all of the sample passed the 3/8" sieve. A representative sample was then collected, which obviously contained a distribution of sizes including some fraction of smaller glass particles. In the TCLP test no restraints are placed on the amount of material less than 3/8", and the samples were not rinsed of adhering fine particles or otherwise modified prior to TCLP leaching. The TCLP leaching tests were conducted by the General Engineering Laboratory (GEL) in Charleston, SC. Each remaining sample was then size reduced to <100 to >200 mesh, to conduct the SRTC PCT leaching tests and determine the total constituent concentrations.

Two glass samples (Nominal 5 and Low Silica 5) were separated into various size fractions to determine the affect of size on the TCLP test results. The sizes ranged from 3/8" to the fine size used for the PCT tests. TCLP testing was conducted by the GEL.

OTD Tests

The M-Area sludge samples (Tanks 7, 8, and 10) were calcined and analyzed at VSL. SRTC provided glass formulations, using Li, B, Al, and Na as additives to the sludge samples. The SRTC glasses were intended to have ~80% dry weight loading, to demonstrate the large volume reduction achievable by vitrification. The glass feed was modified to the SRTC formulation at the end of a treatability study conducted for GTS Duratek by the VSL. The VSL treatability study was being conducted to provide Duratek with additional design data for the VTF melter and off-gas system. Approximately 10 kgs of glass were produced with the SRTC tank 8/10 formulation, while ~49 kgs were produced with the tank 7/10 blend. All of the SRTC formulation glasses were returned to SRTC, and size reduced for TCLP, PCT, and total constituent concentrations. All of the samples were sent to GEL for TCLP analyses. Two of the glass samples sent to GEL were tested for the entire EPA hazardous waste Appendix 8 constituents. These two samples were also tested by the EPA's Multiple Extraction Procedure (MEP). The Appendix 8 analyses and the MEP results were used to prepare an "Upfront Delisting Petition" for the Vitrified M-Area sludges [7].

VTF Glass Analyses

Each feed batch of sludge for the VTF melter was analyzed by the SRTC mobile laboratory, for GTS Duratek. The batch was adjusted, if necessary, with feed additives and then vitrified. Five samples of glass were obtained from the last 10 drums (71 gallon drums) of glass produced from each batch. The glass from the VTF melter was produced as flattened "gems" (ovoids, ~3/4" long and ¼" thick). GTS Duratek then size reduced the gems to pass the 3/8" limit for the TCLP test, and had the glass tested by GEL. Each and every batch passed the contractural limits for the metals of concern, and all of these TCLP results were included in a revised Delisting Petition, which was updated in 2000 [8].

RESULTS AND DISCUSSION

Congruent Dissolution

The PCT and TCLP results for the crucible study glasses (boron, silicon, and uranium) are provided in Table II. The boron PCT results had 3.5 order of magnitude range, while the boron TCLP had a 2.5 magnitude range. The results for the other RCRA hazardous metals are not provided, as both the PCT and TCLP results were usually less than the analytical detection limit. A correlation between the PCT and TCLP boron data in ppm demonstrated a strong correlation between the durability response of these two tests (Figure 1). The strong correlation indicates that the TCLP test exhibits "congruent dissolution" of the glass matrix, and thus the THERMO™ model should be able to predict the TCLP leaching potential of these borosilicate glasses. Previous researchers have attempted to develop a model for predicting the TCLP leaching results, but have

Table II. Comparison of TCLP vs. PCT Durability Test Response

Sample ID	Boron (ppm)		Silicon (ppm)		Uranium (ppm)	
	PCT *	TCLP‡	PCT*	TCLP‡	PCT*	TCLP‡‡
MN-1	6.42	0.28	37.6	0.71	1.15	0.57
MN-2	957	1.72	35.9	1.30	3.98	0.93
MN-3	5270	12.6	54.3	4.2	0.69	2.15
MN-4	8.67	0.24	52.4	0.56	4.10	0.61
MN-5	12.7	1.20	38.1	2.70	1.82	0.79
MN-6	804	6.03	23.6	4.56	2.82	1.70
MN-7	5.29	0.050	58.5	0.50	4.21	0.63
MN-8	642	1.93	33.6	0.72	4.54	0.95
MN-9	279	2.70	34.3	5.60	5.21	1.90
MN-10	0.77	0.088	75.1	0.30	2.21	0.60
MN-11	46.5	0.82	118	2.28	6.82	0.97
MN-12	261	14.9	229	25.5	25.0	5.30
MN-13	19.4	0.97	150	3.79	7.82	1.20
MN-14	108	12.9	349	49.4	31.1	8.10
MN-15	5.90	0.23	68.6	1.75	3.54	0.72
MN-16	18.5	0.88	60.3	2.62	3.21	0.60
MHSI-5A	17.2	1.65	49.2	3.49	3.60	1.35
MLSI-4	12.1	0.81	49.0	2.68	6.87	1.00
MLSI-5A/5L	134	1.48	41.2	3.27	6.65	1.10
MLSI-7	12.9	0.56	59.6	3.61	8.04	1.30
MLSI-8	557	4.13	24.5	6.38	1.00	2.15
MLSI-9	397	4.76	22.0	10.80	1.00	3.60
MLSI-11A	74.3	2.42	114	8.01	5.08	3.10

* Average of 3 replicate PCT analyses
‡ Total constituent analysis by SRTC on TCLP solution returned from GEL
‡‡ TCLP leach by GEL, uranium by SRTC Chem-check

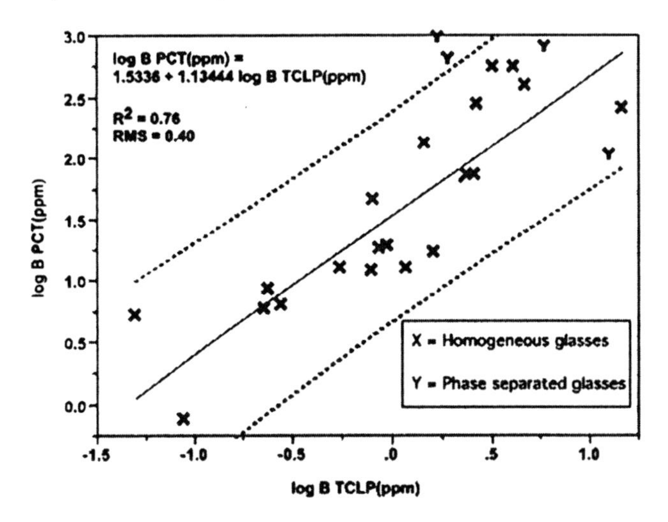

Figure 1. Correlation of boron released from the TCLP and PCT leach tests.

not been successful [9]. Part of this is due to fact that unless a constituent is present in the glass at a substantial concentration ($\geq$ 0.5 wt.%), the TCLP leachant concentrations will be too low to develop an applicable model. The data in Table II indicate that the much more aggressive PCT leaching procedure always resulted in a higher leachant concentration than the TCLP test. The only metals that could be correlated in these studies were Ni and U, due to their relatively high concentrations in the M-Area glasses.

THERMO™ Model for Predicting TCLP for Borosilicate Glasses

The regression analyses for the predicted durability using the THERMO™ model (ΔG_f, kcal/mole) are plotted vs. the normalized elemental concentrations of Ni and U and the TCLP leaching results in Figures 2 and 3. The TCLP data is given in Table III. Although the R^2 for the regressions is somewhat low at 0.65, this scatter can be attributed to lack of size control for the TCLP test procedure, e.g., the TCLP has a noisy durability test response. The SRTC crucibles were used to develop the correlation; the VTF production scale melter data and the OTD pilot scale melter data were used to validate the model. This model can be used to predict the durability of a glass that will need to meet a TCLP leaching limit. For example, the RCRA limit for delisting a mixed waste is 10 ppm TCLP, for nickel (100 x the EPA primary drinking water standard of 0.1 mg/L). The 10 ppm TCLP limit is overlain on Figure 2, and indicates that any glass with a durability better than –25 kcal/mole will pass, at up to 3 wt. % nickel oxide. If the proposed EPA drinking water limit of 0.020 mg/L for uranium were required to be met for a delisting petition, then the TCLP limit would be 2 ppm TCLP. The overlay in Figure 3 indicates that the durability would have to be improved to ~-10 kcal/mole, if the glass contained 5 wt. % U_3O_8.

Glass Size vs. TCLP Leaching Results

The data for the TCLP leaching results for various size fractions is shown in Figure 4. As expected, the smaller the particle size, and the greater the surface area, the greater the TCLP leachant concentration. The difference for a 3/8" size material vs.a < 100 to > 200 mesh material can be a great as 3X to 4X (Figure 4). The data also indicate that the TCLP results are fairly similar until the particle size is reduced to 16 - 30 mesh size. This means that some small amount of material less than the nominal 3/8" size should not affect the TCLP result significantly. But the lack of size control and presence of adhering fines will always affect the reproducibility and accuracy of the TCLP test vs. the PCT test.

CONCLUSIONS

The TCLP test response for boron correlates well with the PCT test response from the same borosilicate glasses. This indicates that the durability of the glass, as expressed by each test response, is controlled by similar mechanisms, e.g., congruent dissolution. The TCLP test response can be adequately modeled using the THERMO model currently employed for PCT durability test response in the SRS DWPF facility. In addition, the TCLP test was performed on glasses sized from 3/8" to 200 mesh and shown to be a strong function of particle size once the size was reduced below ~16 mesh.

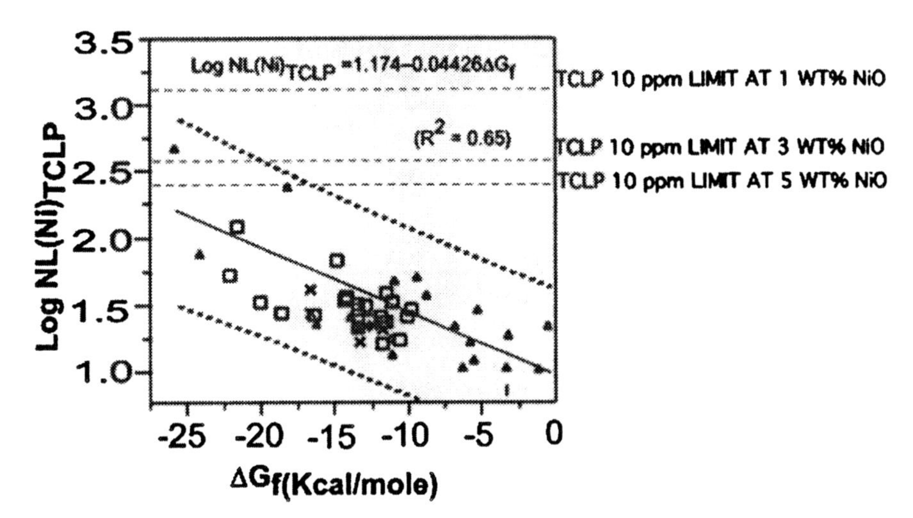

Figure 2. TCLP Ni leaching as a function of the hydration free energy (ΔG_f) and Ni concentration in the glass. The Ni leaching response has been normalized to the mass fraction of Ni in the glass.

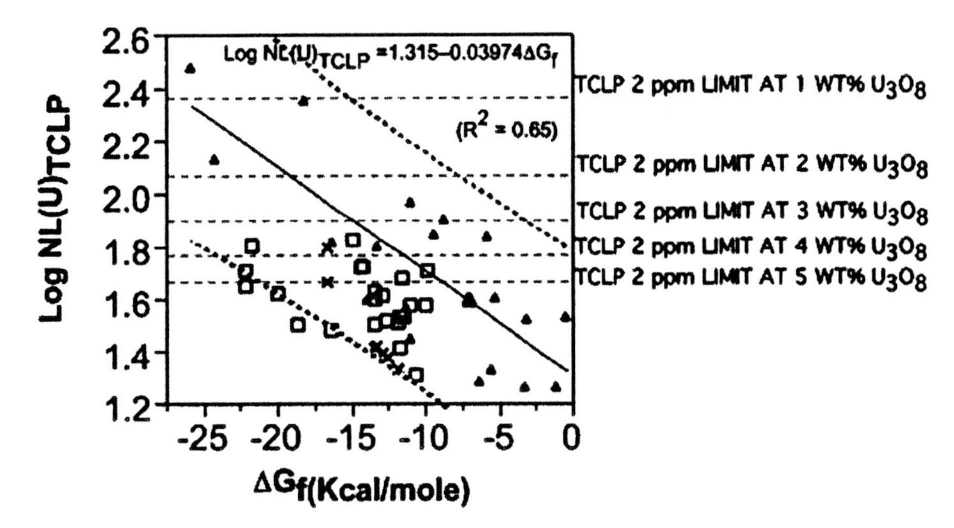

Figure 3. TCLP U leaching as a function of the hydration free energy (ΔG_f) and U concentration in the glass. The U leaching response has been normalized to the mass fraction of U in the glass.

Table III. Uranium and Nickel TCLP Data and Calculated ΔG_f Values

Sample Id	U_3O_8 (wt%)	GEL U TCLP	SRTC U TCLP	NiO (wt%)	GEL Ni TCLP	ΔG_f	Observations
OTD TEST							
MIC4-144	2.85	1.12		0.64	0.20	-16.51	
MIC4-149C	3.53	1.87		0.75	0.34	-16.48	XLS
MIC5-9A	2.53	0.53		0.67	0.17	-12.80	XLS
MIC5-13A	1.75	0.35		0.63	0.16	-12.29	XLS
MIC5-21A	1.50	0.27		0.58	0.14	-11.62	XLS
MIC5-42A	1.42	0.31		0.61	0.15	-13.31	XLS
MIC5-42C	1.35	0.30		0.62	0.12	-13.11	XLS
SRTC CRUCIBLE STUDY							
MN-1	4.98	0.77	0.57	0.63	0.08	-0.98	
MN-2	4.60	1.23	0.93	0.55	0.13	-3.43	ϕ Sep
MN-8	4.59	1.35	0.95	0.56	0.14	-2.64	ϕ Sep
MN-3	4.20	3.00	2.15	0.5	0.43	-4.61	ϕ Sep
MN-4	4.99	0.78	0.61	0.62	0.08	-3.12	
MLSi-4	5.24	1.48	1.00	0.69	0.15	-2.99	
MHSi-5A/5B	3.39	1.98	1.35	0.62	0.12	-5.62	
NM-5 (or 5A)	3.88	1.11	0.79	0.48	0.12	-0.30	
MLSi-5A/5B	4.55	1.55/1.99	1.10/ -	0.68	0.22	-5.17	
MN-6	3.81	2.26	1.70	0.49	0.27	-9.31	ϕ Sep
MN-7	5.00	0.91	0.63	0.62	0.09	-5.41	
MLSi-7	5.45	1.81	1.30	0.68	0.17	-6.72	
MLSi-8A/8B	3.57	1.99/2.69	1.40/2.15	0.78	0.20	-16.19	
MN-9A	3.52	2.42	1.90	0.49	0.20	-8.57	
MLSi-9	3.88	4.55	3.60	0.51	0.41	-24.06	
MN-15	4.69	1.11	0.72	0.63	0.10	-10.93	
NM-16	2.95	0.93	0.60	0.59	0.12	-11.12	
MN-10	4.94	0.81	0.60	0.62	0.08	-6.14	
MN-11	4.20	2.28	0.97	0.52	0.13	-13.11	
MLSi-11A/B	5.25	4.45/4.20	2.95/3.10	0.75	0.39	-10.81	
MN-12	3.72	7.19	5.30	0.47	1.08	-18.12	
MN-13	4.42	1.48	1.20	0.59	0.17	-13.80	
MN-14	4.07	10.50	8.10	0.52	2.26	-25.71	ϕ Sep
VTF DRUMS							
44	1.61	0.89		0.33	0.42	-21.64	
68	1.92	0.74		0.29	0.17	-22.05	
128	1.94	0.86		0.51	0.30	-22.14	
158	1.86	0.67		0.53	0.20	-20.01	
168	1.90	0.52		0.51	0.16	-18.58	
226	4.05	1.10		0.72	0.18	-13.38	
290	6.88	1.77		0.90	0.28	-16.38	
325	5.71	1.28		0.93	0.18	-11.68	
640	3.44	0.97		0.5	0.14	-12.67	
715	3.17	0.92		0.65	0.18	-11.45	
965	2.98	0.83		0.76	0.23	-11.83	
1650	2.42	1.40		0.91	0.66	-14.83	
1780	2.97	1.35		0.94	0.39	-14.23	
1820	3.72	1.27		1.12	0.38	-13.42	
1923	4.39	1.43		1.27	0.48	-11.06	
2014	4.33	1.58		1.49	0.54	-13.45	
2128	4.87	1.99		1.1	0.48	-11.53	
2225	4.38	2.01		0.69	0.27	-14.35	
2256	5.84	1.89		0.87	0.26	-9.99	
2475	4.77	1.69		0.78	0.28	-12.94	
2558	4.51	1.98		0.91	0.31	-9.76	
VTF1636	4.19	1.23		1.24	0.34	-11.70	
VTF1735	0.63	0.11		0.24	0.05	-10.59	

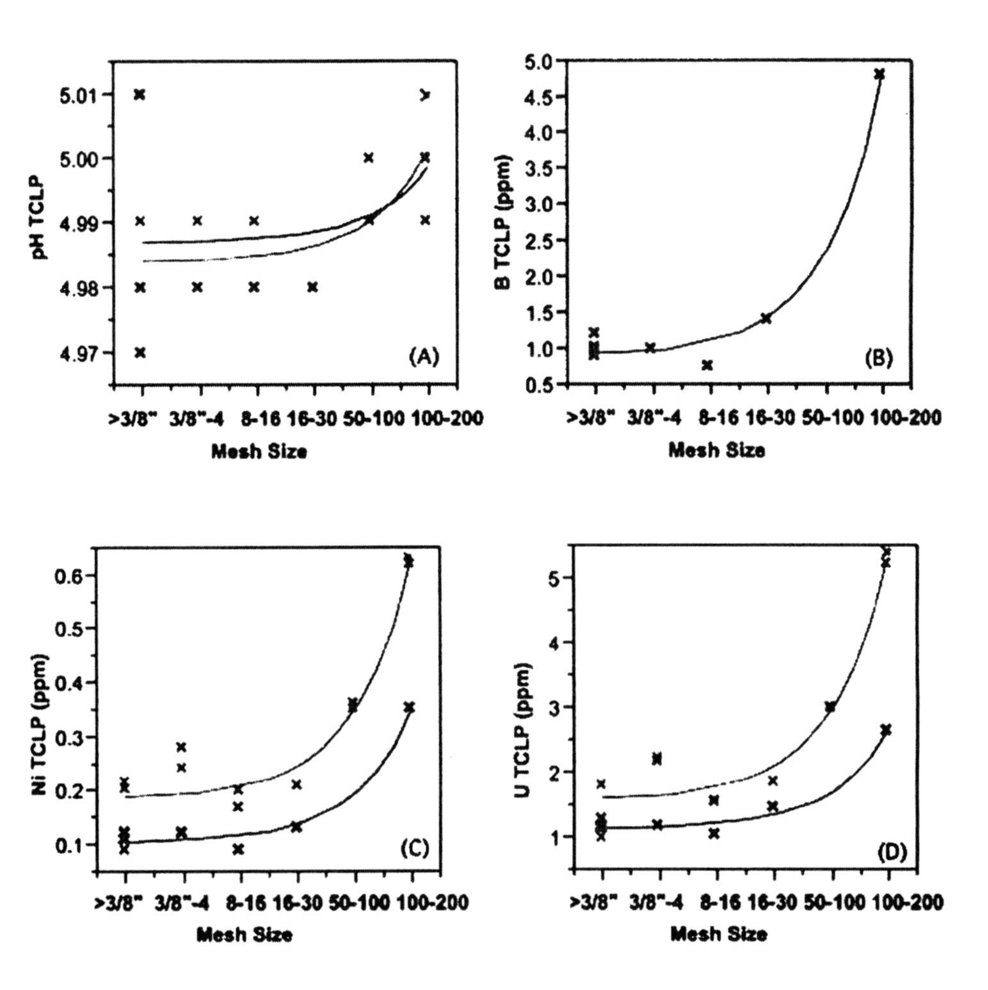

Figure 4. TCLP leaching response as a function of mesh size for glasses MN-5 and MLSI-5 from Table IV

Table IV. Leaching of B, Ni, and U in TCLP as a Function of Mesh Size

	Mesh Size (in. or sieve#)	Mesh Size (mm)	Final pH By GEL	Ni, ppm by GEL	U (ppm) by GEL	U (ppm) by SRTC	B, ppm SRTC	Si (ppm) SRTC
MN-5 (or 5A)	<3/8 inch	<9.5	4.98	0.12	1.11	0.79	1.20	2.70
MN-5B	<3/8 inch	<9.5	4.97	0.11	1.14	0.73	0.89	1.60
MN-5- AL-B*	<3/8 inch	<9.5	5.01	0.11	1.16	0.79	0.97	1.02
MN-5-AL-A*	<3/8 inch	<9.5	5.01	0.11	1.27	0.88	1.02	1.29
MN-5C	<3/8" - >4	<9.5- >4.74	4.98	0.12	1.17	0.80	0.98	1.51
MN-5E	<8 - >16	<2.36- >1.18	4.98	0.09	1.04	0.66	0.75	1.35
MN-5G	<16 - >30	<1.18- >0.60	4.98	0.13	1.46	0.93	1.38	2.00
MN-5K	<100 - >200	<0.15- >0.075	5.00	0.35	2.65	2.05	4.79	4.28
MLSi-5A	<3/8 inch	<9.5	4.99	0.17	1.55	1.10	1.48	3.27
MLSi-5B	<3/8 inch	<9.5	4.98	0.22	1.99			
MLSi-5-AL-B*	<3/8 inch	<9.5	4.98	0.20	1.93			
MLSi-5-AL-A*	<3/8 inch	<9.5	4.98	0.089	2.10	1.50	1.83	3.33
MLSi-5C	<3/8 - >4	<9.5- >4.74	4.99	0.28	2.23			
MLSi-5D	<3/8 - >4	<9.5- >4.74	4.99	0.24	2.16			
MLSi-5E	<8 - >16	<2.36- >1.18	4.98	0.20	1.53			
MLSi-5F	<8 - >16	<2.36- >1.18	4.98	0.17	1.57			
MLSi-5G	<16 - >30	<1.18- >0.60	4.99	0.21	1.85			
MLSi-5I	<50 - >100	<0.30 - >0.15	5.00	0.36	3.02			
MLSi-5J	<50 - >100	<0.30 - >0.15	4.99	0.35	2.99			
MLSi-5K	<100 - >200	<0.15- >0.075	5.01	0.62	5.23			
MLSi-5L	<100 - >200	<0.15- >0.075	5.01	0.63	5.41	3.95		

* Small chunks of $Al(OH)_3$

REFERENCES

1. U. S. Environmental Protection Agency, "Land Disposal Restrictions for Third Third Scheduled Wastes, Final Rule", *55 Federal Register 22627* (1990).
2. C. M. Jantzen, J. B. Pickett, and W. G. Ramsey, "Glassification of Hazardous and Mixed Wastes", *Presentation at the American Chemical Society Symposium on Emerging Technologies for Hazardous Waste Mgt, 9/21-23/92 ,Atlanta, GA.* WSRC-MS-92-261, Westinghouse Savannah River Co. Aiken, SC 29808 (1992).
3. J. B. Pickett, J. C. Musall, and H. L. Martin, "Treatment and Disposal of a Mixed Waste Plating Line Sludge at the Savannah River Site," *Proceed. Second Intl Mixed Waste Symp.; 9/17-20/1993,* A. A. Moghissi, R. K. Blauvelt, G. A. Benda, and N. E Rothermich, (Eds), U. of Maryland, Baltimore, MD (1993).
4. C. M. Jantzen, J. B. Pickett, and W. G. Ramsey,"Reactive Additive Stabilization Process for Hazardous and Mixed Waste Vitrification", 1993. *Proceed. of the Second Intl Mixed Waste Symp., 9/17-20/1993*, A. A. Moghissi, R. K. Blauvelt, G. A. Benda, and N. E Rothermich, (Eds), U. Maryland, Baltimore, MD (1993).
5. C. M. Jantzen, J. B. Pickett, W. G. Ramsey and D. C. Beam, "Treatability Studies on Mixed (Radioactive and Hazardous) M-Area F006 Sludge; Vitrification via the Reactive Additive Stabilization Process (RASP)", *Proceedings of International Topical Meeting on Nuclear and Hazardous Waste Management SPECTRUM '94, Atlanta, GA., 8/14-18/94*, American Nuclear Society, La Grange Park, IL 60525, 737-742 (1994).
6. C. M. Jantzen, "Vitrification of M-Area Mixed (Hazardous and Radioactive) F006 Wastes: I. Sludge and Supernate Characterization", 2001. WSRC-TR-94-0234, Savannah River Technology Center, Westinghouse Savannah River Co., Aiken, SC 29808, (2001).
7. J. B. Pickett, 1996. "Upfront Delisting Petition for Vitrified M-Area Plating Line Wastes", WSRC-TR-96-0244, Rev. 0, Westinghouse Savannah River Co., Aiken, SC 29808, (1996).
8. J. B. Pickett, 2000. "Delisting Petition for Vitrified M-Area Plating Line Wastes", WSRC-TR-96-0244, Rev. 2, Westinghouse Savannah River Co., Aiken, SC 29808, (2000).
9. C. A. Cicero, 1996. "The Effect of Compositional Parameters on the TCLP and PCT Durability of Environmental Glasses", Proceedings of American Ceramic Society Symposium on Environmental issues and Waste Management Technologies in the Ceramic and Nuclear Industries I, April 30-May 3, 1995, Ceramic Transactions Volume 61, Edited by V. Jain and R. Palmer, American Ceramic Society, Westerville OH, 43081, (1996).

EFFECT OF GLASS COMPOSITION ON THE LEACHING BEHAVIOR OF HLW GLASSES UNDER TCLP CONDITIONS

Hao Gan and Ian L. Pegg
Vitreous State Laboratory
The Catholic University of America
Washington, DC 20064

ABSTRACT

Glasses have been developed for the vitrification of high-level wastes from the Hanford tanks. Such glasses are required to meet a variety of property constraints, one of which is performance on the EPA toxicity characteristic leaching procedure (TCLP), which plays an important role in regulatory compliance and product delisting. TCLP data have been collected on 95 glasses (from a total of 146 under investigation). The TCLP leachates have been analyzed not only for the toxic elements but also for the major glass constituents. Correlation analysis shows that the normalized TCLP releases fall into three groups with elements in each group being released at the approximately the same normalized rates. Elements in the "advanced" group (alkalis, alkaline earths, divalent transition metals, B, Ag, and U) are released at about four times the rate of those in the "retarded" group (Si, Tl, Se, Sb, Pb), and generally more than ten times the rate of those in the "slow" group (Al, Fe, Zr, Cr, and As). Based on these observations, a model is described that relates the TCLP release of each element to the glass composition.

INTRODUCTION

The River Protection Project Waste Treatment Plant (RPP-WTP) will separate Hanford tank wastes into HLW and low-activity waste (LAW) streams and each stream will be vitrified separately. Glasses formulations for both of these streams are being developed at the Vitreous State Laboratory[1-4]. Acceptable formulations must meet a variety of processability, product quality, and waste loading requirements that are dictated either by the RPP-WTP contract or by the characteristics of the particular treatment processes that have been selected. These requirements amount to constraints on the acceptable ranges of certain glass properties. These properties are determined first and foremost by the composition of the glass or glass melt. Thus, while there is no *direct* way of controlling the

To the extent authorized under the laws of the United States of America, all copyright interests in this publication are the property of The American Ceramic Society. Any duplication, reproduction, or republication of this publication or any part thereof, without the express written consent of The American Ceramic Society or fee paid to the Copyright Clearance Center, is prohibited.

glass properties of interest during production, there are simple and extremely effective methods of achieving the same result by instead controlling the glass composition. The determination of quantitative relationships between the glass properties that must be controlled and the glass composition can play an important role in the development of such waste treatment facilities. For the RPP-WTP HLW vitrification facility, one of the potential applications of such models is for permitting and regulatory compliance, including LDR compliance and HLW glass delisting. Given the regulatory role of the Toxicity Characteristic Leaching Procedure (TCLP)[5], this paper focuses on the leaching behavior of various glass constituents under TCLP conditions and development of quantitative relationships between TCLP response and glass composition ("models").

EXPERIMENTAL

The glasses used for the present work include a combination of "actively-designed" formulations and "statistically-designed" matrices of formulations. The former are formulated to meet specific constraints by applying past knowledge and experience, whereas the latter are formulated to provide coverage of defined composition regions. The majority of the glasses (one hundred) were statistically designed using the D-optimal method[6] to cover the compositional space specified by a series of single and multiple component constraints using the program Design Expert™ (version 5, Stat-Ease, Inc.). The simulated HLW glasses formulations typically contained about 40 different constituents. To reduce the composition variables considered in the statistical design, single variables (e.g., SiO_2, Al_2O_3, etc.) and combinations of variables were used with multiple component constraints. The details of the experimental design and the glass formulation development work are presented elsewhere[1,2]. A further 46 glasses were "actively formulated" on the basis of existing knowledge to meet specific property constraints[2].

Crucible melts were prepared by melting the appropriate combination of well-mixed chemicals at 1150°C to 1250°C in platinum crucibles for at least two hours. Stirring of the melt was performed with a platinum mechanical stirrer beginning after the first half-hour. The melt was then poured onto a cold graphite plate and quenched. The glass was then distributed for chemical analysis and TCLP testing.

The TCLP[5] consists, in summary, of extracting crushed glass (<3/8") in a sodium acetate buffer solution (pH = 4.93) for 18 hours at room temperature with constant end-over-end agitation. The surface-area-to-volume ratio is about 20 m^{-1}. The leachate concentrations were measured by direct current plasma atomic emission spectroscopy (DCP-AES). The overall uncertainty associated with this test is estimated to be about ±15 %.

Glass samples were subjected to microwave-assisted total HF/HNO_3 acid dissolution in Teflon vessels. The resulting solution was analyzed by DCP-AES using the acid mixture as the matrix blank.

Data Analysis and TCLP Modeling

Despite the lower temperature, shorter duration, and lower ratio of the sample surface area to the leachant volume, the TCLP is much more effective than is the Product Consistency Test (PCT; ASTM C 1285-97), which is the benchmark test for HLW repository acceptance, in leaching toxic elements from the sample, as a result of the lower pH in the TCLP. While only the concentrations of certain toxic elements are normally measured in the TCLP, the data used in the present work also included all of the major glass constituents[1,2]. The inclusion of these elements provides valuable information on the way in which the glass dissolves under TCLP conditions. Following our earlier work on a different glass composition region[7], the first step in the present work involved an analysis of correlations between all of the measured elements; that information was then used to develop an appropriate model form. Statistical correlation analysis and simple data plotting was used to identify these correlations. However, the correlations between TCLP leachate elements are much stronger when the concentrations are first "normalized." The normalization used here is performed by dividing the molar concentration of element i in the TCLP leachate by the ratio of the number of moles of element i in the glass to the number of moles of oxygen in the glass, which gives an approximation to the ratio of the volumetric concentration of that element in the leachate to that in the glass. However, similar correlations are evident by normalizing by simply dividing the mass fraction of element i in the solution by the mass fraction of element i in the glass. In either case, pair-wise correlation analysis reveals simple relationships between different elements in the leachates such that they fall into three distinct groups as follows:

Group 1: Advanced Elements

This group includes all alkalis (sodium release could not be measured), alkaline earths, divalent transition metals, boron, silver, and uranium. Cadmium was selected as a convenient representative of this group against which to display the correlations of other elements in this group. Thus, Figure 1 shows the normalized release of the selected elements in this group plotted against the normalized release of cadmium; the lines in these figures represents a one-to-one relationship (i.e., equality of normalized releases), which most elements follow quite well. This type of dissolution behavior is often referred to as "congruent" and would be characteristic of a simple interface-controlled dissolution process in which the glass simply dissolves "layer-by-layer." Ultimately, differences in solubilities between elements would be expected to lead to departures from this trend at higher concentrations. However, for this group of elements in the pH range of the TCLP, solubility constraints would be expected to be relatively unimportant, as seems to be the case for the present data set.

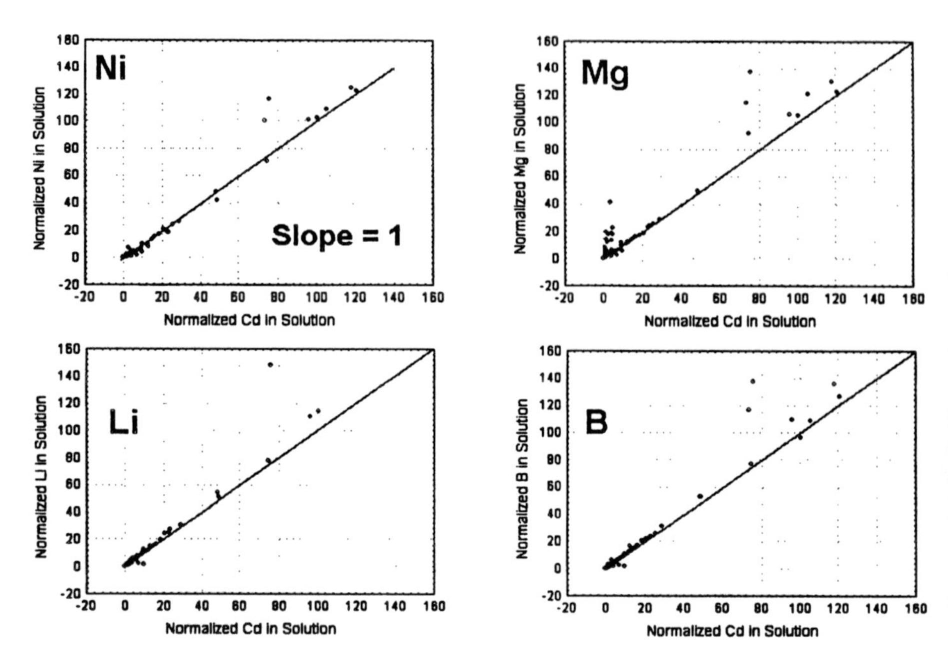

Figure 1. Correlations of the normalized leachate concentrations (mol/l): Ni vs. Cd, Mg vs. Cd, Li vs. Cd, and B vs. Cd.

<u>Group 2: Retarded Elements</u>

This group includes several p-block elements, i.e., Si, Tl, Se, Sb, and Pb. Within this group, the normalized elemental releases were found to be approximately equal to each other. Interestingly, however, the normalized release of these elements is about one-fourth that of the normalized release for the "advanced" group of elements; Figure 2 shows the relationships between the normalized releases of selected Group 1 and Group 2 elements.

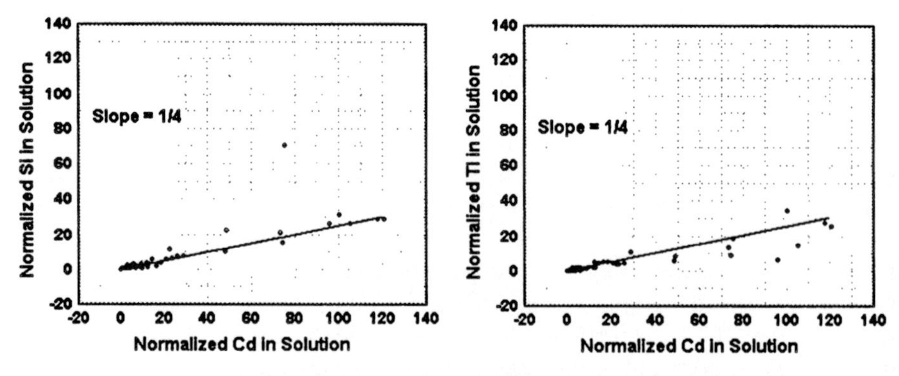

Figure 2. Correlations of the normalized leachate concentrations (mol/l): Si vs. Cd and Tl vs. Cd.

Group 3: Slow and Irregular Elements

Elements in this group include Al, Fe, and Zr, which have low solubilities at pH 5, as well as As and Cr[8]. As illustrated in Figure 3, the normalized releases of these elements are much lower than those of the Group 1 or 2 elements and generally show poor correlations to them.

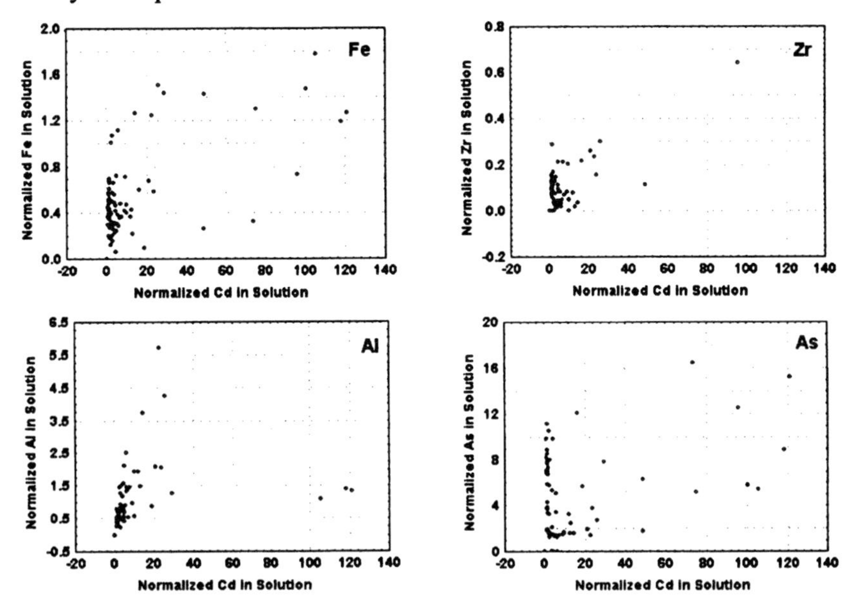

Figure 3. Correlations of the normalized leachate concentrations (mol/l): Fe vs. Cd, Zr vs. Cd, Al vs. Cd and As vs. Cd.

These general observations can be used to motivate a possible form for a model to correlate TCLP release to the glass composition. However, several important conclusions can already be drawn. These simple element groupings can be used as a general guide to assess the relative "difficulty" of complying with each of the elemental concentrations limits from a knowledge of those limits and the expected concentrations of the elements in the glass. Thus, for example, while cadmium (Group 1) is likely to be a "difficult" element for some HLW tanks, as pointed out earlier[1,2], chromium (Group 3) is not. Furthermore, the TCLP behavior of elements that were not included in the present data set can be inferred based on their chemical characteristics. Thus, while beryllium was not included in the present data set, it is very likely to behave as a Group 1 element, such that its normalized release from a given glass, and therefore its TCLP concentration, can be reliably estimated from the normalized cadmium release for that glass. In addition, a simple bounding model based on the Group 1 elements may be adequate for some purposes.

The simple correlations observed between the normalized elemental TCLP releases for elements of Groups 1 and 2 suggest that the TCLP leaching process for these glass elements is an apparent zero-th order kinetic process. If we can, as suggested by the data, ignore solution saturation effects, the important factor becomes the rate constant for the leaching reaction and its dependence on the glass composition. It is reasonable to argue that the dissolution of the major glass constituents would have a significant impact on the release rate of the generally much-less-concentrated toxic elements of concern. For example, the dissolution of B^{3+}, a major component of the borosilicate network, is likely to be necessary for access to and continued release of minor elements. Conversely, higher concentrations of relatively insoluble components such as Al, and Fe would likely tend to hinder that process. Furthermore, a more highly polymerized network is expected to dissolve more slowly than a less polymerized network; the ratio of alkali to silica provides a simplistic measure of this effect. Finally, it is reasonable that the concentration of an element in solution would, to first-order, be proportional to its concentration in the glass and to the amount of glass reacted, hence the relationships between the normalized releases.

Assigning a rate constant K to the apparent zero-th order rate equation for the reaction of glass with surface area A, with a large volume of solution V, the rate of change of the concentration of element i in the solution is:

$$\frac{dC_i}{dt} = K\frac{A}{V}$$

and, therefore,

$$\frac{V}{A}\frac{dC_i}{dt} = K.$$

For the discrete TCLP data over the time interval of the test Δt we then have:

$$\frac{V}{A}\frac{\Delta C_i}{\Delta t} = K$$

where ΔC_i is equal to the change in the TCLP leachate concentration of element i over the test interval Δt, and V/(AΔt) is a constant for all TCLP tests. Consequently,

$$\ln \Delta Ci = -\ln(\frac{V}{A\Delta t}) + \ln K$$

The rate constant K typically follows an Arrhenius dependence on the absolute temperature T,

$$\ln K = \ln K_o + \frac{E}{RT},$$

where E is the activation energy. Since it is often the case that energetic terms are roughly additive with respect to the contributions of the constituents of the glass, the activation energy term and the logarithm of the pre-exponential term (which can be viewed as the limiting case of the energy term) can be approximated by a function of the glass composition $f(X_1, X_2, \ldots X_n)$:

$$\ln \Delta C_i = M + f(X_1, X_2, \ldots X_n),$$

where M is a combination of constants.

At this point we could simply adopt a linear or higher-order mixture model[6] for the function $f(X_1, X_2, \ldots X_n)$ but we prefer to incorporate our earlier reasoning with respect to important compositional factors. Thus, we define MPS_i, the "matrix partial dissolution" parameter for element *i* to reflect the dependence on the dissolution of boron as a major structural component; the inhibiting effect of less-soluble Group 3 components, specifically Al and Fe; the dependence on the extent of polymerization of the glass matrix; and the observed consistency between the normalized releases. In contrast to boron, the other two major trivalent cations in these waste glasses, Al^{3+} and Fe^{3+}, in spite of their similar structural roles (i.e., they tend to reside in tetrahedral sites co-polymerized with silica as boron would), dissolve little with no clear sign of correlation with B in the leachate. We therefore take the ratio of boron to the sum of the three trivalent cations (R_B) as one factor of MPS. The second factor is taken to be the ratio of the total alkali to silica ($R_{M/Si}$), which is approximately a measure of the extent of polymerization of the glass. The third factor of MPS_i is simply the concentration of element *i* in the glass (X_i). Thus, we define

$$MPS_i = R_B R_{M/Si} X_i =$$

$$[(B_2O_3/0.7)/(B_2O_3/0.7 + Al_2O_3 + Fe_2O_3/1.6)][(0.67K_2O + Na_2O + 2Li_2O)/SiO_2)]\ X_i\ .$$

Figure 4 shows a simple monotonic dependence of $\ln \Delta C_{Cd}$ on the MPS_{Cd} parameter. Because of the non-linear nature of the curve, a second-order polynomial in MPS is used to describe the TCLP release. Consequently, the function $f(X_1, X_2, \ldots X_n)$ is expressed as:

$$f(X_1, X_2, \ldots X_n) = B + d_1\, MPS + d_2\, MPS^2$$

and

$$B = \sum_i b_i X_i$$

where a linear composition dependence is assumed for B to account for the variation in Figure 4 that is not accounted for by the MPS parameter. The final form of the model is then:

$$\ln \Delta C_i = M + \sum_i b_i X_i + d_1 MPS + d_2 MPS^2$$

where M, b_i, and d_i are parameters to be determined by multiple linear regression, since this equation is linear in its parameters.

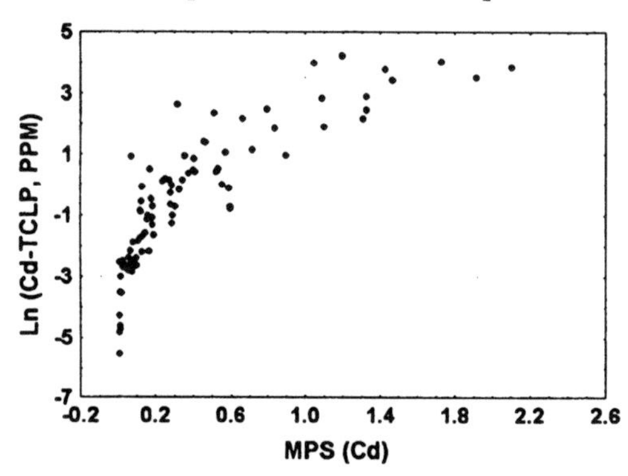

Figure 4. The dependence of the natural logarithm of the Cd TCLP release on the MPS parameter.

Table I. Regression results and component coefficients for the Cd TCLP model; $R^2 = 0.97$, Standard Error = 0.468

Cd TCLP	Coefficient	Standard Deviation
Intercept	1.615142	1.65
Al_2O_3	-0.17733	0.035
BaO	-3.5476	0.75
CaO	0.032233	0.030
CdO	2.977582	0.44
Fe_2O_3	-0.11291	0.028
K_2O	1.771699	0.36
Li_2O	0.039409	0.051
MgO	1.341914	0.46
MnO_2	0.228439	0.070
Na_2O	0.056504	0.025
P_2O_5	-0.39732	0.081
SiO_2	-0.12659	0.023
SrO	-0.15889	0.046
ThO_2	-0.10751	0.088
TiO_2	2.660005	2.08
ZnO	1.100737	0.26
ZrO_2	-0.18003	0.029
MPS	3.227513	1.23
MPS^2	-0.87071	0.46

The above model was fitted to TCLP data for Cd, Ni, and Tl using 85 glasses for Cd, 95 glasses for Ni, and 61 glasses for Tl (the balance of the glasses did not contain Tl). These elements were selected for illustration because they are representative of the two faster-leaching groups of elements and because of their significance for HLW glasses. It is of course, straightforward to generate similar models for all of the measured elements.

However, as discussed above, the normalized TCLP release for elements with each group was found to be approximately equal. On that basis, these models can be used to predict the TCLP release of any Group 1 or Group 2 elements.

The model variables and the regression results for the TCLP model of Cd are listed in Table I. Overall, the models represented the data reasonably well with R^2 values of greater than 0.95 for the Cd and Ni models and around 0.90 for Tl. The calculated and observed TCLP results for Cd are compared in Figure 5. The relative deviations of the model predictions are generally less than about ± 50 %. Except for a few of the glasses with very low Ni releases, all of the calculated TCLP releases for the three elements deviate from the observed values by less than a factor of two. The values of the coefficients indicate, as was expected, that the inert and retarded elements play important roles in controlling the TCLP leaching rate of the selected toxic elements.

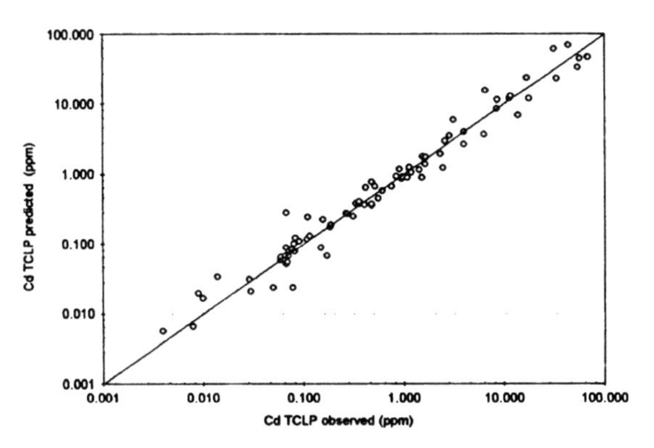

Figure 5. Comparison between predicted and observed values for Cd TCLP release model.

CONCLUSIONS

The TCLP leaching rates of many toxic elements, such as Cd, Ni, and Tl, are strongly dependent on the glass composition. Correlation analysis shows that the normalized TCLP releases fall into three groups with elements in each group being released at the approximately the same normalized rates. Elements in the "advanced" group (alkalis, alkaline earths, divalent transition metals, B, Ag, and U) are released at about four times the rate of those in the "retarded" group (Si, Tl, Se, Sb, Pb), and generally more than ten times the rate of those in the "slow" group (Al, Fe, Zr, Cr, and As). In a qualitative sense, these simple element groupings can be used as a general guide to assess the relative "difficulty" of complying with each of the elemental concentrations limits from a knowledge of those limits and the expected concentrations of the elements in the glass. More quantitatively, based on these observations, a model has been developed that relates the TCLP release of each element to the glass composition. The determination of quantitative relationships between the glass properties, such as TCLP response, that must be controlled during production and the glass composition can play an important role in the development of waste vitrification facilities. One of the potential applications of TCLP models is for permitting and regulatory compliance, including LDR compliance and HLW glass delisting. Ongoing work in this area is investigating the time dependence of glass leaching under TCLP conditions and the relative roles of solution saturation and surface layer formation in controlling the reaction progress[9].

ACKNOWLEDGMENTS

The authors would like to thank the staff of the VSL, particularly Chu-Fen Feng, for support with glass preparation, testing, and analytical work and M.C. Paul for manuscript preparation. This work was support in part by the Department of Energy, Office of River Protection through Duratek, Inc.

REFERENCES

[1]S.S. Fu and I.L. Pegg, "Glass Formulation and Testing with TWRS HLW Simulants," VSL Final Report, January 1998.

[2]W.K. Kot and I.L. Pegg, "Glass Formulation and Testing with RPP-WTP HLW Simulants," Final Report, VSL-01R2540-2, February 16, 2001.

[3]I.S. Muller and I.L. Pegg, "Glass Formulation and Testing with TWRS LAW Simulants," VSL Final Report, January 1998.

[4]I.S. Muller, A.C. Buechele, and I.L. Pegg, "Glass Formulation and Testing with RPP-WTP LAW Simulants," Final Report, VSL-00R3560-2, February 23, 2001.

[5]US Environmental Protection Agency, SW-846, Method 1311, Toxicity Characteristic Leaching Procedure.

[6]J.A. Cornell, "Experiments with Mixtures: Design, Models, and the Analysis of Mixture Data," Second Edition, John Wiley and Sons, New York, 1990.

[7]S.S. Fu and I.L. Pegg, "The Effect of Composition on TCLP Leach Rates for Glasses with Similar Melt Viscosities," Ceramic Transactions, Eds. G.T. Chandler and X. Feng, vol. 107, pp. 261, 1999.

[8]M. Pourbaix, "Atlas of Electrochemical Equilibria in Aqueous Solutions," National Association of Corrosion Engineers, Houston, Texas, USA, (1974).

[9]H. Gan and I.L. Pegg, unpublished work.

CHEMICAL AND PHYSICAL CHARACTERIZATION OF THE FIRST WEST VALLEY DEMONSTRATION PROJECT HIGH-LEVEL WASTE FEED BATCH

Ronald A. Palmer
West Valley Nuclear Services Co.

Harry Smith, Gary Smith, Monty Smith, Renee Russell, and Gert Patello
Pacific Northwest National Laboratory

ABSTRACT

To support the West Valley Demonstration Project's (WVDP) Waste Form Qualification Report (WQR) and data needs associated with the process flowsheet and waste form qualification, the Pacific Northwest National Laboratory (PNNL) was asked to characterize a sample of the first batch of high-level waste (HLW) slurry transferred to the concentrator feed make-up tank (CFMT) identified as CW-H. Cation, anion, and radionuclide concentrations, as well as the slurry physical properties including density, total solids, and suspended solids, were measured. Also, laser ablation-inductively coupled plasma/mass spectroscopy (LA-ICP/MS) results for cation analysis are compared to cation analysis results from inductively coupled plasma/atomic emission spectroscopy (ICP-AES). Radionuclide analysis methods were developed during this work for iodine-129, selenium-79, actinium-227, and neptunium-236, and are also presented.

INTRODUCTION

The former nuclear fuel reprocessing facility located at the Western New York Nuclear Services Center (WNYNSC) in West Valley, New York, has been the focus of a radioactive waste management demonstration project, the WVDP. The WVDP, operated by West Valley Nuclear Services Company (WVNS), is currently vitrifying waste generated from former spent nuclear fuel reprocessing operations. Pacific Northwest National Laboratory was contracted to perform analysis of a HLW slurry sample (CW-H) to support the waste form qualification report (WQR). Approximately 40 mL of the first radioactive HLW slurry batch (Batch 10) transferred from Tank 8D-2 to the CFMT was shipped from the WVDP to PNNL for characterization of various cations, anions, and radionuclides, as well as certain physical properties.

To the extent authorized under the laws of the United States of America, all copyright interests in this publication are the property of The American Ceramic Society. Any duplication, reproduction, or republication of this publication or any part thereof, without the express written consent of The American Ceramic Society or fee paid to the Copyright Clearance Center, is prohibited.

APPROACH AND SAMPLE CHARACTERISTICS

Four 10-ml radioactive HLW slurry specimens from Batch 10 were combined into one homogenized specimen. Since the waste slurry is composed of solids in various degrees of suspension in a liquid, care was taken to ensure the composite HLW slurry sample was well homogenized by vigorously stirring it just prior to the analytical samples being removed . All portions of the HLW slurry sample that were not consumed or significantly altered by testing or analysis were archived in the container in which the four original parts were combined.

The total solids and total oxides of the HLW slurry sample mean value, plus or minus one standard deviation, in weight percent are 33.2 ± 0.2 and 25.4 ± 0.3 wt%, respectively. Based on these results, the mass of the oxides is 77% of the mass measured as solids; therefore, most of the mass of the solids (77%) will be in the vitrified product. A small amount of the solids are in solution (4.34 ± 0.11 weight percent of the sample is dissolved solids), but the majority of the solids are suspended (28.8 weight percent of the sample is attributable to suspended solids). The density of the homogenized waste slurry as received was 1.09 g/mL and 1.08 g/mL in two independent measurements. The pH was 7.46 and 7.51 in two independent measurements.

HLW Waste Slurry Analyses (Chemical Analyses)

Cation analyses on the HLW slurry (Batch 10) were obtained by ICP-AES. Anions were determined by ion chromatography (IC), and carbon in the waste slurry was evaluated by determining the total inorganic carbon (TIC) or carbonate and total organic carbon (TOC) using the Hot Persulfate Method. These results are summarized in Table I. The Hot Persulfate Method oxidizes all the organic (reduced) carbon to carbonate, which is driven out of the sample because of an acid pH. In addition, laser ablation-ICP mass spectrometry results are presented in the second and third columns.

RESULTS AND DISCUSSION

Waste Analyses Results by ICP-AES, IC, TIC, TOC

The radioactive slurry sample, prepared in triplicate by the Shielded Analytical Laboratory (SAL) using HF/HNO_3 acid digestion and KOH caustic fusion sample preparation procedures, was analyzed using ICP-AES. The HF/HNO_3 acid digestion aliquot sample size was about 1 gram of slurry; the aliquot sample size for the KOH caustic fusion procedure was about 0.2 grams of slurry. Both sample preparation procedures appeared to solubilize the entire slurry sample aliquots. The sample contained high concentrations of Fe, Si, Na, and Th at levels greater than 1% by weight. Analyte concentrations reported (µg/g slurry) have been corrected for dilution from sample preparation and for dilutions performed during processing for analysis by ICP-AES.

Table I. Final ICP-AES Cation Analysis, IC Anion Analysis, and Carbon Analysis Results*

Element	Average µg/g	LA-ICP/MS $LiBO_2$ Fusion µg/g	LA-ICP/MS Dry Powder µg/g	Element	Average µg/g	LA-ICP/MS $LiBO_2$ Fusion µg/g	LA-ICP/MS Dry Powder µg/g
Ag	16		8.6	Ru	185	21	163.5
Al	4112	4903	5201.7	Sb	<80		
As	32			Se	<40		
B	1682			Si	14,950		
Ba	145	326	246	Sn	1057		
Be	4.6			Sr	117	119	138.6
Bi	46	1.3	2	Te	<178		
Ca	1112			Th	11,333	13,261	3969.8
Cd	13	1	2	Ti	304		284
Ce	150	174	163.7	Tl	<80		
Co	18	18	17.1	U	1021	967	1429.1
Cr	158	257	354.3	V	<8.9	277	
Cs	164	111	64	W	<356	2.4	
Cu	167	181	288.8	Y	20	54	70.3
Dy	<4			Zn	152	124	200.3
Eu	4.7			Zr	1053	1503	1734.9
Fe	27,150	36,368	40,252.				
K	1997						
La	75	146	112.3				
Li	608			F	240		
Mg	473	711	650.6	Cl	<91		
Mn	2110	2524	2578	NO_2	10,467		
Mo	11		10.5	NO_3	13,567		
Na	11,500			PO_4	<180		
Nd	298	511	400	SO_4	500		
Ni	713	888					
P	1061						
Pb	117	131	128.9	TIC	473		
Pd	238	4.3	10.9	TOC	2247		
Rh	61	30	35.5	TC	2720		

* Data provided by the WVDP for a similar waste slurry sample.

Relatively good agreement in analyte concentration was found between the two sample preparation procedures used. The percent relative standard deviation among the triplicate measurements for concentrations above the estimated quantitation limit (EQL) were typically less than 5%. The HF/HNO_3 acid digestion procedure produced concentration results somewhat higher than those found in the KOH caustic fusion digestion procedure, which are believed to be due to the fact that the concentrations for Si, Th, and Zr were noticeably different between the two digestion procedures. Thorium was about ten times higher, silicon about 25% higher, and zirconium about two times lower in the caustic fusion-prepared sample compared to the HF/HNO_3 acid- prepared sample. Since the concentrations of Si and Th are each about ten higher than the Zr, the results for the other elements may have been depressed in the KOH fusion as a result. The hydrofluoric acid used in the HF/HNO_3-prepared sample may have caused precipitation of thorium as a fluoride compound, although no precipitate was observed and may have volatilized silicon as silicon tetrafluoride. The HF/HNO_3 digestion procedure is more efficient in solubilizing Zr than the KOH fusion procedure because fluoride strongly complexes zirconium. HF is frequently used in preparing stable Zr solutions.

LASER ABLATION - INDUCTIVELY COUPLED PLASMA/MASS SPECTROMETRIC METHOD APPLIED TO THE WVDP WASTE SLURRY ANALYSES

Because of difficulties in obtaining good noble metals data, an analysis of the waste slurry solids was performed by laser ablation - inductively coupled plasma/mass spectrometric (LA-ICP/MS) analysis specifically for palladium, rhodium, and ruthenium. The standard inductively coupled plasma - atomic emission spectroscopy (ICP-AES) technique has difficulty measuring small amounts of noble metals in the presence of iron, thorium, uranium, and other heavy elements which produce a multitude of atomic emission lines. Some of the secondary lines from these metals coincide with or are very near the major noble metals emission lines. This interference increases the ICP-AES measurement uncertainty for the noble metals to an amount equivalent to the entire quantity of the metal present in the waste. The LA-ICP/MS technique uses mass spectrometry to separate and count (by atomic weight) the atoms ablated (vaporized) by the laser beam pulse and does not depend on atomic emission lines.[1,2]

RESULTS AND DISCUSSION

Waste Analyses Results by LA-ICP/MS

Results for the LA-ICP/MS technique are also presented in Table I. The analyses of two types of sample preparation are presented: a dried waste powder and the lithium metaborate fusion. The expected advantage of the lithium metaborate fusion is that the waste elements will be uniformly distributed and dispersed in a nonhygroscopic glass matrix. This assumes that all the different oxides will completely dissolve in the molten lithium metaborate and none of the constituent elements volatilize during the fusion process. As with any analytical technique, there are interferences that can potentially cause misinterpretations. With respect to the LA-ICP/MS technique, these interferences include the formation of molecular ions, the presence of different elements with the same isotope mass, multiple isotopes for a single element, and laser and inductively coupled plasma stability. The values in Table I are an interpretation of the ion count data, correcting for known interferences and knowledge of the expected isotopes produced by the nuclear fuel burn, and any other isotopes that might be there because of the chemicals used in processing the spent fuel.

Based on the results, it was determined that ruthenium volatilized during the fusion process as the ruthenium concentration is depressed relative to the other noble metals. This was not expected because ruthenium was fully recovered in the lithium metaborate fusion of a ruthenium standard made prior to the lithium metaborate fusions including waste. In hindsight, this appears to be explained by the presence of nitrate in the waste. As indicated by Figure1, ruthenium should be present in quantities greater than the other noble metals. Ruthenium oxide (RuO_4), however, is extremely volatile (boiling point of 110°C) and would form during the fusion step because nitrate was available. Odoj, et al.[3] stated that from 50 to 100% of the ruthenium in nuclear waste not containing any reductant may be volatilized during calcining, which would explain the low ruthenium numbers for the lithium metaborate fusion beads in this study. There was some nitrate in the WVDP waste (see Table I) and approximately 6 to 7 times less ruthenium was found in the lithium metaborate glasses than expected. Addition of a reductant e. g., sugar) to the lithium metaborate powder prior to fusion should destroy the nitrate, thereby keeping the ruthenium in the +4 state and would significantly reduce ruthenium volatilization. (However, this was not studied.) Note also that the LA-ICP/MS analysis of the dried waste sample did not show a depletion of ruthenium.

RESULTS AND DISCUSSION

Radiochemical Analyses

The results of the radionuclide analyses are given in Table II. They were performed on the HLW slurry (Batch 10) by techniques that included total alpha, total beta, gamma energy analysis (GEA), and mass spectrometry. Additionally, each of the previous methods was combined with chemical separations to isolate specific elements. Total uranium was determined by laser fluorimetry. The analysis for radioisotopes Se-79, Tc-99, I-129, Ac-227, Np-236, and Am-243 are discussed

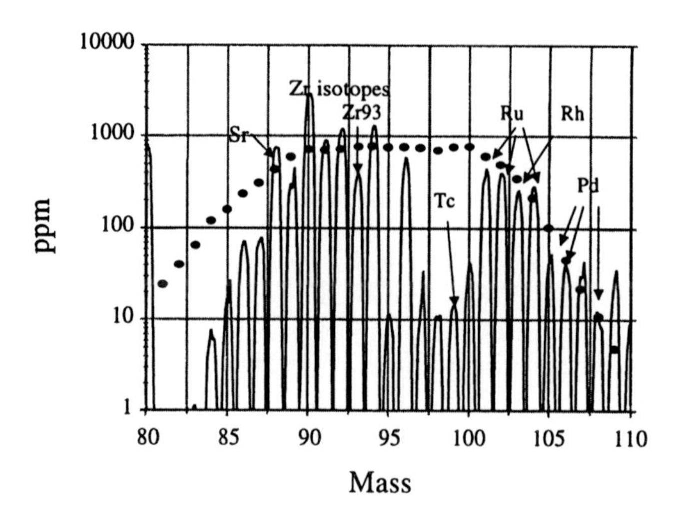

Figure 1.1. Low Mass Fission Product Spectra of West Valley Demonstration Project Slurry Obtained by Direct Analysis by Laser Ablation ICP/MS. Solid symbols represent the relative fission yield expected from reactors employing ^{235}U rich fuel.

individually in some detail below because they were given additional attention for such reasons as the initial detection limit was not low enough to support characterization efforts or the value obtained with the initial technique was clearly inconsistent with the expected value based on previous analytical analyses of related waste.

Table II. Radiochemical Analysis Results*

Element	Date of Analysis Report	Average µCi/g	WVDP-CB10WI.DAT µCi/g*	Element	Date of Analysis Report	Average µCi/g	WVDP-CB10WI.DAT µCi/g*
C-14	6/16/97	0.00049		Th-232	6/26/97	0.0015	
Ni-59	7/8/97	0.07		U-233	6/26/97	0.0036	
Co-60	6/11/97	0.884		U-234	6/26/97	0.0013	
Ni-63	7/8/97	7.36		U-235	6/26/97	0.000038	
Se-79	12/18/97	<0.00002		U-236	6/26/97	0.00011	
Sr-90	7/2/97	2753	2900	U-238	6/26/97	0.00034	
Nb-93m	7/16/97	0.058		Np-236	7/17/97	<0.00003	
Zr-93	12/19/97	0.12		Np-237	6/26/97	0.02**	
Tc-99 (ICP-MS)	6/26/97	14.5		Pu-236	7/2/97	0.022**	
Tc-99 (beta count)	11/10/98	8.45E-2		Pu-238	7/2/97	3.96	
Pd-107	6/26/97	0.0003		Pu-239	7/2/97	1.09	
Sn-126	7/2/97	0.003		Pu-240	7/2/97	0.77	
Cs-134	10/27/97	0.24		Pu-241	7/2/97	34.3	
Cs-135	6/30/97	0.0322		Pu-242	7/2/97	0.00085	
Cs-137	6/11/97	2847	3111	Am-241	6/11/97	32.1	
Sm-151	7/16/97	11.7		Am-243	10/27/97	0.36	
Eu-154	6/11/97	23.4		Cm-242	6/11/97	0.23	
Ac-227	1/22/98	0.012		Cm-243+244	6/11/97	6.98	
Pa-231	6/26/97	0.029					

* Data provided by the WVDP for a similar waste slurry sample
**Large error bars

Se-79 - Based on the results obtained from the chemical separation, a more sensitive determination of Se-79 was required. Additional samples were prepared in triplicate in the hot cells according to a modified acid digestion procedure that accomodates a larger sample size. The modification involved performing an initial separation in the hot cells so that a much larger sample could be analyzed without excess radiation exposure. Approximately 85 mL of this digestion was delivered to the radiochemistry laboratory for analysis as compared to about one mL the first time. As before, selenium was chemically separated from the samples after adding stable selenium to each of the triplicate samples, as well as the hot cell and laboratory blanks. Yields were lower than expected, ranging from 1 to 46%. The low yields may be due to the very large sample volumes used in this procedure. The separated selenium fractions, which were still considerably larger than the initial analytical samples, were counted by liquid scintillation. Detection efficiency was determine using C-14 as before. The triplicate samples, hot cell preparation blank, and laboratory blank showed no measurable activity above background. As a result, all results were reported as less than the minimum detectable activity level of 2.0E-5 μCi/g; a factor of 100 lower than determined previously.

Tc-99 - Technetium-99 was determined by inductively coupled plasma mass spectrometry (ICP-MS) initially, but the value obtained was two orders of magnitude higher than expected, suggesting possible interference from Ru-99. A second analysis was completed using a standard radiochemical procedure followed by beta counting. This analysis provided values consistent with previous analyses made on other samples of the WVDP waste for Tc-99. Both values are included in Table II.

I-129 - Dried HLW material was pulverized to particles <1 mm in diameter. At this point, the dried, pulverized waste sample was divided into three subsamples and each subsample was placed in a zirconium crucible. An iodide carrier solution (a sodium iodide solution that adds enough nonradioactive iodine to the sample to make the iodine easy to follow during the analytical process) was then added to each sample. Each sample was fused for 30 minutes with potassium hydroxide in a furnace between 450 and 500°C. In the absence of silver, bismuth (III), or mercury (II) ions, iodide stays in solution. Therefore, by centrifuging, the iodide is separated from precipitates containing radionuclides that interfere with the iodine measurement.

The solutions were then acidified with concentrated nitric acid after centrifuging. Silver nitrate solution was then added to the solution causing silver iodide to precipitate. The precipitated silver iodide then went through a series of methylene chloride extractions to purify the silver iodide. The filter containing the final silver iodide was mounted on a solid backing and covered with Mylar™ for low-energy photon spectrometry (LEPS) counting for iodine-129.

Iodine-129 was measured in the extracted silver iodide product by LEPS counting. Gamma energy was detected only at 29.8 keV. The results for the concentration of I-129 are shown in Table III. The results are right at the detection limit of the instrument, 1 pCi per gram ± 35% counting error at 1 standard deviation. The iodine-129 concentration in these samples was very low and required counting for 96 hours to detect the iodine-129. The final counting preparations were free of radiochemical interferences.

The original mass of the iodine carrier was accurately known, so by weighing the final silver iodide product, the overall gravimetric recovery could be calculated. The final recovery was about 35%, which is slightly low, but still adequate and about what is expected for samples handled in a hot cell. The gamma counting data were corrected for the gravimetric recovery. It is believed that the gravimetric recovery is accurate as the chemical separation used to isolate the iodine gives a very pure product.

The spikes averaged only about 69% recovery. It is suspected that this is due to the final counting mounts being inadvertently stored in the open under normal room lighting conditions instead of in the dark. Silver iodide is light sensitive and light causes iodine to be lost. The final iodine-129 results were corrected for the low spike recovery under the assumption that the samples and spikes all lost the same amount of iodine. The contribution to the uncertainty from this assumption (69% recovery) was negligible compared to the high counting uncertainty of ± 35%.

The sample results provided in Table III for the original CW-H sample were calculated from the dried sample concentration using the measured weight percent total solids and density. The weight percent total solids measured in the original CW-H sample was 33 wt%. The density of the original CW-H sample was 1.09 g/mL.

Table III. Measured Activities of Iodine-129 in the Sample CW-H with 1-Sigma Total Uncertainties

Sample ID	I-129 concentration in dried sample (pCi/g)	I-129 concentration in original CW-H sample (pCi/g)	I-129 concentration in original CW-H sample (μCi/m³)
CW-H-1	1.63 ± 35%	0.538	0.586
CW-H-2 duplicate	0.742 ± 35%	0.242	0.263

Ac-227- The Ac-227 analysis sample prepared in triplicate from the Na_2O_2 fusion in a zirconium crucible was used as the stock material. An aliquot of 0.2 mL was taken from the 100 mL dilution of the fusion. For the Ac-227 analysis, the samplewas diluted in a 0.095M HNO_3 matrix and then loaded onto a LN-Resin® (Eichrom Industries). Most fission products were removed in this wash, i.e., did not load on the resin. The Ac was then eluted with 0.35M HNO_3 and the elution concentrated to 2M HNO_3. Now the sample was loaded onto a TRU-Resin® (Eichrom Industries) where Ac is poorly retained, thus effecting an Ac-Am separation. The actinium was then precipitation plated and counted by alpha spectrometry immediately following separation and again about 45 days later. The Rn-219 (Ac-227 daughter) region at

6.819 MeV was integrated for both counts. Each sample was analyzed in duplicate; one set was spiked with an Ac-225 tracer, since no Ac-227 was available, the other set remained unspiked. The Ac radiochemical yields on the spiked samples were measured relative to Ac-225 tracer standards. Yields were nominally 96% on five spiked samples and 61% on the reagent blank. The yield corrections found for the spiked samples were applied to the unspiked samples.

Np-236 - The upper limit for Np-236 activity was evaluated in three different ways for this waste material. First, its activity was implied by alpha energy analysis (AEA) based on Pu-236, which is a daughter product of Np-236. Second, a thermal ionization mass spectrometry (TIMS) analysis was run on a separated Np fraction. And third, the concentration was calculated from the ICP-MS data. The TIMS analysis provided the lowest detection limit.

Am-243 - Initially, Am-243 was measured by gamma-counting the sample. However, a large amount of Eu-154 and Eu-155 were present, complicating the gamma spectra. A second approach using alpha spectroscopy gave an excellent resolution of the alpha spectra such that tailing of the Am-241 peak into the lower energy Am-243 peak was insignificant. The hot cell preparation blank and the laboratory blank did not result in any detectable Am-243. No analytical problems were noted. A small amount of Am-243 (0.355 ± 13% μCi/g) was found. The Am-241 present in the sample matrix and previously quantified, was used as an internal tracer for the preparation blank, laboratory blank, and blank spike.

CONCLUSIONS

In summary, the WVDP waste slurry sample CW-H analysis is consistent with the expected analysis based on earlier analytical work performed by the WVDP both in terms of oxides and radionuclides (Sr-90, Cs-137). ICP-AES analytical interference problems were addressed by the application of alternative methods such as LA-ICP/MS for the noble metals. Radiochemical analytical sensitivity and interference problems were overcome by techniques as simple as extending the counting time to changes of the detection method (e.g., using radiochemical separation and counting to measure Tc-99 instead of mass spectrometry).

The direct analysis of dried waste sludge by LA-ICP/MS was the most successful technique for the analysis of the noble metals. The thousands of points sampled per analysis appear to mitigate any problems of inhomogeneity of the sample at the 50 μm spot size of the laser beam. The noble metal isotope distribution is consistent with the fission product levels expected for power reactor spent fuel. The noble metals were found in quantities (e. g., Pd - 82 μg/g, Rh - 231 μg/g, and Ru - 1063 μg/g) that were consistent with previous waste analyses. The LA-ICP/MS analysis was able to quantify the amount of each noble metal better than ICP-AES analyses.

Other sample preparation methods for LA-ICP/MS were found to be faulty as applied. The salt fusion sample preparation was unsuccessful due to the extremely hygroscopic nature of the fused material. Lithium borate fusion beads (fused at 905 °C) provided a clean stable medium for doing LA-ICP/MS analysis. The lithium borate fusion bead prepared from the standard glass (PNL-246) gave very consistent results in line with the expected composition. However, in the present case, ruthenium was selectively volatilized during the production of the lithium metaborate fusion beads while Pd and Rh are measured at levels consistent with their expected levels for this kind of waste. It is believed that the ruthenium was oxidized to ruthenium +8 by the nitrate in the waste during the fusion cycle. Note that the PNL-246 glass did not appear to lose any ruthenium when fused into the lithium metaborate glass at 905 °C. This suggested that the presence of some nitrate in the dried waste caused the observed loss of ruthenium.

REFERENCES

[1]Smith, M. R., J. S. Hartman, M. L. Alexander, A. Mendoza, E. H. Hirt, T. L. Stewart, M. A. Hansen, W. R. Park, T. J. Peters, and B. J. Burghard. "Initial Report on the Application of Laser Ablation - Inductively Coupled Plasma Mass Spectrometry for the Analysis of Radioactive Hanford Tank Waste Materials," (PNNL-11449) Pacific Northwest National Laboratory, Richland, Washington, December 1996.

[2]Smith, M. R., J. S. Hartman, M. L. Alexander, A. Mendoza, E. H. Hirt, T. L. Stewart, M. A. Hansen, W. R. Park, T. J. Peters, B. J. Burghard, J. W. Ball, C. T. Narquis, D. M. Thornton, and R. L. Harris. "Laser Ablation-Inductively Coupled Plasma Mass Spectrometry: Analysis of Hanford High-Level Waste Materials," Science and Technology for Disposal of Radioactive Tank Wastes, Edited by W. W. Schultz and N. J. Lombardo, Plenum Press, New York, 1998.

[3]Odoj, R., E. Merz, and R. Wolters. "Effect of Denitration on Ruthenium Volatilization," Scientific Basis for Nuclear Waste Management, Vol. 2, ed. Clyde J. M. Northrup, Jr., Plenum, New York, 1980.

Alternative and Innovative Waste Forms

AQUEOUS BASED POLYMERIC MATERIALS FOR WASTE FORM APPLICATIONS

Liang Liang,[1] Harry Smith,[1] Renee Russell,[1] Gary Smith[1]
and Brian J.J. Zelinski[2]

[1]:Pacific Northwest National Laboratory, 902 Battelle Blvd.,
Richland, WA 99352
[2]:Dept. of Materials Sci. & Eng., University of Arizona, Tucson, AZ 85721

Abstract

Organic based polymers, which are applicable to waste encapsulation generally, require the use of volatile, flammable organic solvents in the initial steps of fabricating the waste form. This work describes the initial investigations of changing over to an aqueous based system. Beginning with aqueous solutions of latex and modified latexes and making a series of chemical modifications to these based solutions, materials with good resistance to water and organic solvents were identified. The more promising materials were then loaded with sodium nitrate and their ability to hold these salts while immersed in de-ionized water was measured. The ability of a material to isolate soluble salt wastes while submerged in water was observed to be a function of the degree and rate of crosslinking of each polymer based material.

Introduction

It is estimated that the Department of Energy (DOE) contains about 200 million kilograms of solid waste in its inventory. These solid wastes may be treatment residues, fly ashes and sludges with high concentration of salts (>15% by weight). The traditional method of cement stabilization of the solid waste is expensive because the basic Portland cement formulation can only load about 15 wt.% of salts. This process inefficiency and its related high disposal cost offsets any benefits. In addition, the salts can affect the setting rate of cements and can react with cement hydration products to form expansive and cement damaging compounds (1).

Several material strategies have been proposed to attempt loading more waste salts into a final waste form. Waste vitrification is one way to produce a favorable waste form. However, if sulfate exists in the waste salts, the lower solubility of sulfate in the glass melt may cause a molten sulfate pahse to

To the extent authorized under the laws of the United States of America, all copyright interests in this publication are the property of The American Ceramic Society. Any duplication, reproduction, or republication of this publication or any part thereof, without the express written consent of The American Ceramic Society or fee paid to the Copyright Clearance Center, is prohibited.

form, which is highly corrosive to the melter refractory materials (1).

Macro-encapsulation technology encapsulates waste salts with polyethylene (PE), which is melted, mixed with waste, and extruded (2). After cooling, the polyethylene forms a low-permeability barrier between the waste and the leaching media. This technology is relatively intolerant to the presence of free liquids and organics, which limits its application to waste treatment.

Crosslinked polysiloxane can be used to encapsulate waste salts. After curing at room temperature, and the waste form can be taken directly to a permitted disposal facility (3). However, polysiloxane has poor mechanical properties and the material is limited to non-aqueous solid materials. The generation of hydrogen gas during the curing process of forming polysiloxane is an additional limitation for this technology.

Magnesium oxide (MgO) and mono-potassium phosphate (KH_2PO_4) react to produce $MgKPO_4$, a hard, insoluble, stable and dense phosphate loaded waste matrix (4). However, it takes more than 15 days for the complete curing of the waste matrix, also a highly porous matrix is formed, which results in a higher leaching rate for the salt containing waste form. Although fly ash can be added to decrease the porosity of the form, phase separation may occur, which decreases the strength of the waste form and its leach resistance.

A waste matrix can also be synthesized by cross linking polyester. The polyester is generated by condensation between an organic alcohol (ROH) and organic acid (ROOH) (5). Then, a reaction between styrene and polyester is initiated by methyl ethyl ketone peroxide to form the crosslinked polyester. One major drawback of this technique is the use of toxic and flammable styrene at concentration as high as 60% in the polyester solution.

A polyceram is a hybrid organic-ceramic material. One version made using a modified polybutadiene and silane as the waste matrix produced a material with strong, tough, and water-resistant characteristics (6). Compressive strength and leachability tests confirm that polyceram-based salt waste forms have the potential for disposing of salt-containing mixed wastes. However, the hazardous and flammable organic solvent tetrahydrofuran is used in this process to dissolve the polybutadiene and TEOS so they could be reacted together.

Eliminating the need to use a toxic and flammable organic solvent during the polyceram fabricating process is very important in making these materials applicable to a wide variety of waste problems. A potential alternate processing route would use organic reactants emulsified in water as the starting materials. As a preliminary step towards developing a water-based procedure for fabricating polycerams, this work focused on developing polymer composites using aqueous composites. These polymer composites may well be useful for encapsulating salt waste in their own right. In latex emulsions, the polymer droplets (or micelles) are stabilized by a surfactant (7). Polymers can be emulsified in aqueous solutions by choosing a suitable surfactant. Epoxy resin, for instance, has excellent mechanical and chemical resistance properties (8) and the simple method used to cure epoxy resin makes it possible to combine it with a latex polymer. The addition of the latex component increases the ability of the composite to resist chemical and water corrosion. By combining the polymer latex and epoxy resin emulsions, a new type of waste form precursor is generated, which is water based. This paper reports a method to synthesize composite waste forms polybutadiene-styrene latex and epoxy resin emulsions, and polyacrylic acid latex and epoxy resin emulsions and the basic properties of salt loaded composites.

Experimental

Starting Materials - The following materials were used in this work: Polystyrene-butadiene latex (BSAF, Styronal ND 656, PSB), acrylic latex (Paranol T-6300, Para Chem., Ac), epoxy resin (R-100, Composite Material Inc., Ep), N, N, N',N'-tetramethylenediamine (TMEDA, 99%, Aldrich), diethylenetriamine (DETA, 99%, Aldrich), benzoyl peroxider (BP, 99%, Aldrich), acetone (99%, Aldrich) and dodecylsulfonic acid sodium salt (DSAS, 98%, Aldrich) used without purification. De-ionized water with conductivity of 18 S/cm was used in the experiments.

Preparation of Composites – A typical procedure to prepare a composite from polystyrene-butadiene latex (or polyacrylic acid latex) and epoxy resin can be described as follows: 1.25 g of epoxy resin was mixed with 1 g of DSAS at room temperature by fast stirring (30 min). The mixing solution was added drop wise into 10g of polystyrene-butadiene latex at stirring for 1 hour. Then, 0.1 g of TMEDA (or DETA) as catalyst of epoxy resin was added in solution at room temperature. BP (0.01 g) as the crosslinking agent for polybutadiene was dissolved in 0.2 ml of acetone first, then added into the above solution. The final solution was cured in an oven at 80°C overnight. The salt, $NaNO_3$ was loaded into the solution before the solution was transferred to oven.

Determination of the Water Uptake (swelling) of the Composites - The swelling of the composite in water is measured by immersing 0.5 g of the composite in water for a different lengths of time. After the sample weight stabilized, the composite was moved from water and the free water on the surface of the sample was wiped off by tissue paper. The swelling ratio was calculated from following Equation.

$$S = \frac{W_{wet} - W_{dry}}{W_{dry}} \quad (1)$$

where W_{wet} (after exposure) and W_{dry} (initial) are the weight of the sample after and before exposure to water.

Determination of the Soluble Polymer - The soluble polymer in the composite was extracted with tetrahydrofuran using Sohext extractor over a time period for 24 hours. The soluble polymer in the composite, P can be calculated from the weight of the composite before and after extraction.

$$P(\%) = \frac{W_1 - W_0}{W_1} 100 \quad (2)$$

where W_1 and W_0 are the weight of the composite before and after extraction.

Determination of Released Salt - The conductivity of water containing $NaNO_3$ was determined using a conductivity meter (JENCO 1671) and a standard curve was plotted of the solution conductivity against the amount of $NaNO_3$ dissolved in the solution. The leached $NaNO_3$ from the composites can be calculated by the change of conductivity of the water. Stirring was kept during the measurement and the test temperature is room temperature.

Results and Discussion

1. Preparation of composites

Composites were fabricated from polymer latex emulsions and epoxy resin emulsions to produce composites with enhanced mechanical properties and chemical durability. Commercially available epoxy emulsion proved to be ineffective. In-house techniques for emulsifying epoxy resin, using surfactants produced stable emulsions containing as much as 50% epoxy. These emulsion were then mixed with the polystyrene-butadiene or polyacrylic acid latex emulsions. The flow sheet used to fabricate the composite polymer waste form loaded with salt is shown in Figure1. The epoxy resin was emulsified using DSAS as a surfactant. Although final emulsions appear to be stable at room temperature, the settling of epoxy resin can be observed after one month. It is best to cross link the epoxy matrix with the polymer latex as soon as possible after it is generated because the emulsion is thermodynamically unstable. Sodium nitrate (salt waste surrogate) is loaded into the emulsion along with the proper crosslinking agents to cause both the epoxy resin and polymer latex in the emulsion to form a crosslinked material. After curing at 80° C, a tough and hard material was obtained. The high degree of crosslinking prevents damage to the composite matrix by subsequent attack from organic solvents and water.

2. Mechanisms producing crosslinked composites

It is known that the crosslinking of epoxy resin can be carried out either through the epoxy groups or hydroxyl groups (9). Two types of curing agent, N, N-tetramethylenediamine and diethylenetriamine were used in this study. The former is a tertiary amine that worked as a catalyst and the latter is a polyfunctional amine that worked as a crosslinking agent. In the catalytic system, the epoxy ring is readily attacked by tetramethylenediamine, (Scheme1).

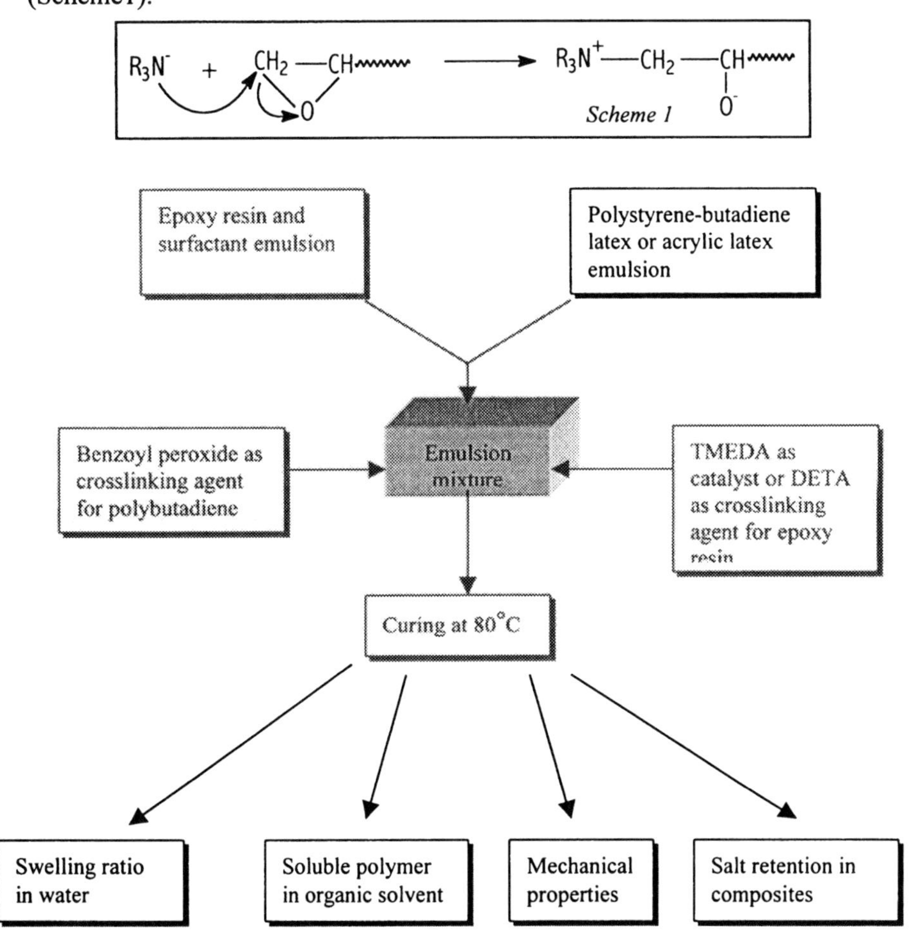

Figure 1. Flow sheet to prepare salt loaded composites from polystyrene-butadiene latex (or acrylic acid latex) and epoxy resin.

Since this reaction can occur at both ends of TMEDA, a three dimension network will built up (Scheme 2). The crosslinking by tetramethylenediamine is described by Scheme 3. It is known that the crosslinking by a polyfunctional amine is much faster than that by a tertiary amine. The faster crosslinking reduces phase separation when the emulsions are mixed with waste salt. Homogeneous composites made by faster crosslinking exhibited better mechanical properties and a higher salt retention.

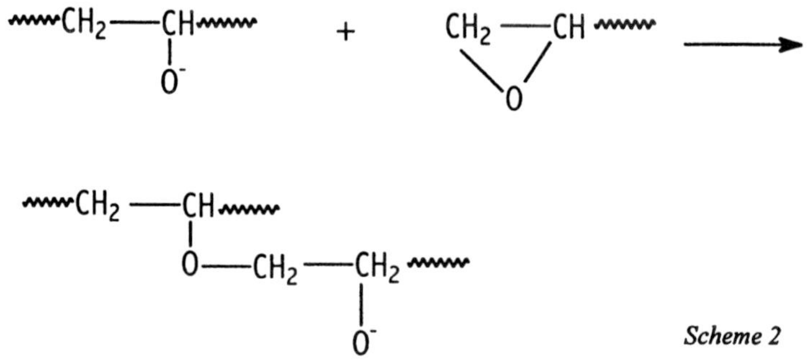

$$\sim CH_2-CH(O^-)\sim \; + \; CH_2-CH\sim \text{ (epoxide, O)} \longrightarrow \sim CH_2-CH(-O-CH_2-CH_2(O^-)\sim)\sim$$

Scheme 2

$$\sim CH(O)CH_2 \; + \; H-N(R)-H \; + \; CH_2(O)CH\sim \longrightarrow \sim CH(OH)-CH_2-N(R)-CH_2-CH(OH)\sim$$

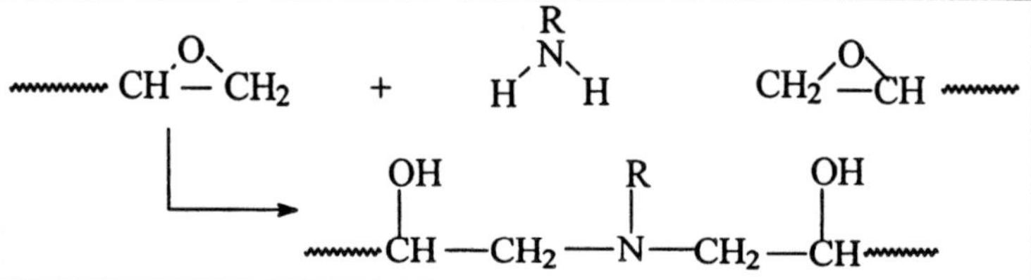

Scheme 3

When polyacrylic acid was used in the polymer-epoxy emulsion, the epoxy ring can react with carboxyl groups on the polyacrylic acid chains (Scheme 4). The reaction will also result in the crosslinking between epoxy resin and polyacrylic acid.

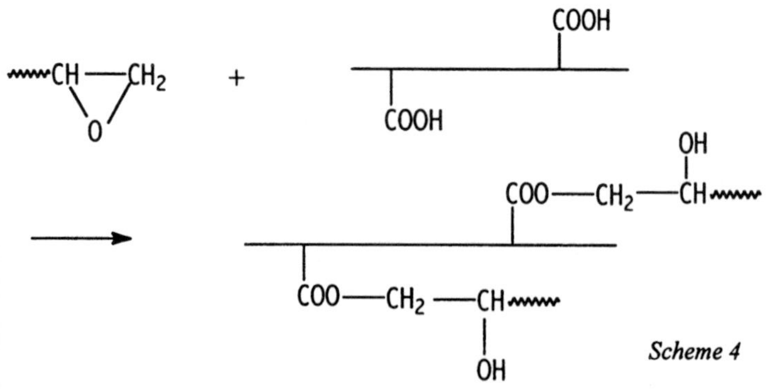

$$\sim CH-CH_2 \text{ (epoxide, O)} \; + \; \text{polyacrylic acid (COOH, COOH)} \longrightarrow \text{chain with } COO-CH_2-CH(OH)\sim \text{ and } COO-CH_2-CH(OH)\sim$$

Scheme 4

On the other hand, polystyrene-butadiene latex can be crosslinked by benzoyl peroxide (BP) initiator [9]. The benzoyl radicals formed by the initial bond rupture may decompose by the reaction shown in scheme 5. The final free radicals then attack the double bonds on the polybutadiene chains to generate the crosslinking materials. It has been found that polystyrene-butadiene cannot be dissolved by toluene after introducing BP, which confirms the presence of crosslinked materials.

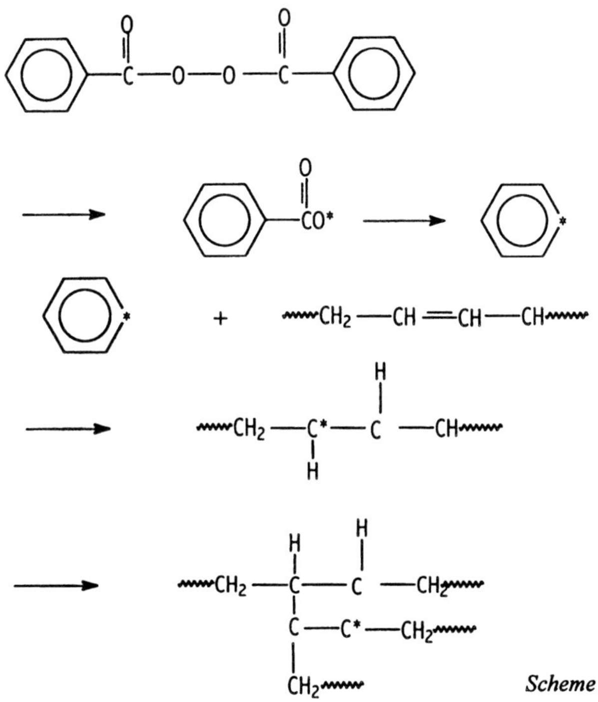

Scheme 5

3. Water resistance and chemical durability of composites

Table 1 lists the weight change of composites after solvent extraction by tetrahydrofuran. As indicated by the results from Table 1, sample PSB-Ep-50 is resistant to solvent extraction by THF. These results indicate that all of the polymer composites are cross-linked into the network in this sample. Because of incomplete crosslinking, soluble polymer in other samples is extracted by the solvent as indicated by appreciable weight loss.

Figure 2 shows the water uptake (swelling ratio) of the composites in water. For all samples the swelling ratio is lower than 0.1 water (g)/polymer (g) i.e., only about 9 wt.% water was absorbed by the composite. Even through polyacrylic acid itself has hydrophilic properties, the cross linking along the polyacrylic acid chain effectively reduces the amount of carboxyl groups and results in the low water uptake (swelling ratio) of the composite polymer material.

Table 1. Soluble polymer in composites

Sample	PSB (wt. %)	Ac (wt. %)	Ep (wt. %)	Extracted (wt.%)
PSB-Ep-50	50	-	50	~ 0
PSB-Ep-80	80	-	20	41.8
Ac-Ep-50	-	50	50	14.8
Ac-Ep-80	-	80	20	13.3

Tetrahydrofuran as solvent; Extraction temperature: 66°C; Extraction time: 24 hours.

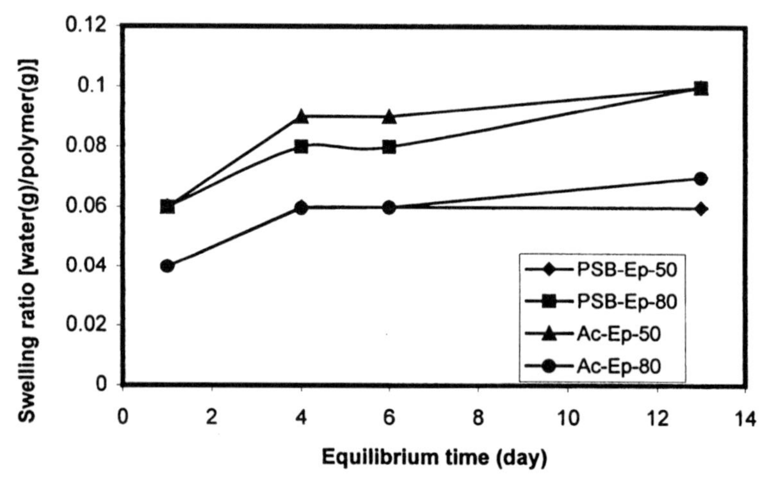

Figure 2. Swelling ratio of different composites in water.

4. Composites after loading with salt

Figure 3 shows the composite made from a polystyrene-butadiene and epoxy resin emulsion loaded with 20% $NaNO_3$. The maximum loading with $NaNO_3$ in such a composite is about 60%, for which mechanically sturdy composite can still be produced. The most highly loaded composites do not display good leach resistance probably because of the inhomogeneous nature of the cured materials. Use of DETA, a faster cross-linking agent, results in improved homogeneity and a significant reduction in the leaching rate. Figure 4 shows the quantity of salt retained in composites immersed in water as a function of time. Salt retention was calculated from the conductivity of the immersion solution. The samples used for the test contained 20% $NaNO_3$ and were the same geometry as those used for the water resistance test. The data indicate that using a faster crosslinking agent (i.e., APTS) reduces the initial leaching rate by a factor for about 50. It is speculated that increasing the crosslinking rate resistant a tendency towards phase separation during the curing of sample containing high concentration of salt.

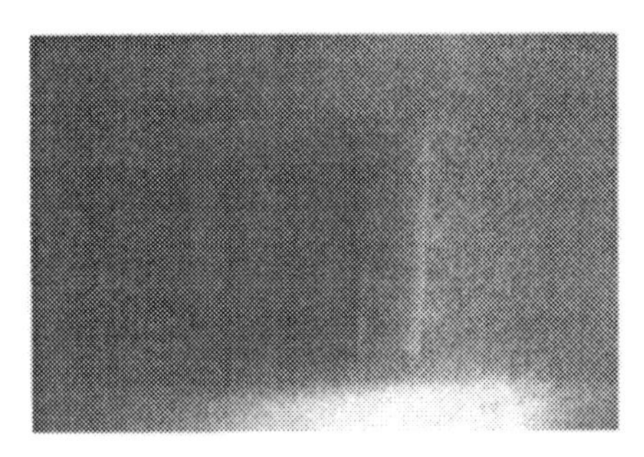

Figure 3. Composite from PSB-Ep (PSB/Ep = 50:50) loaded with 20 wt.% of $NaNO_3$.

Conclusions

Composites of PBS or Ac latex polymers modified by epoxy resin can be fabricated from emulsions of these polymers components. These physically tough composites form monolithic bodies when loaded with waste salt (up to 60% $NaNO_3$). Good mechanical property and high resistance to water and organic solvent were observed for those composites. It appears that the low leach resistance of these waste forms is due to the formation of open porosity in the waste form during the curing step. Acceleration of the curing reaction of the composites significantly enhances the leach resistance of the salt waste matrix. These results represent an important step in developing aqueous-based polymer or polyceram waste matrices. The water-borne polymer emulsions show promise for use as a simple, low cost, environmental benign precursor material for fabricating highly loaded salt waste form.

Acknowledgements

This research work was supported by a grant provided by Environmental Technology Division at Pacific Northwest National Laboratory. Pacific Northwest National Laboratory is operated for the US Department of Energy by Battelle under Contract no. DE-AC06-76RLO 1830.

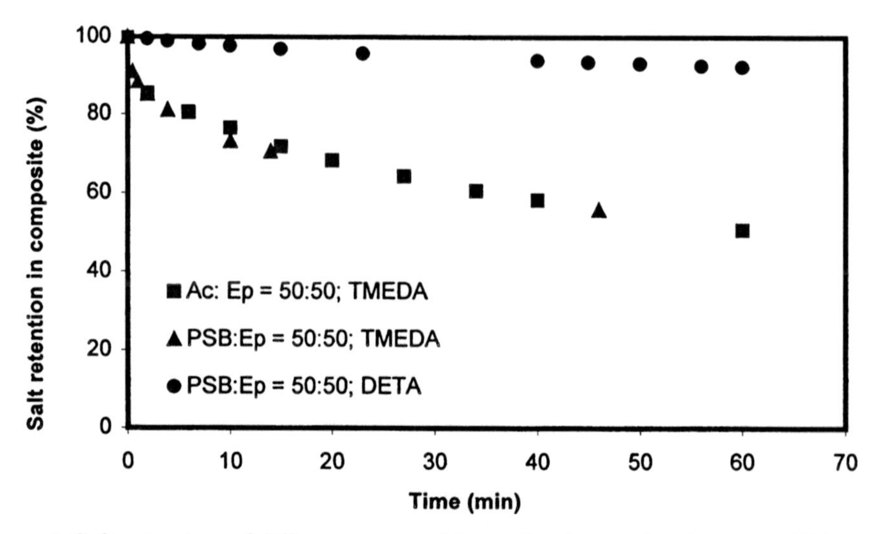

Figure 4. Salt retention of different compositions after immersion in water (All samples were loaded 20 % of NaNO3).

References

1. A. S. Wagh and D. Singh, Stabilize high salt content waste using sol-gel process, Summary report, DOE/EM-0473, 1999.
2. T. E. Williams and P. Owca, Polyethylene macroencapsulation, DOE/OST reference #30, 1998.
3. G.G. Loomis, S. Prewett, and C. Miller, Stabilize high salt content waste using polysiloxane stabilization, DOE/EM – 0474, 1999.
4. R. D. Spence, Stabilization of high salt waste using a cementitious process, DOE/EM-0500.
5. R. K. Biyani, W. Owca and V. Maio, Mixed waste encapsulation in polyester resins, DOE/OST-1658, 1999.
6. Stabilization of salt wastes using polybutadiene-based polycerams, G.L. Smith and B.J.J. Zelinski, DOE/EM, 1999.
7. Q. Wang, S. Fu, and T. Yu, Prog. Polym. Sci., 19 (1994) 703.
8. M.L. Berins, Plastic Engineering Handbook, Chapter 3, Charpman & Hall, New York, 1991.
9. J.A. Brydson, Plastics Materials, Chapter 26, Butterworth-Heinemann Ltd, Oxford, 1995.
10. G. Odian, Principles of Polymerization, Chapter 3, John Wiley & Sons, New York, 1991.

PERFORMANCE OF HYDROCERAMIC CONCRETES ON RADWASTE LEACH TESTS

Darryl D. Siemer
Bechtel BWXT Idaho, LLC
P. O. Box 1625
Idaho Falls, ID 83415-7111

ABSTRACT

Hydroceramics (HC) are geopolymeric-type concretes made by curing modeling clay-like mixtures of calcined waste, calcined clay, vermiculite, Na_2S, NaOH, plus water under hydrothermal conditions. The resulting "rocks" are comprised of feldspathoid (alkali aluminosilicate) mineral assemblages rather similar to the zeolitized tuff indigenous to the USA's current proposed HLW repository site[1]. This paper compares the leach test performance of these concretes with those of a variety of radwaste-type glasses and points out the advantages of this approach to preparing defense-type reprocessing waste for disposal.

INTRODUCTION

Several independent reviews of the US federal government's radioactive waste management programs have concluded that a cementitious solidification technology would probably be more appropriate than vitrification for much of its nominally "high level *" (1-100 watt/m^3) reprocessing waste[2-3]. The latest such opinion[4] points out that the Idaho National Engineering and Environmental Laboratory's (INEEL's) calcined waste is especially well suited for such an alternative. The main reason for this is that it contains no organic materials and a lesser proportion of intrinsically-difficult-to-fix salts than does that stored at DOE's other major nuclear fuel reprocessing facilities. Hydroceramic concretes

* "Historic" reprocessing wastes exhibiting up to 500 watt/m^3 worth of radioactivity and 27,000 nCi/g of TRU have recently been rendered road-ready by grouting facilities at Sellafield. Great Britain reserves vitrification for a low-volume, homogeneous, waste stream more than two orders of magnitude "hotter" than is US defense-type HLW.

To the extent authorized under the laws of the United States of America, all copyright interests in this publication are the property of The American Ceramic Society. Any duplication, reproduction, or republication of this publication or any part thereof, without the express written consent of The American Ceramic Society or fee paid to the Copyright Clearance Center, is prohibited.

differ from conventional radwaste "grouts" in that they are comprised of alkali aluminosilicate minerals rather than hydrated calcium silicates (C-S-H in cement-chemistry shorthand). Such minerals (cancrinites, sodalites, and some zeolites) form "cages" around individual salt molecules when salt-containing caustic solutions are reacted with clays under hydrothermal conditions[5]. The process would implement the chemistry of Hanford's "Clay Reaction" process[6] via ORNL's FUETAP autoclaved-concrete technology[7] to produce a canistered monolith. The resulting concrete chemically fixes not only the traces of "easy" waste constituents (polyvalent cations - most fission products & RCRA metals plus all TRU) that most grouts perform adequately on but also the far larger fraction that they "immobilize" only by serving as a gross physical encapsulant.

Since many decision makers and some scientists consider the relatively poor leach resistance of conventional grouts to be the most compelling argument for vitrification, this paper will be devoted to demonstrating that it is indeed possible to make concrete waste form materials that perform well on radwaste leach tests.

EXPERIMENTAL

Two previously published papers describe how the specimens discussed in this paper were made[8-9]. To recapitulate, a stiff (modeling clay-like) grout is made from a mixture of crude metakaolin ("Troy Clay", *ASHGROVE CEMENT Co.*), about 10% as much powdered raw vermiculite (*W. R. Grace*, fine grade), 20-40 wt% of a waste simulant consisting either of a mix of salts or pilot plant INEEL calcine, ~1 wt% Na_2S, plus a concentrated (25-40 wt%) aqueous solution of NaOH. The proportions of alkali metals, aluminum, silicon and "heteroanions"* in the formulations are adjusted to approximate the composition of sodalite; i.e., atom-wise molar ratios are: $(Na+K+Cs)_a:Al_b:Si_c:X_d$ where $b \geq a$, $c \geq a$, & $d < 0.25a$. The raw "dough" is usually rolled into a ball, wrapped with tin foil or put into a Teflon® bottle† and then autoclaved for an hour or so (typically at ~ 200°C).

LEACH TEST RESULTS

There are four reasons why conventional grouts usually don't perform as well as glasses on most radwaste leach tests. The first is that vitrification is usually reserved for a "volume reduced" fraction from which the readily soluble materials have been removed via "sludge washing". Since these washings end up

---

* "heteroanions" = X= all anions other than oxide, hydroxide, silicate, and aluminate. Fluoride present as fluorite (CaF_2), a major constituent of INEEL calcines, is essentially inert and therefore doesn't count. Calcination enhances waste loading because it reduces the heteroanion content of the waste.

† The grout-balls are wrapped or bottled to prevent preleaching by liquid water during the curing step.

in the grout, it has to deal with the toughest-to-immobilize (most soluble) fractions. Second, conventional grouts* are not well suited for that job because they do not form water-insoluble minerals with the predominant salts in DOE reprocessing waste. Third, radwaste leach protocols (short term, small size, etc.) obviate a key advantage of cementitious solidification, i.e., that it would be easy/safe/cheap to make *large* waste forms. Because all inorganic concretes are intrinsically porous, specimens expose much more surface area to the leachant than do equal sized pieces of glass†. This especially enhances *initial* leach rates – usually the only characteristic measured. The fourth reason is that decision-makers apparently do not consider the leach resistance of "low" waste forms to be very important. This is reflected by the promulgation of "easy" waste acceptance criteria (WACs)‡, over-emphasis of waste loading, and refusal to implement pretreatment technologies which would produce better-quality grouts§.

MCC-1: The MCC-1 (presently embodied as ASTM C1220-98[10]) was the primary leach test when the final decision was made about how to deal with the Savannah River Site's reprocessing waste (circa 1981). It exposes a small monolith (a 1 cm^3 cube would be typical) of known composition to a relatively large volume ($A_{specimen}/V_{water} = 0.1 cm^{-1}$) of hot (90°C) distilled water for 28 days. The fractions of its various constituents in the leachate are then used to derive normalized leach rates for each in units of $gram/m^2/day$. Table I compares MCC-1 leach performance of two HCs, several radwaste-type glasses, and a hot-isostatically-pressed (HIPed) ceramic ("ANLW cer")[11]. The HCs contained ~30 wt% of two representative INEEL pilot-plant calcines along with "Troy clay", powdered vermiculite (raw, not the expanded form), ~0.5 wt% Na_2S, plus NaOH. Please note that the leach rates of those elements responsible for most of the radioactivity of DOE-type reprocessing waste (Cs & Sr) are considerably lower from the HCs than from the vitrified materials. Of special relevance is the fact that the MCC-1 leachability of Cs from HCs is about 2 orders of magnitude better than it was from the original FUETAP concretes[7].

---

* Grouts based upon magnesium phosphate (e.g., Ceramicrete™) and calcium aluminate cements behave similarly (or worse) than do C-S-H forming systems when applied to salt-wastes.

† Hydroceramic concretes crushed to the extent specified by the PCT protocol (75-149 micron screen size) exhibit BET surface areas on the order of 15 $m^2/gram$. Perfectly impervious particles with the same size and bulk density (~1.8 g/cm^3), would possess a surface area of ~0.033 m^2/g.

‡ Today's more stringent grout WACs specify a ANS/ANSI 16.1 leach index of ≥6.0 plus the ability to meet UTS criteria on the TCLP. In other instances, the sole requirement is that the material pass a "paint filter test" – in other words, not exhibit too much standing water.

§ DOE is shutting down its existing calcination and incineration facilities, not building new ones. While virtually every conceivable way of implementing "thermal treatment" has been successfully demonstrated at one time or another in the DOE complex, only a small fraction of the waste which would benefit from it has been so-treated.

Table I: A comparison of 28-day MCC-1 leach rates ($g/m^2/day$) of two HCs, four radwaste-type glasses and a hot-isostatically-pressed glass ceramic[a]

Material\Component	Na	Cs	Sr	Al	Si	Ca
EA GLASS	1.4	1.6	0.20	0.57	1.1	0.23
SRL-131	1.1	0.89	0.14	0.50	0.86	0.0034
WV-39-2	4.7	5	2.0	3.85	3.8	4.1
JSSA	0.39	0.1	0.27	0.18	0.31	0.29
ANLW (glass ceramic)	0.39	0.39	0.14	0.57	0.14	0.075
HC1	13	0.096	0.012	1.6	0.54	0.0014
HC2	15	0.07	0.024	0.72	0.75	0.013
[a] *Glass & glass ceramic leach data taken from reference 11*						

PCT: Most of today's WACs specify the Product Consistency Test (PCT)[12]. For DOE's high-level glasses, these WACs decree that fractional loss of bulk constituents (sodium, boron, lithium, and silicon) must not exceed those from a benchmark, Environmental Assessment (EA) glass. While the WACs for DOE's proposed low-level glasses are not yet finalized, a typical proposal would limit PCT leachability of sodium to under $1 g/m^2/day$*. Because the PCT exposes 200-300 times more of the material's surface area to a given volume of hot (90°C) water than does the MCC-1, it generates an estimate of its gross solubility under conditions that encourage back reaction/saturation. [This is a much more realistic "repository failure" scenario† than that assumed by the MCC-1 or the ANS/ANSI 16.1 tests.] Table II compares several HCs with "representative"[13] radwaste-type glasses with regards to PCT-leachability of their most readily-solubilized common component, sodium‡. The HCs would have outperformed these glasses by a greater margin for the "easy" constituents of defense-type reprocessing wastes (e.g., ^{90}Sr and TRU).

---

* $1 g/m^2/day$ corresponds to a fractional PCT sodium dissolution rate of ~2% per day – about the same as that exhibited by EA glass.

† Most of the "failure" scenarios modeled in DOE's performance assessments assume that liquid water somehow manages to reach the waste forms - the system can't fail otherwise. These scenarios vary primarily in how and when this contact occurs.

‡ To make this test both more relevant (calcine-containing HCs are inhomogeneous at this scale) and "tougher", after the HC has been ground to a fine powder, *everything* that will pass through a 100 mesh (150 micron) screen is leached – not just a prewashed 75-150 μm size fraction.

Table II: Comparison of HCs and glasses on the PCT

MATERIAL	% Na_2O	mg/l Na in leachate	% Na dissolved
EA GLASS	16.9	1720	13.7
PUREX GLASS	12.1	941	10.4
SRL-131	12.9	931	9.7
HC#1 $NaAlO_2$/NaOH/TROY clay	16.7	718	5.8
HC#2 NaOH, $NaNO_3$ (25% of Na)/TROY clay	12.6	513	5.5 (2.6% of the NO_3^- had also leached)
HC#3 38% alumina calcine/NaOH/DEA/TROY clay	13.1	554	5.7
HC#4 46% zirconia calcine/NaOH/TROY clay	12.4	558	6.1
HC#5 30% sugar-calcined SBW/TROY clay[a]	12.6	925	9.9
HC#6 NaOH/ Englehard Metakaolinite, 9-hr cure @ 200 °C	16.3	229	1.9 (ANSI 16.1 LI_{Na}=11.6

[a] This HC violated the "sodalite composition" rule of thumb - too much carbonate

ANS/ANSI 16.1: Table III lists results of an ANS/ANSI 16.1 leach test[14] performed on one of the calcine-containing HCs in Table II. This test measures the mobility* of individual components *within* a monolithic specimen immersed in fresh water†. Back reactions/saturation are prevented by replacing the water at regular intervals. Note both that the most readily soluble bulk constituents of US radwaste (sodium and nitrate) evinced diffusivities ~four orders of magnitude lower (better) than the usual grout WAC (10^{-6} cm^2/sec) and that those of "easy" components (Zr, Sr, etc.) were several orders of magnitude lower still. Note in

---

* The mobilities of individual constituents are expressed in terms of their bulk diffusion constants, D in units of cm^2/s. A component's "leach index" = $-\Sigma \log_{10} D)/n$ where n is the number of leach intervals.

† ANS/ANSI 16.1 results are immediately useful for predicting the consequences of dropping a "naked" waste form (no canister) directly into a river.

particular that cesium's diffusivity was over seven orders of magnitude lower (better) than the usual standard. The substantial drop in nitrate mobility after the first leach interval indicates that ~90% of it had been microencapsulated by the "cage minerals" – the rest leached as it would from a conventional grout.

Table III: ANS/ANSI 16.1 Leach Performance*

Interval (hrs)	Sodium		Cesium		Zirconium		Strontium		Chromium		Nitrate	
	ppm	-logD	ppb	-logD	ppb	-logD	ppb	-logD	ppm	-logD	ppm	-logD
2.83	32	9.55	<2	>13.8	33	15.3	3	14.1	0.28	10.6	23	8.3
5.7	20	9.7	<2	>13.5	68	14.4	2	14.1	0.21	10.6	5	9.3
15.3	28	9.8	<2	>13.9	140	14.1	7	13.5	0.076	11.8	2.3	10.4
19.5	13	10.3	<2	>13.7	<10	>16.3	1	15.0	0.05	11.1	0.9	11.1
22	22	9.8	<2	>13.9	100	14.2	6	13.4	0.03	12.4	1	10.9
35.8	21	10.1	<2	>13.5	<10	>16.4	1	15.2	0.02	13.0	1	11.1
25.5	15	9.9	<2	>13.7	<10	>16.0	3	14.2	0.02	12.6	0.4	11.5
36	14	10.2	<2	>13.7	<10	>16.2	1	14.9	0.01	13.4	0.9	11.0
LI	9.9		>13.7		>15.4		14.3		12.0		10.4	
Total % Leached	8.26		<0.0099		<0.0025		0.015		1.2		10.5	
< > figures are based upon the detection capabilities of the analytical instrumentation: ICPAES for all metals except Cs, graphite furnace AAS for Cs, and ion chromatography (IC) for nitrate												

TCLP: The environmental impact of DOE's reprocessing wastes is apt to be dominated by their chemical toxicity, not their radioactivity. One of the most "impactful" components is apt to be nitrate because: 1) with the exception of sodium and water, it is the most plentiful component; 2) it has an infinite "half life"; 3) it is a plant food*; 4) it is not retained by natural soil minerals; and, 5) it is somewhat toxic. The ~10% of INEEL's reprocessing waste which has not already been calcined (denitrated) is called "sodium bearing waste" (SBW). Such wastes (~1-2 M free acid, ~1.5 M Na^+, ~0.5 M Al^{+++}, ~5 M NO_3^-, plus trace levels of many other things†) cannot be efficiently processed by INEEL's existing calcination facility unless a reducing agent (e.g., sugar) is added because $NaNO_3$

* The reason why this is important is that "low" waste forms are apt to end up in shallow graves. Plant roots feeding from them are apt to bring radionuclides to the surface.

† Traces add up: INEEL's ~10^6 gallons of SBW contains about 8 kg of plutonium (SBW has a higher concentration of TRU than did the already-calcined waste) and ~2 tonnes of mercury.

does not thermally decompose at an otherwise appropriate* temperature. Table IV gives the results of a TCLP leach test[15] of a HC specimen made with a sugar-calcined INEEL SBW simulant doped with several RCRA-characteristic metals. The raw simulant was pretreated as follows: After 38 grams of sucrose per mole of nitrate had been dissolved in it, the liquid was added dropwise to a stainless steel beaker seated on a maximum-temperature hotplate. The beaker was then placed into a 500°C muffle furnace for a few minutes to burn out residual elemental carbon. The HC was made with 30 wt% of the resulting calcine, ~1% sodium sulfide, sufficient household lye to provide the free hydroxide† required to reform the clay, plus enough water to make a stiff dough which was then autoclaved for two hours @ ~200°C. Table IV lists regulatory limits along with the concentrations of RCRA metals in both the calcine and the TCLP leachate. Hydroceramic waste forms do not exhibit the "characteristic of toxicity".

TABLE IV: TCLP Results: Sugar-calcined sodium bearing waste specimen[a]

ANALYTE	Found(μg/g)	UTS Limit (μg/g)	Calcine (μg/g)
As	<0.002	5	10.8
Ba	0.35	7.6	48
Cd	0.13	0.14	1372
Cr	0.023	0.86	950
Pb	<0.1	0.37	1500
Se	<0.002	0.16	6.9
Ag	<0.1	0.3	1510

[a]Mercury was not added to the liquid simulant because it would have been lost during calcination.

* "Appropriate" means a temperature high enough to decompose organic materials and *most* nitrate/nitrite salts but low enough to minimize volatilization of toxic metals (Cd), fission products (^{99}Tc and ^{137}Cs), and corrosives (F and Cl). Consequently, radwaste calciners typically operate at ~500°C – about 600 degrees lower than do glass melters. "Sugar calcination" also reduces NO_x emissions because most of the nitrate ends up as N_2.

† Approximately one third of the sodium in the calcine itself was present as the aluminate - most of the rest was Na_2CO_3. A better-quality calcine (and HC) would have been produced if some clay had been added along with the sugar - less carbonate (a "heteroanion") would have been retained.

VHT: Because its developers made the most reasonable assumption about how water will interact with waste forms in a vadose*-zone repository (e.g., DOE's Yucca Mountain and "Greater Confinement Disposal" facilities[16]), today's most realistic glass durability ("weathering") test is the Vapor Hydration Test (VHT)[17]. It is also most realistic in that it quickly demonstrates the ultimate fate of glasses in such environments; i.e., more or less quantitative transformation to "alteration products". The VHT is performed by suspending a wafer of the sample above the bottom of a stainless steel vessel with a platinum wire†. After the addition of sufficient water to saturate the gas phase at the test temperature (typically 200°C), the vessel is hermetically sealed and put into a preheated oven for a time typically ranging from several days to several weeks. The thin film of liquid water which forms on the surface of the specimen is maintained by condensation of steam. In practice, the corrosion rate of the glass is low until the concentrations of ions leaching into that film reach a point that initiates the precipitation of stable (under those conditions) minerals, primarily felspathoids (analcime, sodalite, cancrinites, etc.), zeolites, and silicates. Because hydroxide ion tends to remain in solution, the overall corrosion (alteration) rate accelerates to a level determined by its ability to release the rate-limiting component of the product-forming reactions (usually aluminate‡) from the glass. After the test is over, that rate is determined by measuring the thickness of the remaining (unaltered) glass layer with a microscope.

Since VHT "leach" (alteration) conditions are essentially identical to those used to cure HCs, a properly made example doesn't change much when so-tested. Furthermore, since the application of successively more rigorous hydrothermal curing conditions tends to improve their leach performance[8], any changes that do occur are apt to be for the better. Since HC-type concretes are manufactured under conditions generally considered to represent a "worst case" repository scenario (hydrothermal), it is reasonable to conclude that they would prove to be more rugged than glasses if those conditions were to actually be realized. A fundamental difference between the formulations of HCs and radwaste-type

* Vadose" doesn't mean dry. Subterranean air is almost always saturated with water vapor regardless of whether or not individual soil particles are surrounded with a continuous film of liquid water. An essentially infinite reservoir of warm, wet, air will provide near-ideal alteration conditions for glasses at Yucca Mountain.

† PNNL's Procedure No. GDL-VHT, Rev. 1, is a "proceduralized" version of the VHT.

‡ This is the reason why some radwaste management experts apparently believe that neither the glass waste form itself nor its immediate surroundings at the repository should contain "available" forms of aluminum; e.g., soils, clays, grouts, etc.. Protecting glass from dirt in a subterranean repository will prove to be an expensive proposition.

glasses is that the latter contain insufficient aluminum* to form insoluble minerals with 100% of the alkali metals in them. This means that a natural corrosion process cannot convert these glasses to mineral assemblages comprised entirely of low solubility minerals.

CONCLUSIONS

HC concretes match the leach-test performance of radwaste glasses because their intrinsically lower water solubility compensates for their greater porosity and microscopic surface area. A key advantage of HC-type concrete over glass as a disposal form is that waste forms could be easily/safely/cheaply stabilized in situ by surrounding them with any desired amount of a pumpable barrier-buffer grout (a.k.a. "backfill" or "overpack"). The process itself is attractive relative to vitrification because it would be both cheaper and safer to implement and because it would create much less "incidental" waste. Reasons why it is especially well suited for INEEL include: 1) that site has already calcined 90% of its waste (and could quickly calcine the rest); 2) it recently produced all of the caustic "activator" needed to so-render its calcines road-ready by reacting ~175,000 gallons of "waste" metallic sodium reactor coolant with water; 3) it's situated close to a cement plant that could produce/provide calcined "Troy Clay" at low cost, and, 4) its decision-makers have made no tangible commitment to any other solidification technology.

ACKNOWLEDGMENT

The author would like to thank Professors Della Roy, Barry Scheetz, and Michael Grutzeck, of the Pennsylvania State University for their support of the research outlined by this paper. Without their help, the hydroceramic alternative could not have been developed.

REFERENCES

[1]J. Winograd, "Radioactive Waste Disposal in Thick Unsaturated Zones", *Science,* 212, pp. 1457-1464 (1981).

[2]Solidification of High-Level Radioactive Wastes", doc. NUREG/CR-895, prepared by the National Research Council (NRC) for the U. S. Nuclear Regulatory Agency, 1979, pp. 116-118.

[3]Jimmy Bell, "Alternatives to HLW Vitrification: The Need for Common Sense", *Nuclear Technology*, Vol 130, April 2000, 89-97.

---

* A primary goal of sludge washing is to remove aluminum – the rationale for doing so is that it simultaneously reduces the size of the "high" fraction and makes it easier to melt.

[4]Letter from Michael Corradine, Chair NRC Board of Radioactive Waste Management, to Carolyn Huntoon, Undersecretary DOE-EM, 2Nov00: see http://emsp.em.doe.gov/pdfs/nrc_hlw.pdf

[5]R. M. Barrer, "Zeolites and Clay Minerals as Sorbants and Molecular Sieves", Academic Press, NY 1978, pp. 30-31.

[6]G. S. Barney, "Fixation of Radioactive Wastes by Hydrothermal Reactions with Clays", *Advances in Chemistry Series #153*, American Chemical Society, Wash. D. C., pp. 108-125, 1976.

[7]Dole et. al., "Cement -Based Radioactive Hosts Formed Under Elevated Temperatures and Pressures (FUETAP Concretes) for Savannah River Plant High-Level Defense Waste", ORNL/TM-8579, March 1983 (ORNL's last and most comprehensive report).

[8]D. D. Siemer, Johnson Olanrewaju, Barry Scheetz, N. Krishnamurthy, and Michael Grutzeck, "Development of Hydroceramic Waste Forms," in Proceedings of the International Symposium on "Environmental Issues and Waste Management Technologies in the Ceramic and Nuclear Industries VI," ed. by Dane R. Spearing, Gary L. Smith, and Robert L. Putnam, Ceramic Transactions, Vol.119, pp. 383-390, The American Ceramic Society, Westerville, Ohio, 2001.

[9]D.D. Siemer, Johnson Olanrewaju, Barry Scheetz, and Michael Grutzeck, "Development of Hydroceramic Waste Forms for *INEEL* Calcined Waste," in Proceedings of the International Symposium on "Environmental Issues and Waste Management Technologies in the Ceramic and Nuclear Industries VI," ed. by Dane R. Spearing, Gary L. Smith, and Robert L. Putnam, Ceramic Transactions, Vol.119, pp. 391-398, The American Ceramic Society, Westerville, Ohio, 2001.

[10]American Society for Testing and Materials, ASTM C1220-98 "*Standard Test Method for Static Leaching of Monolithic Waste Forms for Disposal of Radioactive Waste*," West Conshohoken, Pennsylvania.

[11]R. W. Benedict & H. F. McFarlane, "EBR-II Spent Fuel Treatment Demonstration Project Status", *Radwaste Magazine*, July 1998, pp. 23-28.

[12]C. M. Jantzen, et. al., "Characterization of the Defense Waste Processing Facility (DWPF) Environmental Assessment (EA) Glass Standard Reference Material", doc. WSRC-TR-92-346, Rev 1, June 1, 1993.

[13] Shi-Ben Xing and Ian L. Pegg, "Effects of Container Materials on PCT Leach Test Results for High-Level Nuclear Waste Glasses", *Mat. Res. Soc. Symp. Proc. Vol. 333*, pp. 557-564, 1994 (Scientific Basis for Nuclear Waste Management XVII)

[14]American National Standards Institute, Inc., American Nuclear Society, ANSI/ANS - 16.1 - 1986, "*American National Standard Measurement of the Leachability of Solidified Low-Level Radioactive Wastes by a Short-Term Test Procedure*," La Grange Park, Illinois.

[15]Toxicity Characteristic Leaching Procedure (TCLP), 1992, SW-846, Method 1311, Rev. 2 in *Test Methods for Evaluating Solid Waste*, Volume 1C: Laboratory Manual Physical/Chemical Methods, U.S. Environmental Protection Agency, Office of Solid Waste and Emergency Response, Washington, D.C..

[16]E. J. Bonano, M. S. Y. Chu, S. H. Conrad (SANDIA) and P. T. Dickmann (DOE-NEV), "The Disposal of Orphan Wastes Using the Greater Confinement Disposal Facility", Waste Management '91, Vol. 1, Post & Wacker Eds., pp. 861-868.

[17]W. L. Ebert, K. K. Bates, and W. L. Bourcier, "The Hydration of Borosilicate Glass in Liquid Water and Steam at 200°C", *Waste Management,* 11, pp. 205-221 (1991).

CERIUM AS A SURROGATE IN THE PLUTONIUM IMMOBILIZED FORM

James C. Marra, Alex D. Cozzi, R. A. Pierce, John M. Pareizs, Arthur R. Jurgensen and David M. Missimer
Westinghouse Savannah River Company
Aiken, SC 29808

ABSTRACT

The Department of Energy (DOE) plans to immobilize a portion of the excess weapons useable plutonium in a ceramic form for final geologic disposal. The proposed immobilization form is a titanate based ceramic consisting primarily of a pyrochlore phase with lesser amounts of brannerite, rutile, zirconolite, vitreous phases and/or other minor phases depending on the impurities present in the feed. The ceramic formulation is cold-pressed and then densified via a reactive sintering process. Cerium has been used as a surrogate for plutonium to facilitate formulation development and process testing. The use of cerium vs. plutonium results in differences in behavior during sintering of the ceramic form. The phase development progression and final phase assemblage is different when cerium is substituted for the actinides in the ceramic form. However, the physical behavior of cerium oxide powder and the formation of a pyrochlore-rich ceramic of similar density to the actinide-bearing material make cerium an adequate surrogate for formulation and process development studies.

INTRODUCTION

The U.S. Department of Energy (DOE) has determined that at least a portion of the excess weapons-useable material will be immobilized in a titanate-based ceramic for final disposal in a geologic repository [1]. The technology to be employed will involve immobilizing the Pu in ceramic "pucks", placing the pucks in cans and then encasing these cans in radioactive high-level waste glass (i.e. can-in-canister) [2]. The can-in-canister configuration, including the high radiation field afforded by the waste glass, will provide the necessary proliferation resistance.

The proposed immobilization formulation contains chemical precursors to produce a durable ceramic form as well as sufficient neutron absorbers to preclude criticality. The baseline formulation shown in Table I was designed to produce a ceramic consisting primarily of a highly substituted pyrochlore with minor amounts of brannerite and hafnia-substituted rutile. Many of the prospective feed streams contain large amounts of uranium in addition to plutonium. The pyrochlore phase was shown to readily accommodate large quantities of actinides [3]. In addition, given that pyrochlore has a cubic structure, it is anticipated that the isotropic nature of the mineral phase will minimize radiation damage [3]. When impurities are present in the Pu feed-streams other minor phases such as zirconolite, perovskite, glassy phases, and/or other minor phases may be present in the ceramic form. The composition was designed such that all of the phases that contain actinides also have significant amounts of neutron absorbers (with the exception of unreacted actinide oxide phases - present at less than 1 vol %). Chemical formulas (showing the atomic substitutions) for the various mineral phases observed in the ceramic are shown in Table II.

To the extent authorized under the laws of the United States of America, all copyright interests in this publication are the property of The American Ceramic Society. Any duplication, reproduction, or republication of this publication or any part thereof, without the express written consent of The American Ceramic Society or fee paid to the Copyright Clearance Center, is prohibited.

Table I. Baseline composition for Pu immobilized form

Oxide	wt %	Role
CaO (added as $Ca(OH)_2$)	9.5	Precursor (mineral formation)
TiO_2 (anatase form)	37.7	Precursor (mineral formation)
HfO_2	11.1	Precursor (mineral formation/neutron absorber)
Gd_2O_3	7.6	Precursor (neutron absorber)
UO_2	23.3	Feed material/additive to stabilize pyrochlore
PuO_2	10.8	Feed material

Table II. Mineral phases in the Pu immobilized form

Mineral Phase	Stoichiometry	Observed range
Pyrochlore	$(Ca,Gd)(Hf,U,Pu,Gd)Ti_2O_7$	62 - 90%
Brannerite	$(U,Pu,Hf,Gd,Ca)Ti_2O_6$	0 - 22%
Zirconolite	$(Ca,Gd)(Hf,U,Pu,Gd)Ti_2O_7$	0 - 25%
Rutile	$(Ti,Hf)O_2$	0 - 16%
Perovskite	$CaTiO_3$	0 - 6%
Glassy phase	silica glass	0 - 6%
Other minor phases (e.g. monazite)	various	0 - 6%
Unreacted actinide oxide	$(Pu,U)O_2$	0 - 1%

Concurrently with formulation development efforts, a ceramic fabrication process is being developed based on cold pressing and sintering. In summary, the current baseline process consists of:

- milling the actinide oxide feeds to less than 20 μm in an attritor mill
- mixing/milling the milled actinide oxides with a commercially supplied precursor batch in a second attritor mill
- granulating the conditioned powders
- cold pressing the granulated powders into "pucks"
- sintering the pressed "pucks."

Nominal sintering conditions have been established for the ceramic form consisting of heating at 3° C/min to 300° C, 2 hour hold at 300° C for binder burnout, 5° C/min to 1350° C, hold at 1350° C for 4 hours, and cooling at 5° C/min to room temperature [3]. The objective of the sintering process is to produce a monolithic product with a phase assemblage that will ensure product durability. Unlike solid state or single phase sintering processes, the sintering of the Pu immobilized ceramic form is "reactive" in nature, where the oxide precursors react to form highly substituted mineral phases. Cerium has been used as a surrogate for plutonium to facilitate formulation development and process testing. The use of cerium vs. plutonium results in differences in behavior during sintering of the ceramic form. The objectives of this work are to quantify these behavioral differences to determine the adequacy of cerium as a surrogate for the actinides in the ceramic form.

EXPERIMENTAL

Precursor Batch and Ceramic Formulation Preparation

The precursors listed in Table I (without the actinides) were weighed and mixed in the amounts to form the desired precursor batch size. The batch was added to a polyethylene bottle filled approximately half full with 0.64 mm (0.25 inch) cylindrical calcium stabilized zirconia grinding media. Deionized water was added to cover just the batch and grinding media. The mixture was then ball milled for about 20 hours. The milled slurry was poured through a #40 sieve opening to catch the grinding media and allow the slurry to collect in a drying pan. The solution was then dried at 105° C so that no moisture was retained. The dried material was removed from the pan and calcined at 750° C for 1 hour in a shallow alumina tray. After cooling, the precursor batch was set aside for later mixing with PuO_2 and UO_2 (or CeO_2 as a surrogate for the actinides) to form the baseline composition.

For the baseline formulation, PuO_2 and UO_2 powders were weighed out and added to the appropriate amount of precursor batch to coincide with the baseline composition (Table I). This formulation was referred to as Hf-Pu-U. For the surrogate baseline formulation, CeO_2 as a surrogate for both PuO_2 and UO_2 on a one-to-one molar basis was added to the appropriate precursor batch amount. The surrogate formulation was referred to as Hf-Ce-Ce. Hf-Pu-U and Hf-Ce-Ce batches were added to a polyethylene bottle filled approximately half full with 1/4 inch cylindrical calcium stabilized zirconia grinding media. Deionized water was added to cover just the batch and grinding media. The batches were milled for approximately 8 hours. The slurry was screened to remove the grinding media and dried on a hot plate. The resulting cake was crushed and forced through a #40 sieve screen. The powder was then used for thermal analysis or for pressing into pellets and subjected to isothermal heat treatments. Pellets (approximately 1 g) were pressed using uniaxial compression at a pressing pressure of 13.8 MPA (2000 psi) in 12.7 mm (0.50 inch) dies. After pressing, the diameter, thickness and mass of each green pellet were measured.

Thermal Analysis

Differential thermal analysis (DTA) and thermogravimetric analysis (TGA) were run (in duplicate) on samples of the Hf-Ce-Ce batch to identify reaction temperatures. Typical scan rates of 10° C/min were used in all of the testing. The DTA scans were run to the nominal sintering temperature of 1350° C. Due to instrument limitations, the TGA scans were run only to 1200° C in these initial tests. The results were also evaluated using the first derivative to aid in the identification of reactions. For the Hf-Pu-U composition, TGA and differential scanning calorimetry (DSC) were used in an attempt to quantify reaction temperatures. DSC runs were made only to 700° C due to equipment limitations.

Toward the end of this study, a simultaneous DTA/TGA was available and samples of Hf-Ce-Ce baseline powder were run at 10° C/min to 1350° C in air, oxygen and helium atmospheres.

Isothermal Heat Treatments

Once potential reaction temperatures were determined, a series of isothermal heat treatments at temperatures slightly above those depicted in the thermal analysis scans were run on the material. In areas in the thermal analysis scans that were difficult to discern (perhaps due to multiple reactions occurring), several additional temperatures were selected for testing. Two pellets of both Hf-Ce-Ce and Hf-Pu-U compositions were heat-treated for two hours at the prescribed temperature and then quenched. Pellets were placed on reticulated yttria stabilized zirconia setters for the heat treatments to preclude interactions with the setter.

Sample Characterization

Following heat treatment, the diameter, thickness and mass for each pellet was measured. From these measurements, the geometric density was calculated for each pellet. X-ray diffraction scans were run on a sample from each of the prescribed heat-treatments to determine the phases present.

RESULTS AND DISCUSSION

The phases present in the Hf-Pu-U isothermally heat-treated samples are shown in Table III. The results of the isothermal heat treatments for the Hf-Ce-Ce samples are shown in Table IV.

Upon heating, both compositions exhibited similar behavior with respect to the $Ca(OH)_2$ decomposition, in-growth of $CaCO_3$ and decomposition of $CaCO_3$. This behavior in the Pu immobilization ceramic has been previously described [4].

Hf-Pu-U System

In the 600° C heat treatment in the Hf-Pu-U system, a calcium uranium oxide phase ($CaUO_4$) was evident. This phase continued to increase in relative concentration with additional heat treatments until 950° C and then disappeared from the phase assemblage in the 1100° C heat treatment sample. Examination of the CaO-UO_2 phase diagram (which indicated extensive solid solubility between CaO and UO_2) [5] and subsequent heat treatments in the CaO-UO_2 binary system confirmed the formation of this phase. Not surprisingly, UO_2 was also found to oxidize upon heating in air to form U_3O_8. However, by 900° C, no uranium oxide phase was detected in the samples. An additional low temperature uranium bearing phase, brannerite, also appeared to form at relatively low temperatures (750° C) and then be consumed in further development of the

Table III. Phase Assemblages of Hf-Pu-U Isothermally Heated Samples

Temperature (°C)	Phases Present/Relative Concentration Trend
As-received	Precursors
325	Precursors, $CaCO_3$, $Ca(OH)_2$
400	Precursors, $CaCO_3$ ↑, $Ca(OH)_2$ ↓
500	Precursors, $CaCO_3$ ↑, no $Ca(OH)_2$
600	Precursors, no $CaCO_3$, no $Ca(OH)_2$, *$CaUO_4$*
725, 800, 850	TiO_2 (a), HfO_2, Gd_2O_3, U_3O_8, $CaUO_4$, *brannerite*, PuO_2
900	TiO_2 (a), HfO_2, Gd_2O_3, *no U_3O_8*, $CaUO_4$↑, *no brannerite*, *CaU_2O_7*, PuO_2
950	TiO_2 (a), HfO_2, Gd_2O_3, $CaUO_4$↑, PuO_2, *zirconolite*
1025	TiO_2 (a), TiO_2 (r), HfO_2, $CaUO_4$, *no zirconolite*, *pyrochlore*, PuO_2
1100, 1130, 1150	Pyrochlore (major), TiO_2 (r), HfO_2, *no $CaUO_4$*, PuO_2
1200	Pyrochlore (major), TiO_2 (r), *no HfO_2*, PuO_2, *brannerite*
1250	Pyrochlore (major), TiO_2 (r), *no PuO_2*, *brannerite*
1300, *1350*	Pyrochlore (major), TiO_2 (r), brannerite

↓ - relative concentration appears to be decreasing compared to lower temperature

↑ - relative concentration appears to be increasing compared to lower temperature

(a) – anatase phase, (r) – rutile phase

Table IV. Phase Assemblages of Hf-Ce-Ce Isothermally Heated Samples

Temperature (°C)	Phases Present/Relative Concentration Trend
As-received	Precursors
325	Precursors, $CaCO_3$, $Ca(OH)_2$
400	Precursors, $CaCO_3$, $Ca(OH)_2$ ↓
500	Precursors, $CaCO_3$ ↑, no $Ca(OH)_2$
600	Precursors, no $CaCO_3$, no $Ca(OH)_2$
725, 800, 850	TiO_2 (a), HfO_2, Gd_2O_3, CeO_2, *perovskite*
900	TiO_2 (a), HfO_2, Gd_2O_3, CeO_2, perovskite, *pyrochlore*
950	TiO_2 (a), TiO_2 (r), HfO_2, Gd_2O_3, CeO_2, perovskite, pyrochlore↑
1025	TiO_2 (a) ↓, TiO_2 (r) ↑, HfO_2, CeO_2, perovskite, pyrochlore↑
1100, 1130, 1150	Pyrochlore (major), *no perovskite*, HfO_2, TiO_2 (r), CeO_2
1200	Pyrochlore (major), *no perovskite*, TiO_2 (r), *zirconolite*, CeO_2
1250	Pyrochlore (major), *no perovskite*, TiO_2 (r), *zirconolite, no* CeO_2
1300	Pyrochlore (major), *perovskite*, TiO_2 (r), zirconolite, no CeO_2
1350	Pyrochlore (major), perovskite ↑, TiO_2 (r), zirconolite

↓ - relative concentration appears to be decreasing compared to lower temperature
↑ - relative concentration appears to be increasing compared to lower temperature
(a) – anatase phase, (r) – rutile phase

microstructure at 900° C. The appearance and disappearance of the brannerite phase was surprising since it was observed to reappear at higher temperatures (1200° C) and is always present (at about 10 vol %) in the Hf-Pu-U baseline ceramic upon sintering at 1350° C. It is quite possible that the composition of the "low temperature" and "high temperature" brannerite phases differ. Future microscopic and chemical analyses will be aimed at examining this hypothesis.

In the 950° C heat treatment, zirconolite was observed as well as the initiation of the anatase to rutile phase transformation. At 1025° C, no zirconolite was evident in the XRD patterns, however, pyrochlore formation was observed. The dimensional measurements of the samples (see below) indicated that densification in the ceramic coincided with the formation of the pyrochlore phase. At 1100° C, all anatase was converted to rutile and the only other oxides from the initial batch that were not reacted completely from the initial batch were HfO_2 and PuO_2. At 1200° C all the HfO_2 had reacted and by 1250° C all the PuO_2 had reacted.

Hf-Ce-Ce System

At 725° C, a perovskite phase formed in the Hf-Ce-Ce system. The relative concentration of a perovskite phase appeared to be constant until it disappeared at 1100° C. A perovskite phase was observed to reappear at 1300° C and increase in relative concentration in the final sintered product at 1350° C. Similar to the behavior with brannerite in the Hf-Pu-U composition, the perovskite behavior was puzzling since it is always observed in the Hf-Ce-Ce composition in discernible concentrations following sintering at 1350° C. The potential for differing compositions of the "low temperature" and "high temperature" perovskite phases will also be examined in future microscopy studies. It is known that Ce readily changes oxidation states from Ce^{4+} to Ce^{3+} at elevated temperatures [6]. Thermal analysis studies (see below) confirmed this behavior in this system. The "high temperature" perovskite phase could very likely result from the reduction of Ce and the "ejection" of the Ce^{3+} from the pyrochlore phase.

At 900° C, the initiation of pyrochlore formation in the Hf-Ce-Ce ceramic was observed. The relative concentration of pyrochlore increased in subsequent heat treatments at 950° C and 1025° C. After the 1100° C heat treatment, pyrochlore was the major crystalline phase in the assemblage. Dimensional measurements again pointed to the formation of pyrochlore coinciding with densification (see below).

Similar to the Hf-Pu-U system, the anatase to rutile transition was observed to initiate in the 950° C heat treatment and the conversion was completed following the 1100° C treatment. The HfO_2 was also observed to all be reacted by 1200° C. Finally, CeO_2 was found to behave very similarly to PuO_2 in that complete dissolution of the CeO_2 occurred by 1250° C.

Zirconolite was also observed to form in the Hf-Ce-Ce ceramic at 1200° C and remain in the ceramic at the final sintering temperature. Similar to the "high temperature" perovskite phase, the reduction of cerium is also very likely responsible for stabilizing the zirconolite phase.

Densification

The densification behavior was examined by performing measurements on the samples before and after heat treatments. Figure 1 shows the fractional diameter vs. the heat treatment temperature. The fractional diameter is the diameter after heat treatment at the prescribed temperature divided by the original diameter.

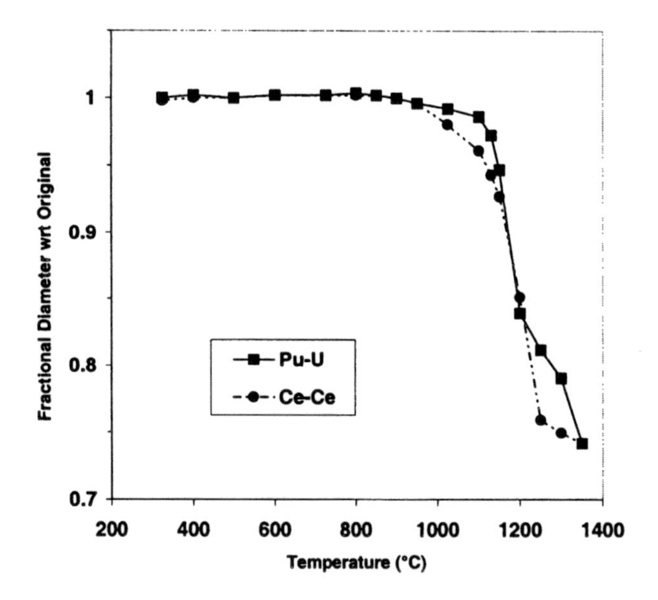

Figure 1. Densification behavior of Hf-Pu-U and Hf-Ce-Ce ceramics expressed as the fraction obtained from the diameter of the pellet divided by the original diameter at the prescribed temperature.

When comparing the phase development results to the densification behavior, densification was observed to coincide with pyrochlore formation in both systems. Densification in the Hf-Ce-Ce began at a significantly lower temperature than in the Hf-Pu-U system (approximately 900° C vs. 1050° C for the Hf-Ce-Ce and Hf-Pu-U systems, respectively). In both systems, pyrochlore formation was initiated near these respective temperatures.

In addition, the phases comprising the final assemblages were present by 1250° C in the Hf-Pu-U system and by 1300° C in the Hf-Ce-Ce systems yet significant densification occurred after these temperatures. This was especially evident in the Hf-Pu-U ceramic. Future microscopic analysis will be used to evaluate the microstucture development at these temperatures.

Interestingly, the dimensional changes of the pellets in both systems following sintering at the baseline sintering temperature of 1350° C were identical. Although the end-points were the same, the shape of the curves indicated that the densification paths were different for the two systems. The formation of brannerite may be retarding sintering in the Hf-Pu-U system. Future dilatometry studies and thermokinetic analysis are planned for these systems with the intention of modeling the sintering behavior in the two systems. This should provide insight into the sintering mechanisms occurring in the two systems.

Thermal Analysis

When it was determined that the reduction of cerium may be playing an important role in the phase development in the Hf-Ce-Ce composition, a few scoping experiments were performed in an attempt to verify the cerium reduction. Thermogravimetric scans were run on samples of the Hf-Ce-Ce composition in air, oxygen and helium atmospheres (Fig. 2).

The weight loss in the TGA scans at the higher temperatures pointed to the reduction of cerium. The fact that this behavior initiated at a lower temperature and the loss was exaggerated in

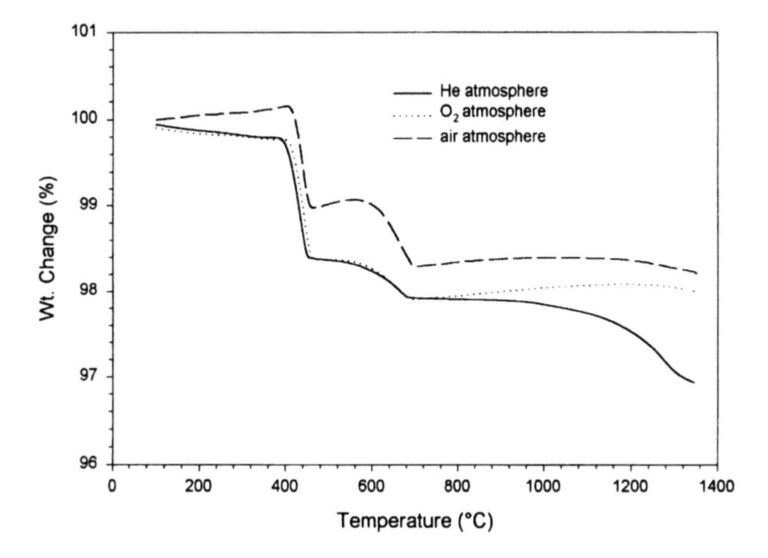

Figure 2. Thermogravimetric analysis scans of Hf-Ce-Ce sample heated in different atmospheres.

the sample run in helium further confirmed the hypothesis. Additional TGA work with a coupled mass spectrometer is planned to quantify this behavior. Efforts are also underway to determine the cerium speciation (i.e. ratio of Ce^{3+} to Ce^{4+}). The weight gain at lower temperatures in the sample heated in air was observed previously and is associated with in-growth of calcium carbonate [4].

CONLCUSIONS

Cerium oxide as a surrogate for the actinide oxides in the Pu immobilized form is adequate from many perspectives and is, thus, a suitable "stand-in" for process development studies. Although not discussed here, cerium oxide performed well as a physical surrogate for batching, powder handling and compaction. In both systems, the major phase formed was pyrochlore and the dissolution behavior of cerium oxide and plutonium oxide into the crystalline assemblage generally coincided. Finally, the overall shrinkage of the two systems under the baseline sintering conditions was essentially identical.

There were, however, significant differences between the behavior of cerium oxide and the actinide oxides that limits the use of cerium oxide as a true chemical surrogate. The response of calcium varied between the two systems. In the Hf-Pu-U system, calcium reacted at low temperatures to form a calcium uranium oxide phase whereas in the Hf-Ce-Ce system calcium reacted at a slightly higher temperature to form perovskite. The formation of pyrochlore and the onset of densification occurred at a lower temperature in the Hf-Ce-Ce composition. The most pronounced limitations appeared to be associated with the reduction of Ce at elevated temperatures. The reduction was undoubtedly responsible for the occurrence of zirconolite and perovskite in the final phase assemblage in the Hf-Ce-Ce formulation. The cerium reduction also appeared to have an effect on the densification behavior in the Hf-Ce-Ce ceramic and may be responsible for different sintering mechanisms occurring in these systems.

ACKNOWLEDGEMENTS

The information contained in this paper was developed during the course of work under Contract No. DE-AC09-96SR18500 with the U.S. Department of Energy.

REFERENCES

1. Fissile Materials Storage and Disposition Programmatic Environmental Impact Statement Record of Decision, *Storage and Disposition Final PEIS*, 62 Federal Register 2014, January 14, 1997.
2. M. E. Smith and E. L. Hamilton, "Plutonium Immobilization Project Phase 2 Cold Pour Test," This Proceedings.
3. B. B. Ebbinghaus, G. A. Armantrout, L. Gray, H. F. Shaw, R. A. VanKonynenburg and C. C. Herman, "Plutonium Immobilization Baseline Formulation Report," UCRL-ID-133089, Rev. 1, Lawrence Livermore National Laboratory, Livermore, CA, September 2000.
4. J. C. Marra, A. D. Cozzi, J. M. Pareizs, A. R. Jurgensen, D. M. Missimer, J. W. Congdon, B. Bukovitz, E. C. Skaar and T. D. Taylor, "Phase Development and Sintering Studies on An Immobilized Plutonium Ceramic Form", Environmental Issues and Waste Management Technologies in the Ceramic and Nuclear Industries V, Edited by G. T. Chandler and X. Feng, Ceramic Transactions, Vol.107, American Ceramic Society, Westerville, OH, 2000, p. 517.
5. Phase Diagrams For Ceramists, Vol. XII, Edited by A. E. McHale and R. S. Roth, American Ceramic Society, Westerville, OH, 1996, p. 79.
6. D. K. Peeler, J. E. Marra, I. A. Reamer, J. D. Vienna and H. Li, "Development of the Am/Cm Batch Vitrification Process," Environmental Issues and Waste Management Technologies in the Ceramic and Nuclear Industries V, Edited by G. T. Chandler and X. Feng, Ceramic Transactions, Vol.107, American Ceramic Society, Westerville, OH, 2000, p. 517.

RELEASE OF URANIUM AND PLUTONIUM FROM THE EBR-II CERAMIC WASTE FORM

Lester R. Morss and William L. Ebert
Chemical Technology Division
Argonne National Laboratory, Argonne, IL 60439.

ABSTRACT

A ceramic waste form (CWF) has been developed to immobilize radioactive electrorefiner salt from the electrometallurgical treatment of spent metallic fuel from the experimental breeder reactor EBR-II. Tests are being carried out to qualify CWF for disposal in a high-level waste repository. There are two major phases in U/Pu-loaded CWF, sodalite ($Na_8Al_6Si_6O_{24}Cl_2$) and glass; and two important minor phases, halite (NaCl) and oxide $(U,Pu)O_2$. The $(U,Pu)O_2$ phase is present as colloid-sized particles in the glass, near sodalite-glass phase boundaries. Tests have been conducted to measure the release behavior of the $(U,Pu)O_2$ particles as the CWF corrodes. The releases of matrix and radioactive elements from CWF samples into water were determined from tests in which crushed material was reacted with water at 90 or 120 °C for durations from 7 to 365 days. Colloids in the test solutions were characterized by sequential filtration, followed by analysis of filtrates for cations. Plutonium is released partially as $(U,Pu)O_2$ colloids similar to the particles in the uncorroded CWF.

INTRODUCTION

Glass-bonded sodalite is the ceramic waste form (CWF) for disposition of radioactive electrorefiner salt resulting from the electrometallurgical treatment of spent sodium-bonded nuclear fuel.[1,2] It is produced by blending electrorefiner salt with dehydrated zeolite 4A and then consolidating the zeolite into sodalite ($Na_8Al_6Si_6O_{24}Cl_2$) with binder glass at 850-915°C. For this study, four CWF samples containing uranium, plutonium, and simulated fission products were prepared by hot isostatic pressing (HIPing) with 1:3 and 3:1 U:Pu ratios, and 0.15% and 1.5% water-content in dried zeolite, that are representative of the CWF expected during production (Table I).[2] Microstructural examination showed halite and $(U,Pu)O_2$ particles 0.02-0.05 μm in diameter in the glass phase near sodalite–glass phase boundaries.[2,3,4] The oxide particles result from reaction of UCl_3 and $PuCl_3$ with residual water in zeolite 4A during blending. This paper reports the release behavior of uranium and plutonium from CWF consolidated by hot isostatic pressing (HIPing) at 850°C.

A pressureless-consolidation (PC) process has been selected for the inventory reduction phase of the EBR-II spent fuel program. The corrosion tests described in this report are being repeated with U/Pu-loaded CWF samples made by the PC

To the extent authorized under the laws of the United States of America, all copyright interests in this publication are the property of The American Ceramic Society. Any duplication, reproduction, or republication of this publication or any part thereof, without the express written consent of The American Ceramic Society or fee paid to the Copyright Clearance Center, is prohibited.

process. From SEM examinations of the non-radioactive and radioactive PC CWF specimens, the uranium and plutonium are also present as $(U,Pu)O_2$, mostly at or near sodalite-glass phase boundaries but sometimes distributed throughout the glass phase. Therefore the Pu behavior measured in the corrosion tests with HIPed CWF should provide useful and timely insight into the expected behavior of Pu during corrosion of PC CWF.

Table I. HIPed (U,Pu)-Loaded Ceramic Waste Form Samples

CWF Sample	U Content (mass %)	Pu Content (mass %)	Water in Zeolite (mass %)
237x	0.44	0.15	0.12
238x	0.15	0.44	0.12
239x	0.44	0.15	3.5
240x	0.15	0.44	3.5

CORROSION TESTING OF (U/Pu)-LOADED CERAMIC WASTE FORM

The objectives of the corrosion tests were (1) to characterize the form(s) in which uranium and plutonium were released during corrosion tests and (2) to determine if the corrosion of the wasteform was altered significantly by presence of uranium and plutonium in different ratios or by formation from low- or high-water-content zeolite. The corrosion test method used was the Product Consistency Test (PCT-B).[5] This test method was selected because it highlights the feedback effect of the solute species, which is expected to be important in a repository environment in which only small amounts of water are likely to contact the waste forms. Tests were conducted in Type 304L stainless steel vessels using demineralized water at CWF/water mass ratios of 1:10 [typically 1 g of −100+200-mesh CWF in 10 g H_2O, yielding (CWF surface area)/(solution volume) = $S/V \approx 2300\ m^{-1}$] at 90 and 120°C, and at mass ratios of 1:20 (yielding $S/V \approx 1150\ m^{-1}$) at 120°C. Tests at 90°C with a CWF/water mass ratio of 1:10 are standard PCT conditions and may be compared with tests conducted with other wasteforms for the same durations. Tests were conducted at 120°C to accelerate the CWF degradation to ensure that some tests would result in a measurable Pu release. Tests at 120°C were conducted at two mass ratios because the 1:10 mass ratio was expected to achieve higher solution concentrations and pH, while the 1:20 mass ratio was expected to produce greater corrosion depth in the wasteform. After each test, the test solution and wasteform were removed; material adsorbed on test vessel walls was dissolved by acid stripping the vessel overnight in 1% HNO_3(aq) at 90°C.

After the test solutions had cooled, sequential filtrations were carried out by passing portions through 0.45-, 0.1-, and 0.005-μm filters. The concentrations of matrix and trace elements were measured in these filtrates by inductively-coupled plasma-mass spectrometry (ICP-MS). To confirm the identification of colloidal particles, 0.005-μm aliquots of the 0.45-μm filtrates were wicked through lacey carbon foils and examined by transmission electron microscopy (TEM).

RESULTS

Normalized elemental mass losses (NLs) in each solution were calculated (g/m^2) as follows: NL(i) = $m_i/(S \bullet f_i)$ where m_i is the mass of element i in the test solution (g), corrected for background; S is the surface area of the test specimen (m^2); and f_i is the mass fraction of element i in the CWF. Because there was no significant difference among NLs of test solutions from the four CWF samples in Table I, the means of NLs from tests on the four CWF samples are presented to compare the releases of matrix elements and actinides.Only the results of the 90°C tests at S/V = 1150 m^{-1} are reported in this paper.

1. Release of matrix elements and actinides versus test duration

a. *Release of matrix elements (B, Al, Si).* As found in other PCTs with CWF, the concentrations of Si and Al saturate but concentration of B increases at a nearly constant rate as test duration increases (Fig. 1).

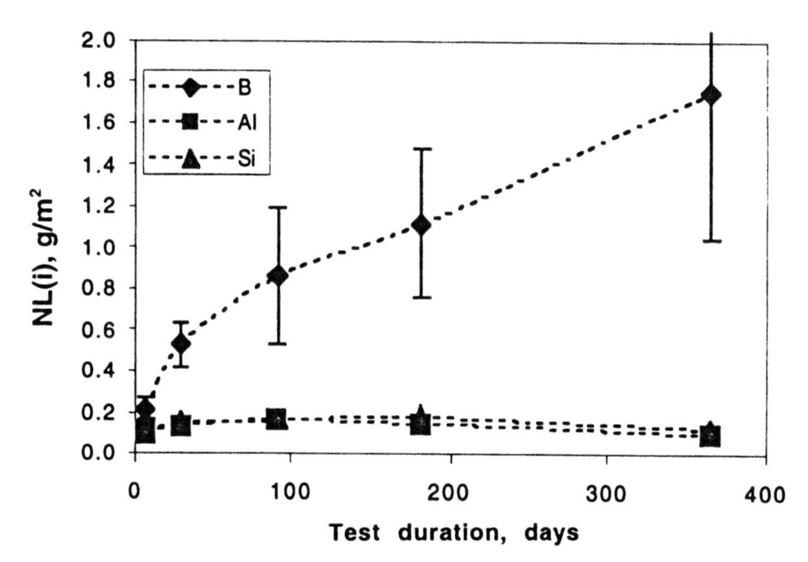

Fig. 1. NL(B, Al, and Si) (sum of NLs from 0.45-μm filtrates and adsorbed fractions) of 90°C PCT test solutions as function of test duration. NL values are averages from tests with the four CWF materials; error bars represent one standard deviation from the four tests. Dashed lines are drawn to guide the eye.

b. *Release of uranium and plutonium:* Each data point in Figs. 1 and 2 is the sum of NL(0.45-μm filtrate) and NL(adsorbed). The NL(adsorbed) for U and Pu have high uncertainties because adsorbed U and Pu are removed along with CWF fine particles during vessel rinsing prior to the acid soak used to removed adsorbed U and Pu. Thus the uncertainty of each datum in Fig. 2 is quite large; it is not possible to conclude that the releases increase as a function of test duration.

Because the $(U,Pu)O_2$ particles in the CWF are within the glass phase, the release of these particles occurs when they are exposed as glass dissolves, and their releases should be limited by the release of boron (Fig. 1) and increase as a function of test duration. Such an increase is not observed in Fig. 2. It may be that as CWF corrodes $(U,Pu)O_2$ particles are retained on slowly-dissolving exposed surfaces of the CWF. The PCT data are inadequate to calculate mass balances of the actinides. Examination of surface microstructure of corroded CWF particles or stainless steel used in these tests is in progress and may help to establish the relative amounts and physical form of actinides retained on these surfaces. The release behavior of these elements is discussed further in the section on sequential filtration.

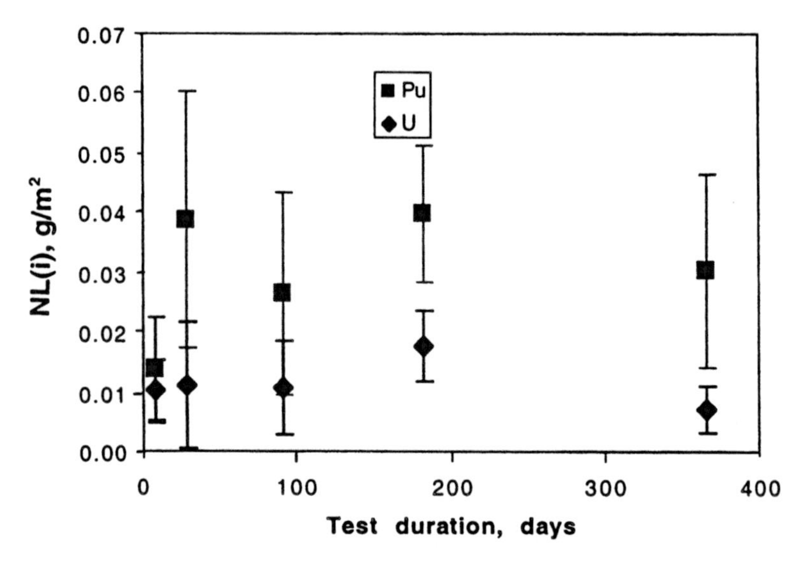

Fig. 2. NL(U) and NL(Pu) (sum of NLs from 0.45-μm filtrates and adsorbed fractions) of 90°C PCT test solutions as function of test duration. NL values are averages from tests with the four CWF materials; error bars represent one standard deviation from the four tests.

2. Releases of matrix elements and actinides as a function of sequential filter size

Sequential filtration was used to determine the presence and size distribution of colloidal particles in PCT solutions. Test vessels were allowed to cool to room temperature, after which solution aliquots were removed and filtered sequentially through 0.45, 0.1, and 0.005-μm filters at room temperature. Elemental analysis of the filtrates provided an indirect determination of the composition of the colloidal particles.

Figure 3 shows that B concentrations do not decrease as the solutions pass through the sequential filtrations. Negligible concentrations of boron were found in acid strip solutions. We conclude that all of the boron released is truly in solution. Alteration products or colloidal particles are not likely to contain significant amounts of boron.

The decreases in Al and Si concentrations in Fig. 3 show that ~25% of Al and Si present in test solutions were retained by the 0.005-μm filter. The retained material is probably aluminosilicate colloids, a conclusion confirmed by TEM identification of amorphous aluminosilicate agglomerates of colloidal (~0.1-μm) size along with smaller $(U,Pu)O_2$ particles on wicked carbon foils.

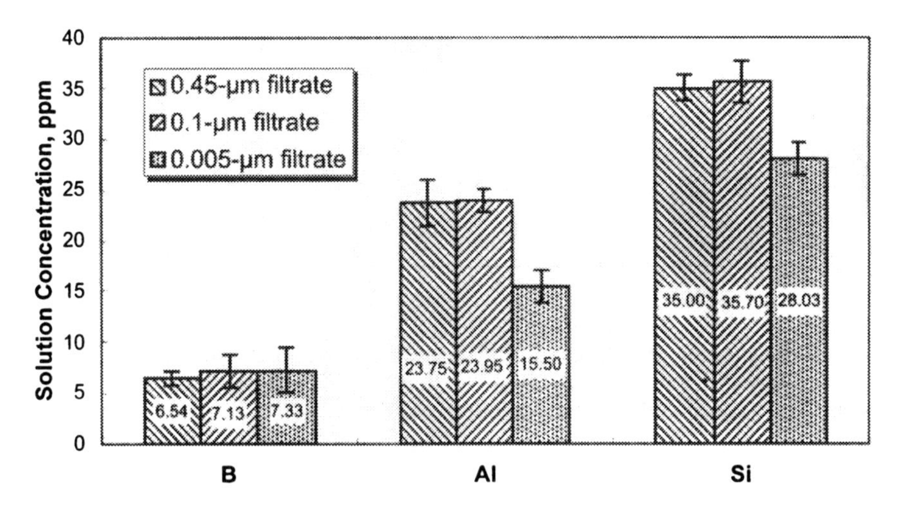

Fig. 3. Solution concentrations of matrix elements from (U/Pu)-loaded HIPed CWF in 90°C 7-day PCTs vs. particle size range. Concentrations are averages from tests with the four CWF materials; error bars represent one standard deviation from the four tests. Fractions adsorbed on vessel walls are not included.

Figure 4 shows the decrease in concentration of neodymium, uranium, and plutonium after each step in the sequential filtrations. The neodymium concentrations are representative of all the rare earths as well as that of americium, all of which are +3 ions in solution. Uranium is expected to be present as the hexavalent uranyl ion UO_2^{2+} in the test solutions. The sequential filtration data for uranium (the difference between uranium concentration in 0.45-μm filtrate and that in 0.005-μm filtrate) are consistent with release of ≈80% of uranium as U^{4+} in $(U,Pu)O_2$ colloids and the remainder as dissolved U^{6+} (as hydroxo- and carbonato complexes).

Plutonium is expected to be present as the tetravalent ion Pu^{4+} in colloids and in the test solutions. The sequential filtration of Pu indicates that >95% of Pu is in $(U,Pu)O_2$ colloids and the remainder is dissolved Pu (approximately at saturation concentration as hydroxyl and carbonato Pu^{4+} complexes at pH 9). The ions Nd^{3+} and Pu^{4+} are extremely insoluble (~0.2 ppb) at the pH region of these tests (~pH 9) in water equilibrated with atmospheric CO_2 at 25°C. This research confirms the expectations that rare earths and actinides, especially the +3 and +4 ions, are released from CWF during corrosion testing mostly as $(R, U,Pu)O_{2\pm x}$, where R represents the +3 elements (rare earths and transplutonium actinides).

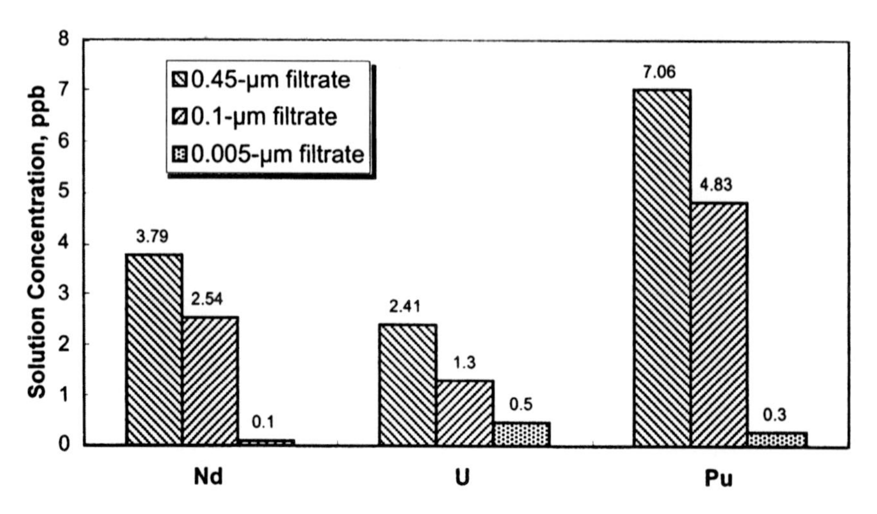

Fig. 4. Solution concentrations of Nd, U, and Pu from (U/Pu)-loaded HIPed CWF in 90°C 7-day PCTs at S/V ≈ 2300 m^{-1} as function of particle size range. Fractions adsorbed on vessel walls are not included.

Figure 5 shows the mean release of Pu from the four 90°C 2300 m^{-1} PCTs with the four HIPed CWF materials as a function of test duration and particle size range. Each bar has four segments. The lowest segment represents Pu in the 0.1- to 0.45-μm fraction and is very small. The second segment represents Pu in the 0.005-0.1-μm fraction. It is the major fraction of Pu in the test solution and represents Pu within $(U,Pu)O_2$ colloidal-sized particles. The third segment represents Pu in the 0.005-μm filtrate; these Pu species are considered to be dissolved rather than colloids. This segment is also very small. The uppermost segment represents the Pu concentration in the acid strip solution; it represents Pu that adsorbed on vessel walls during the tests, either as released $(U,Pu)O_2$ particles or as dissolved Pu that hydrolyzed and became adsorbed as Pu hydrolyzed polymers.

The overall Pu release (the total height of each bar) increases with test duration. The rate of Pu release appears to decrease as a function of test duration.

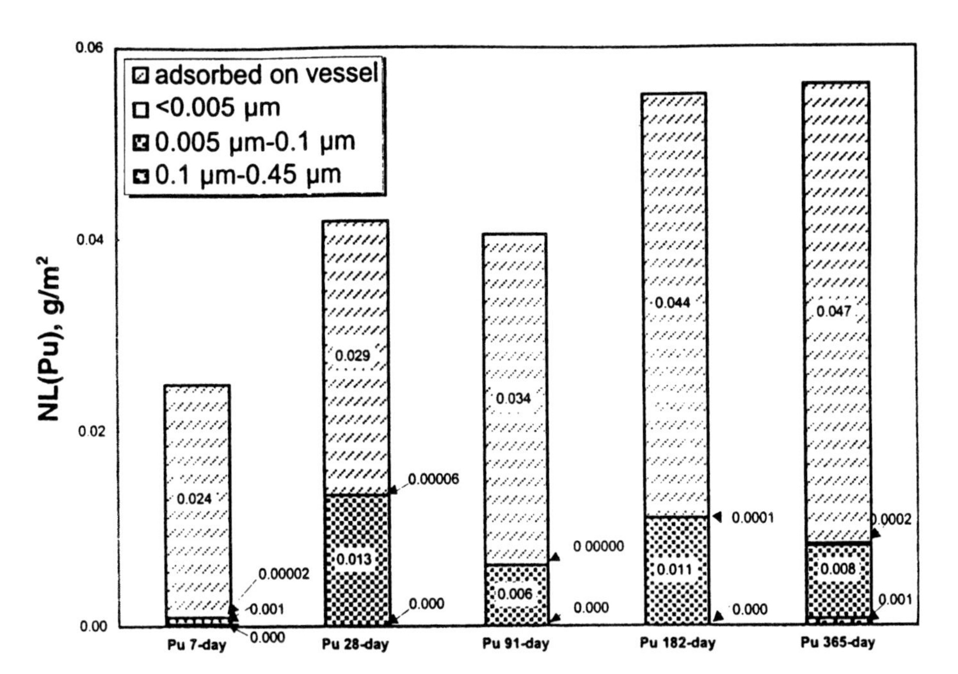

Fig. 5. NL(Pu) from HIPed CWF in PCTs at 90°C and S/V ≈ 2300 m^{-1} with (U/Pu)-loaded HIPed CWF for four test durations vs. particle size range and adsorbed Pu. The 0.1- to 0.45-µm and <0.005-µm bar segments are too small to be visible. NL values are means from tests with the four CWF materials.

CONCLUSIONS

1. Almost all of the plutonium released from (U/Pu)-loaded HIPed CWF is suspended in solution as colloidal-sized $(U,Pu)O_2$ particles or adsorbed on test vessel walls. The presence of filterable 0.005- to 0.1-µm particles is confirmed by evidence of similar particles observed by TEM from PCT solution that was deposited by wicking solution through carbon films.
2. The Pu adsorbed on vessel walls in PCTs is a significant fraction of total released Pu, always larger than that remaining in solution. The fraction of Pu found on vessel walls varies from test to test; thorough wiping of vessel walls after rinsing shows that this adsorbed Pu includes CWF fine particles as well as Pu that was released from the CWF during the test and subsequently adsorbed on the vessel walls. More efficient vessel rinsing and removal of corroded CWF is needed. The nature of the adsorbed elements must also be determined, because fines of corroded CWF adhering to vessel walls are test artifacts that should not be included in normalized mass losses.
3. The releases of matrix elements and trace elements as functions of test duration have been determined.

WORK IN PROGRESS

1. A limited test matrix of PCTs with CWF made by the PC process. The corrosion testing will be carried out with new stainless steel vessels and the sequential filtration procedure will be carried out at specified short intervals.
2. SEM examination of corroded (U/Pu)-loaded CWF particles, to determine if $(U,Pu)O_2$ particles were released and then re-adsorbed on the corroded CWF surface, in particular the sodalite granules.
3. SEM examination and radioanalysis of Type 304L stainless steel coupons placed in some of the PCTs to characterize adsorbed Pu species, particularly to identify $(U,Pu)O_2$ particles, and to determine whether most of the adsorbed Pu is present in fine particles of corroded CWF or as Pu species adsorbed after being released.
4. TEM examination of colloidal particles removed from PCT solutions.

REFERENCES

---

[1] K. M. Goff, L. L. Briggs, R. W. Benedict, J. R. Liaw, M. F. Simpson, E. E. Feldman, R. A. Uras, H. E. Bliss, A. M. Yacout, D. D. Keiser, K. C. Marsden and C. W. Nielsen, "Production Operations for the Electrometallurgical Treatment of Sodium-Bonded Spent Nuclear Fuel," Argonne National Laboratory Report ANL-NT-107, June 1999.

[2] W. L. Ebert, D. Esh, S. M. Frank, K. M. Goff, S. G. Johnson, M. A. Lewis, L. R. Morss, T. Moschetti, T. P. O'Holleran, M. K. Richmann, W. Riley, L. J. Simpson, W. Sinkler, M. M. Stanley, C. D. Tatko, D. J. Wronkiewicz, J. P. Ackerman, K. Bateman, T. J. Battisti, D. G. Cummings, T. DiSanto, M. Gougar, M. C. Hash, K. Hirshe, L. Leibowitz, J. S. Luo, M. Noy, H. Retzer, M. F. Simpson, A. R. Warren, and V. N. Zyryanov, "Ceramic Waste Form Handbook," Argonne National Laboratory Report ANL-NT-119.

[3] S. M. Frank, S. G. Johnson, T. L. Moschetti, T. P. O'Holleran, W. Sinkler, D. Esh, and K. M. Goff, "Accelerated Alpha Radiation Damage in a Ceramic Waste Form, Interim Results," MRS Symposium on Scientific Basis of Nuclear Waste Management, Vol. 608, Edited by R.W. Smith and D.W. Shoesmith, 2000, pp. 469-474.

[4] W. Sinkler, T. P. O'Holleran, S. M. Frank, M. K. Richmann, S. G. Johnson, "Characterization of a Glass-Bonded Ceramic Waste Form Loaded with U and Pu," MRS Symposium on Scientific Basis of Nuclear Waste Management, Vol. 608, Edited by R.W. Smith and D.W. Shoesmith, 2000, pp. 423-429.

[5] "Standard Test methods for Determining Chemical Durability of Nuclear, Hazardous, and Mixed Wastes: The Product Consistency Test (PCT)," ASTM C 1285-97, West Conshohocken, PA, Vol. 12.01, pp. 694-711 (1999).

MONITORING CONSISTENCY OF THE CERAMIC WASTE FORM

M. A. Lewis and W. L. Ebert
Argonne National Laboratory
Argonne, IL 60439

ABSTRACT

Glass bonded sodalite is the ceramic waste form (CWF) being developed to immobilize salt waste generated during electrometallurgical conditioning of spent sodium-bonded fuel. CWF will be prepared using a pressureless consolidation (PC) method. The processing conditions will be selected to optimize waste loading while maintaining physical and chemical durability. The possible use of the Product Consistency Test (PCT) to measure the consistency of CWF products is being evaluated by measuring test repeatability, reproducibility, sensitivity to different compositions and processing conditions, and production consistency. The response of the PCT was the same, within 95% uncertainty limits, in replicate tests with the same product, indicating test repeatability. The response of the PCT was the same for replicate products, within 95% uncertainty limits, indicating production consistency. The response of the PCT was sensitive to PC processing conditions. Releases of Al, B, Na, and Si from the PC CWF in 7-day PCTs were lower than those from Environmental Assessment glass. These observations show that the PCT is an appropriate test for confirming that processing conditions meet specifications during CWF production as well as determining that the CWF has adequate and consistent chemical durability.

INTRODUCTION

A glass-bonded sodalite waste form is being developed to immobilize the waste salt generated during the electrometallurgical conditioning of sodium-bonded spent nuclear fuel for disposal in a federal high-level radioactive waste repository. Standard borosilicate waste forms, such as those being used to immobilize Department of Energy (DOE) tank wastes at the Savannah River and Hanford sites, are not amenable to wastes containing high concentrations of Cl. This waste form is prepared in two steps: (1) occlude waste salt within the micropores of a zeolite and (2) heat the salt-loaded zeolite with a glass binder. During the second step, the zeolite converts to the mineral sodalite, $Na_8(AlSiO_4)_6Cl_2$, and the glass softens and encapsulates the sodalite granules to produce a dense waste form. Small amounts of halite (NaCl) inclusions form within the glass phase of the resulting waste form (the halite content is typically less than 3 mass %). Small amounts of various oxides and silicates also form either when the salt is loaded into the zeolite or during the encapsulation step. These also form inclusions in the glass

To the extent authorized under the laws of the United States of America, all copyright interests in this publication are the property of The American Ceramic Society. Any duplication, reproduction, or republication of this publication or any part thereof, without the express written consent of The American Ceramic Society or fee paid to the Copyright Clearance Center, is prohibited.

phase. The resulting product is glass with mineral inclusions that is referred to as the "ceramic waste form."[1]

In the past, the CWF was densified by hot isostatic pressing (HIP), rather than pressureless consolidation.[1] Preliminary tests have shown that the microstructure, the major phase content, and the chemical durability are virtually the same for HIP CWF and PC CWF.[2] Since the PC method requires less sophisticated equipment in a hot cell and is potentially less dangerous because ambient pressures are used, PC has been chosen as the preferred densification method. HIP remains a back-up method.

In this report, we present the results of tests that measured the consistency and dissolution behavior of the PC CWF. We have chosen to test the PC CWF with the PCT because it is the standard test method used for qualifying standard borosilicate glass waste forms. The DOE Waste Acceptance System Requirements Document (WASRD) specifies acceptance criteria for borosilicate waste glasses in terms of the PCT response to the Environmental Assessment (EA) glass.[2] However, the PC CWF is a heterogeneous material with crystalline phases encapsulated in a glass matrix and it is necessary to evaluate whether the PCT is an appropriate test method and the WASRD requirements are relevant. We are doing that by demonstrating the following:

(1) The PCT can be conducted repeatedly with PC CWF in replicate tests with well-homogenized materials taken from individual PC CWF products.
(2) The PCT method is sensitive to differences in PC CWF resulting from changes in fabrication conditions or composition.
(3) The PCT can be used to measure the consistency of the consolidation process by conducting tests with multiple products prepared identically but at different times.
(4) The repeatability and reproducibility of the PCT with PC CWF and glass are similar.

We also compare the PCT responses of the PC CWF and of EA glass to show that the PC CWF meets that criterion.

EXPERIMENTAL

Materials

Several types of PC CWF products (individual PC CWFs) were prepared as described below:

(1) Baseline PC CWF: Baseline PC CWF products consisted of a mixture of 50 mass % glass and 50 mass % salt-loaded zeolite that was heated at 850°C for 4 h. Samples taken from a 5 kg product were used in replicate 7-day tests to measure the repeatability of the test (intra-laboratory precision). In addition, ten replicate 20-g products were produced, each on a different day, from a mixture of glass and salt-loaded zeolite that was nominally the same as that used for the 5 kg product. Samples of these products were subjected to 7-day PCTs to measure the production consistency.

(2) Temperature-time (T-t) PC CWF: Thirty-six 20-g products were made with 25 mass % glass and 75 mass % salt-loaded zeolite at six processing temperatures (850, 875, 900, 915, 925, and 950°C) and six processing times

(1, 4, 8, 16, 24, and 36 h). Samples from each were used in 7-day PCTs to measure PCT sensitivity.

(3) Advanced PC CWF: These PC CWF products were made by heating a mixture of 25 mass % glass and 75 mass % salt-loaded zeolite at 915°C for 16 h. This set of fabrication conditions was deemed optimal after the temperature-time study. A 500-g product (PC10402) was used in replicate 7-day tests to measure the repeatability of the test. Samples from this product are also being used in an inter-laboratory round robin study to measure test reproducibility.

Test Methods

Preparation of crushed glass for PCTs is described in detail in ASTM C1285.[3] The procedure includes steps to remove fines adhering to the glass prior to testing. In the procedure, it is recognized that some glasses may contain soluble phases that are removed during the washing process. If preferential extraction of such phases is likely to occur when the sample is washed with water, the PCT procedure provides the option of either analyzing the wash solution directly, or omitting the water wash and washing only with ethanol to remove fines. For the PCTs with PC CWF, we have chosen the former approach because it permits determining the abundance of the halite phase and its composition. This information is needed for process control.

The water wash method that is used for PC CWF, referred to as the Rapid Water Soluble (RWS) test, consists of two steps: (1) The crushed CWF is washed with absolute ethanol to remove fines without dissolving halite; and (2) it is washed ultrasonically for 2 min with demineralized water to dissolve the halite exposed at the surface of the sample. The water is then decanted and filtered with a 450-nm pore-size filter and is referred to the "RWS solution." The crushed material is dried at 90°C overnight. After drying, the crushed material is divided into approximately 1-2 g lots for use in the PCT. All tests were performed with demineralized water in Teflon vessels. At the completion of the test, the test solution is removed from the vessel with a pipette and passed through a 450-nm pore-size syringe filter. The filtrate solution is referred to as the "PCT solution." The RWS test and PCT are conducted with the same mass ratio of CWF and demineralized water, i.e., 1:10. This allows the results of the RWS and the PCT to be added directly to determine the total release from all phases of the CWF.

The RWS and PCT solutions were analyzed following the same procedures. Aliquots of the test solutions were analyzed for Cl and I without further treatment. The chloride concentration was measured with a chloride-ion selective electrode, and the iodide concentration was determined with inductively coupled plasma-mass spectrometry (ICP-MS). Solution aliquots taken for analysis of cations were stabilized with nitric acid and then analyzed with either ICP-MS or inductively coupled plasma-atomic emission spectroscopy (ICP-AES). The Cs concentrations were measured with ICP-MS. The solution concentrations were used to calculate the normalized elemental mass loss, NL(i), as defined elsewhere.[3]

RESULTS

Determination of Test Repeatability

Samples from the 5-kg baseline PC CWF product and the advanced PC CWF were used to measure the repeatability of the 7-day PCT with PC CWF. For the 5-kg baseline product, preliminary RWS tests with material from the center, mid-radius, and edge showed that the halite was distributed uniformly within 95% uncertainty limits. Crushed materials from the three sections were consolidated for the RWS procedure and the PCT. The NL(*i*) for the RWS solution was added to the NL(i) for the PCT. The mean values of the total NL(*i*) (i.e., the sum of the releases in the RWS and PCT steps) ± two standard deviations (2s) for three sets of triplicate tests are plotted in Fig. 1a. Also shown are the mean total NL(*i*) ± 2s for the entire set of nine tests.

To show that test repeatability does not depend on composition or processing conditions, similar tests were conducted with the advanced PC CWF. One 500-g advanced PC CWF product was entirely crushed, and random samples were taken for use in the RWS/PCT. The results of three triplicate tests and the entire set of nine tests are shown in Fig 1b. The standard deviation (s) was less than 10% of the mean for all elements in all tests. The mean NL(*i*)s for each triplicate lies within 2s of the mean NL(*i*) for the entire set of 9 tests, indicating that the releases are the same within 95% uncertainty limits.

While the precision in conducting the tests is the same for the advanced and baseline PC CWF, the NL(*i*) values may vary due to differences in composition or in processing conditions. For example, the NL(Cl) is larger for the advanced PC CWF (2.7 g/m^2) than for the baseline PC CWF (0.60 g/m^2). The higher chloride release is believed to be due to the greater generation of halite at higher processing temperatures (915 vs. 850°C) and longer processing times (16 vs. 4 h). The NL(B) is significantly smaller for the advanced PC CWF than for the baseline PC CWF, and is discussed below.

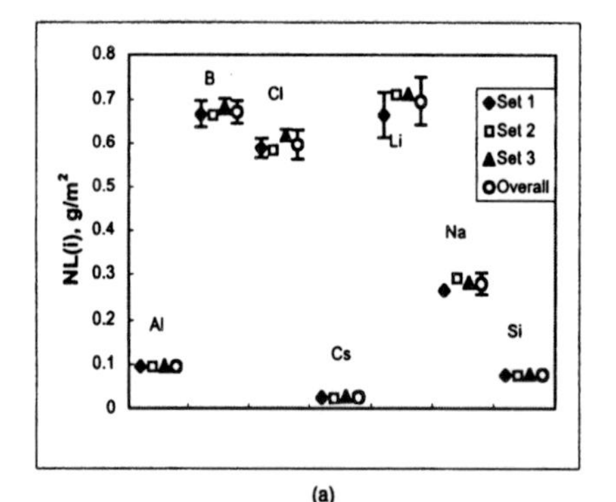

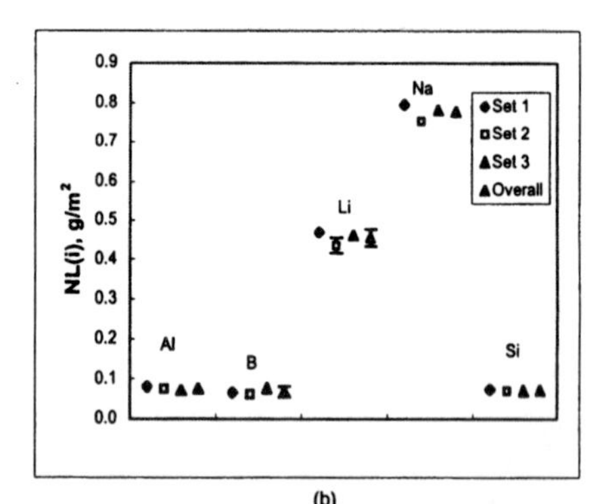

Fig. 1. Total NL(i) in 7-day PCTs for the (a) Baseline PC CWF and (b) for the Advanced PC CWF. For each element, the first three data points represent the mean of the three triplicate tests while the fourth data point represents the mean of the nine tests. The NL(Cl) is off scale in (b) and is not shown for reasons of clarity. Errors bars represent 2s and are within the symbol if not shown explicitly.

Determination of PCT Sensitivity to Product Differences

We investigated the 7-day PCT response of the PC CWF made under different conditions to determine the optimum processing conditions and to measure the sensitivity of the PCT method to processing changes. For each of the 36 products, triplicate RWS tests and triplicate 7-day PCTs were conducted.

The mean NL(*i*) values for the replicate RWS tests and the PCT were summed to give the total NL(*i*). Two types of release behavior were noted. The release behavior of halite is illustrated in Fig. 2, while that for the glass and sodalite phases is shown in Fig. 3. The NL for Cl, Na, and I showed the largest variation with the processing conditions. The NL(Cl) are plotted for the six processing temperatures versus the six times in Fig. 2a. The NL(Cl) increased with both processing time and temperature, though temperature has the greater effect for times > 1 h. Lines are drawn through the 16 h results to guide the eye. Standard deviations (not shown for clarity) for these data are within 10% of the mean. For example, for the product made at 915°C for 16 h, the mean NL(Cl) ± s = 0.32 ± 0.024 in the PCT and 2.64 ±0.085 in the RWS .

In Fig. 2b, the total NL for I and Na are plotted against the total NL(Cl). There is good correlation between the release of Cl and Na and the release of Cl and I. The correlation indicates that constant fractions of I and Na are incorporated into the halite phase. Most of the release of Cl and I occurs in the RWS test.

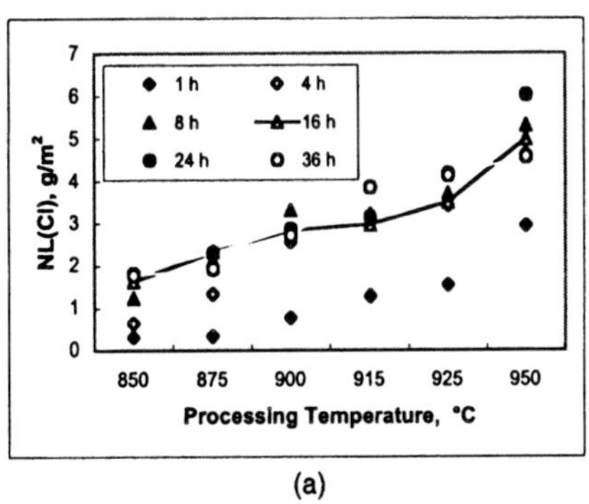

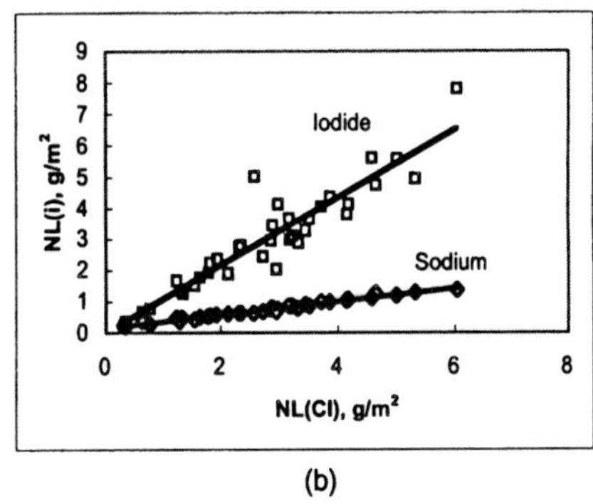

(a) (b)

Fig. 2. (a) Total NL(Cl) for Products Made under Different Processing Conditions and (b) Correlation of Total NL(Cl) with Total NL(Na) and NL(I). The lines are guides for the eye.

Samples of the baseline and advanced PC CWF were analyzed with powder X-ray diffraction (XRD) and compared with reference standards. Most of the peaks for the PC CWF can be assigned to sodalite. Nearly all patterns contained peaks assigned to halite (NaCl) and the intensity of the halite peaks varied with processing conditions. For the products in the T-t study, a reasonable correlation was obtained between the %NaCl determined in a semiquantitative analysis of the XRD peak intensities and the %Cl released in the RWS, which was calculated from the mass of Cl in the RWS solutions and the mass initially present in the solid PC CWF.

The second type of release behavior is illustrated in Fig. 3, which shows the total NL(i) for B and Si. Most of the Al, Si, and B are released from the relatively insoluble glass and sodalite matrices in the PCT, not in the RWS. The total NL(i) values either decreased or were invariant with increasing temperature/time. For example, the NL(B) and the NL(Si) for the six processing times are shown versus the six processing temperatures, in Fig. 3a and 3b, respectively. For products made with a 1-h hold time, the NL(B) decreases with increasing temperature between 850 and 900°C. A similar but much smaller effect is seen with a 4-h hold time. At 900°C and above, the NL(B) is nearly constant. This may indicate that the glass durability depends on processing conditions. We discounted the idea that the lower boron releases resulted from a loss of boron during processing. Analysis of solids from products made at 850 and 915°C showed the same boron fraction within experimental error. The NL(Si) varied similarly but on a more compressed scale. The NL(i) for Al, Cs, and Li (not shown) were similar to NL(Si) in their dependence on processing time and temperature. Cs and Li are released primarily in the PCT. For example, for products made at 915°C for 16 h, the releases constitute about 98 and 88% of the total release, indicating that Cs and Li are immobilized primarily in the glass and sodalite phases.

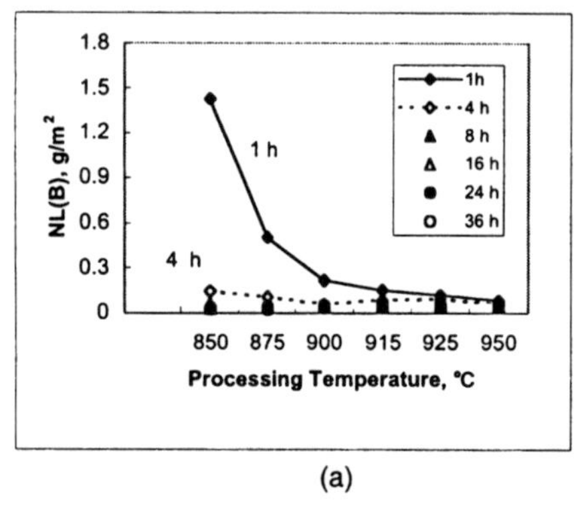

(a)

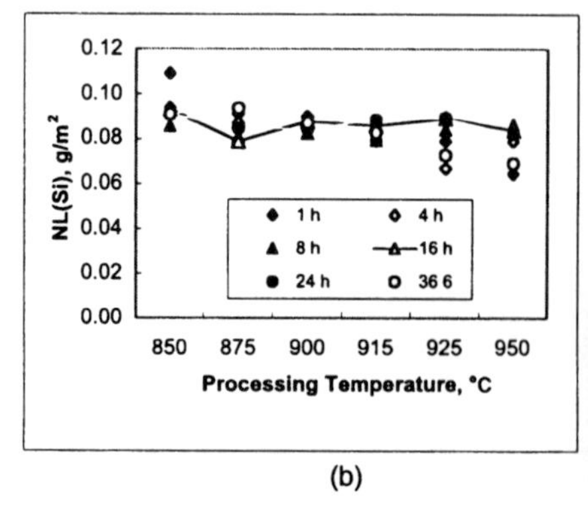

(b)

Fig. 3. (a) Total NL(B) and (b) NL(Si) for Products Made Under Different Conditions. The lines are meant only as a guide for the eye.

Processing conditions of 915°C for 16 h were selected for future production CWF. These conditions represent a compromise between lower halite abundance and a more durable glass phase. A broader range (900-925°C and 8-24 h) will probably be acceptable, especially for larger products.

Determination of Production Consistency

Seven-day PCTs were conducted with baseline PC CWF products (made with 50 mass % glass) that were made on 10 different days using the same source materials and following the same procedure. The objective was to measure the consistency with which PC CWF products could be made.

The NL(i) are given for the RWS test and the PCT in Fig. 4a and 4b, respectively. The data were examined to determine the consistency of the products, i.e., how reproducible the release was in the RWS and the PCT. In the RWS, the NL for Na and Cl were the same for all of the products within 95% uncertainty limits (within 2 standard deviations of the mean). In the PCT, the NL(i)s for all elements were the same within 95% uncertainty limits for all of the products (not shown for clarity). These results indicate that the baseline PC CWF can be made consistently. Further details of these experiments and others described herein are given elsewhere.[5]

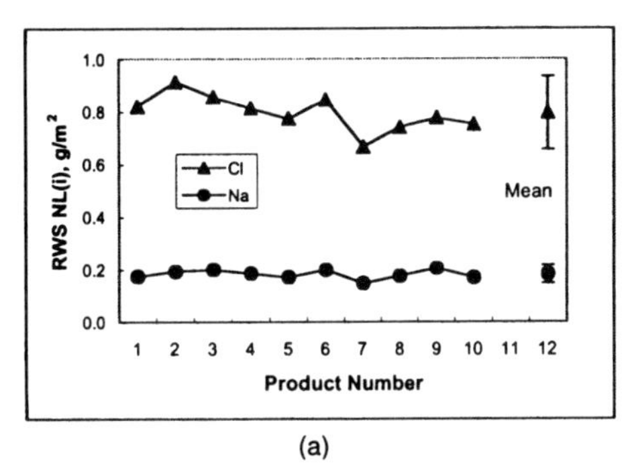

(a)

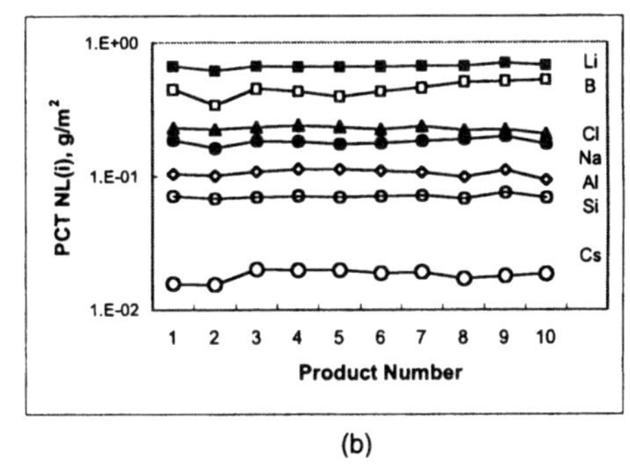

(b)

Fig. 4. NL(i) in 7-day PCTs for Ten Replicate Baseline Products: (a) for the RWS solutions and (b) for the PCT solutions. The mean NL(i) ± 2s for Cl and Na are given in Fig. 4a at the right.

DISCUSSION

Several types of PC CWF were tested with the 7-day PCT. We showed that the 7-day PCT could be conducted repeatedly and that the PCT method is sensitive to differences in composition and fabrication conditions. Having determined repeatability and sensitivity, we were able to determine that products can be made consistently. We have also used the 7-day PCT to compare the release behaviors of the various PC CWF products described above. Among the PC CWF, the most significant differences were (1) the greater releases of Na and Cl in the RWS with PC CWF prepared at high temperatures compared with those prepared at lower temperatures, which reflects the greater amounts of halite that are generated during processing at the higher temperatures, and (2) the smaller release of B from PC CWF prepared at temperatures > 850°C and with smaller amounts of glass. However, these differences are minor when the responses in PCT with the PC CWF are compared with the response of Environmental Assessment (EA) glass. The response of EA glass in the 7-day PCT is specified as a benchmark for high-level waste glasses in the DOE Waste Acceptance System Requirements Document (WASRD)[2] and will probably be used as a benchmark for CWF. Comparison of the NL(i) for B, Li, Na, and Si for baseline and advanced PC CWF and EA glass are given in Table I. All NL(i) in tests with CWF are significantly lower than the NL(i) in tests with EA glass. The highest value for the CWF products tested is NL(Cl), which is 2.7 g/m^2 (for the advanced PC CWF). This is significantly lower than the highest value for the EA glass, which is NL(B) = 8.5 g/m^2. It is likely that the final CWF will be formulated to lower the halite content, so that NL(Cl) for actual waste forms will be less than 2.7 g/m^2. These results show that CWF can be produced to meet the WASRD requirement for the PCT.

Table I. Total NL(i), g/m^2, for B, Cl, Li, Na, and Si for PC CWF and EA glass

	NL(B)	NL(Li)	NL(Na)	NL(Si)	NL(Cl)
Baseline PC CWF	0.67	0.70	0.28	0.080	0.60
Advanced PC CWF	0.068	0.46	0.78	0.072	2.7
EA Glass	8.5	4.8	6.7	2.0	NA[1]

[1]NA = not applicate

CONCLUSIONS

The objective of this series of tests was to evaluate the use of the PCT with the PC CWF. Since the PCT specifies application to homogeneous glass waste forms,[3] it was incumbent for us to show that the attributes of the PCT method can also be applied to heterogeneous, crystalline-glass waste forms. The results showed that the PCT can be conducted repeatedly with PC CWF, within 95% uncertainty limits. A large suite of PCTs conducted with CWF made under different processing conditions showed that the PCT method was sensitive to differences in the abundance of halite and glass content and durability. For a given set of processing parameters and replicate products, the PCT showed production consistency within 95% uncertainty limits. These observations indicate that the PCT method can be used to confirm that proper processing conditions were used during waste form production as well as confirming that the waste forms have adequate and consistent chemical durability.

REFERENCES

[1]W. L. Ebert et al., "Ceramic Waste Form Handbook," Argonne National Laboratory Report, ANL-NT-119, (August 1999).

[2]*Civilian Radioactive Waste Management System Waste Acceptance System Requirements Document*, Rev. 3. DOE/RW-0351, April, 1999.

[3]American Society for Testing and Materials, *Annual Book of ASTM Standards*, **12.01**, "Standard Test Method for Determining Chemical Durability of Nuclear Waste Glasses: The Product Consistency Test (PCT)," C1285-94, pp. 796-813 (1996).

[4]M. K. Andrews and N. E. Bibler, "Radioactive Demonstration of DWPF Product Control Strategy," *Ceram. Trans.*, **39**, Environ. and Waste Manag. Issues in the Cer. Ind., pp 205-213.

[5]M. A. Lewis and W. L. Ebert, Argonne National Laboratory Report, to be published.

IMPURITY EFFECTS IN TITANATE CERAMICS FOR ADVANCED PUREX REPROCESSING

Ewan R Maddrell
British Nuclear Fuels plc
Sellafield
Seascale
Cumbria, CA20 1PG
United Kingdom

ABSTRACT

The effects of small additions of sodium, phosphate and silica on a titanate ceramic phase assemblage tailored for the immobilisation of highly active waste streams containing large quantities of iron, zirconium, chromium and nickel have been studied. Increases in leach rates follow similar trends to those caused by the same impurities in Synroc C. Variations in the abundance of perovskite are noted due to soda and phosphate additions. Coarsening of the waste form microstructure in comparison with undoped samples is also observed.

INTRODUCTION

Alternative dissolution techniques for processing irradiated nuclear fuel can involve significant amounts of the fuel assembly components being taken into solution. This leads to waste streams in which zirconium, iron, chromium and nickel account for 60-70 wt%, on an oxide basis, in addition to the fission products and non-recycled actinides. Economic immobilisation of this waste stream requires the use of a waste form with a high waste loading and can be achieved through the use of a titanate ceramic in which the waste constituents are a functional part of the waste form.[1] In comparison with Synroc C, zirconia can be omitted from the precursor because it is present in the waste, and alumina is also omitted because its charge balancing role is replicated by the iron and chromium in the waste stream.

Initial work concentrated on a specific waste stream composition and the resulting phase assemblage contained zirconolite, $[(Ca,REE)Zr(Ti,Fe,Cr)_2O_7]$ hollandite, $[Ba(Cr,Fe)_2Ti_6O_{16}]$ loveringite, $[Ca(Ti,Fe,Cr,Zr)_{21}O_{38}]$ perovskite $[CaTiO_3]$ and spinel, $[(Fe,Cr,Ti)_3O_4]$. The precursor composition by weight was: 80.8% TiO_2,

To the extent authorized under the laws of the United States of America, all copyright interests in this publication are the property of The American Ceramic Society. Any duplication, reproduction, or republication of this publication or any part thereof, without the express written consent of The American Ceramic Society or fee paid to the Copyright Clearance Center, is prohibited.

7.2% BaO and 12.0% CaO, and a waste loading of 50% was achieved. Leach rates were comparable to those from Synroc C.

Fuel assembly designs, however, vary among reactors and contain differing quantities of Zircaloy, Inconel and stainless steel. After dissolution, this leads to variations in the waste stream composition and the waste form described above was generalised such that it was applicable to a range of fuel assembly designs.[2] Because the throughput of Purex-type plants is based on a fixed daily tonnage of fuel, the generalisation process was designed to allow a constant quantity of precursor material per tonne of fuel, thus facilitating constant operating conditions in the waste immobilisation plant. Consequently there were variations in waste loadings and waste volumes per tonne fuel.

Throughout the above development work, the silica that might be present in the waste stream due to the silicon content of stainless steels was omitted. It was recognised that this was potentially significant due to the known deleterious effect of silica on leach rates from Synroc C.[3] Further, there are routes by which other species of possible concern could be routed to the waste form. Phosphate rich by-products from the solvent extraction process contain fission products and are best routed to the highly active [HA] waste form, and sodium-containing surfactants may prove beneficial during processing giving rise to residual soda in the waste form.

The deleterious effect of silica in Synroc C has been attributed to the formation of pollucite,[4] $CsAlSi_2O_6$, and in the current waste form it is possible that a chromium analogue of this phase might form; the known iron analogue is improbable given that iron is reduced to Fe^{2+} or the metal. The loveringite phase, however, contains two tetrahedral sites per formula unit [5] and an alternative is that silica could be accommodated here. Sodium can be incorporated in a variety of phases such as hibonite, freudenbergite, loveringite,[6] and sodium β-alumina.[4] Sodium can also be accommodated in perovskite.[7] Phosphate additions have been observed to form monazite, however, increased calcium and strontium leach rates suggested the formation of less durable alkaline earth phosphate based phases.[4]

EXPERIMENTAL DETAILS

Simulated waste forms were prepared by wet chemical routes using alkoxides and nitrates as previously described.[1] Four of the eight fuel assembly designs used for the generalisation of the waste form [2] were chosen, representing the Zircaloy rich assembly, the Inconel/stainless steel rich assembly, one other PWR assembly and one of the BWR assemblies. Consistent with this work,[2] the fuel assemblies will be referred to as A, B, C and G. The HA waste arisings, as oxides, from these assemblies are given in Table I, assuming that 85 % of the Zircaloy spalls off as zirconia and is routed to intermediate level waste. To establish the

Table I. Normalised Waste Stream Data per Tonne Fuel Pre-Irradiation (kg)

Fuel Assembly		A	B	C	G
Fe_2O_3		43.64	33.32	71.71	38.35
Cr_2O_3		12.53	12.93	24.31	11.01
NiO		4.98	14.75	20.98	4.37
SiO_2		0.92	0.60	1.41	0.81
MnO		1.11	0.72	1.70	0.98
MoO_3		0.00	0.79	0.91	0.00
Nb/Ta_2O_5		0.00	1.25	1.45	0.00
ZrO_2		74.34	53.02	58.76	65.14
SnO_2		1.07	0.76	0.84	0.94
Fission Products		57.90	57.90	57.90	57.90
Gd_2O_3		9.10	9.10	9.10	9.10
Actinides		3.30	3.30	3.30	3.30
Total		207.97	187.82	250.97	191.09
Mass Precursor		250.97	250.97	250.97	250.97
Waste form mass		458.94	438.79	501.93	442.06
Waste loading (%)		45.3	42.8	50.0	43.2
Impurity	P_2O_5	4.00	4.00	4.00	4.00
	Na_2O	2.50	2.50	2.50	2.50
	SiO_2	0.92	0.60	1.41	0.81

effects of silica, sodium and phosphate it was decided to make one waste form doped with each of the individual impurities and a further waste form doped with all three. The silica content was determined by the level at which it would occur in the waste stream as given in Table I. The phosphate additions were set at the equivalent of 4 kg P_2O_5 per tonne fuel reprocessed and the sodium content set at 2.5 kg Na_2O per tonne fuel.

For each formulation a batch of undoped simulant, corresponding to $1/1000^{th}$ of the waste arisings per tonne, was prepared and divided into four equal portions. Individual and conjoint impurity additions were made to each portion and these were then homogenised and stir dried on a hot plate. Silica was added as a colloidal powder, sodium was added as the nitrate and phosphate was added as $(NH_4)_2HPO_4$. After drying, the powders were calcined at 750 °C in N_2 - 5 % H_2 flowing at 0.5 litre/min for 10 hours and ball milled to improve homogeneity. The

Table II. Summary of leach rates from waste forms ($g \cdot m^{-2} \cdot d^{-1}$)

	Cs	Mo	Ba	Ca	Sr	P_2O_5	Na	SiO_2
Type A Undoped	0.18	0.49	0.09					
Type A + P_2O_5	0.17	1.18	0.12	0.06	0.13	≤0.17		
Type A + Na_2O	0.21	0.85	0.12	≤0.02	≤0.06		≤0.19	
Type A + SiO_2	0.29	1.03	0.10	≤0.02	≤0.07			0.47
Type A + All	0.41	1.84	0.09	0.28	0.61	1.04	2.11	1.46
Type B Undoped	0.29	0.74	0.15					
Type B + P_2O_5	0.19	0.79	0.18	0.03	0.06	≤0.17		
Type B + Na_2O	0.24	0.78	0.17	0.02	≤0.06		≤0.18	
Type B + SiO_2	0.22	0.63	0.18	0.02	≤0.06			0.19
Type B + All	0.14	0.60	0.10	0.15	0.24	0.91	0.63	0.38
Type C Undoped	0.21	0.75	0.04					
Type C + P_2O_5	0.19	0.64	0.08	0.17	0.44	≤0.20		
Type C + Na_2O	0.18	0.37	0.07	0.04	≤0.07		≤0.20	
Type C + SiO_2	0.52	0.58	0.07	0.03	≤0.07			0.99
Type C + All	1.56	1.17	0.08	0.27	0.28	1.58	2.31	2.88
Type G Undoped	0.17	0.34	0.12					
Type G + P_2O_5	0.16	1.78	0.15	0.08	0.16	≤0.17		
Type G + Na_2O	0.20	1.07	0.13	0.03	≤0.06		≤0.20	
Type G + SiO_2	0.25	0.88	0.13	0.03	≤0.06			1.39

powders were blended with 2 wt% metallic titanium and hot isostatically pressed [HIPed] at 1200 °C and 200 MPa for 2 hours. Because of difficulties during processing it was not possible to prepare a waste form corresponding to fuel assembly G with all of the impurities.

The samples were characterised by X-ray diffraction (XRD), using Cu Kα radiation, and scanning electron microscopy with energy dispersive spectroscopy (SEM/EDS). Leach tests were carried out in triplicate on the 75 - 150 μm grains from a ground sample of the waste form for 24 hours at 100 °C in deionised water. The surface area was estimated by assuming that grains were of tetrahedral geometry. Typically a quantity of material was used to provide a surface area of the order of 20 - 30 cm^2 and the volume of leachate was 50 ml.

RESULTS AND DISCUSSION

The means of the normalised leach rates of selected elements for the fifteen waste forms are given in Table II. In cases where the concentration of the element analysed for was below the detection limit, an upper bound leach rate is given. Cs, Mo, and Ba leach rates from the undoped waste forms are also included; Ca and Sr in these leachates were below detection limits. It is noted that in some cases there was significant scatter among the three leach test results. For example, the three values of Sr leach rate for the Type C waste form with phosphate were 1.05, 0.15 and 0.14 $g \cdot m^{-2} \cdot d^{-1}$.

For phosphate additions it can be seen that the Ca and Sr leach rates increase and in fuel assemblies A and G there is also an increase in Mo leach rate. Sodium additions cause no measurable changes in leach rates whilst silica additions increase Cs leach rates, although only for the Type C waste form do leach rates fall outside the range of values observed for the undoped waste forms.[2] The synergistic effect of the impurities is more marked, with barium alone of the analysed elements showing no increase in leach rate. Leach rate increases are most notable in fuel assembly C which gave rise to the highest silica content in the waste form.

X-ray diffraction traces were in good agreement with the previous work

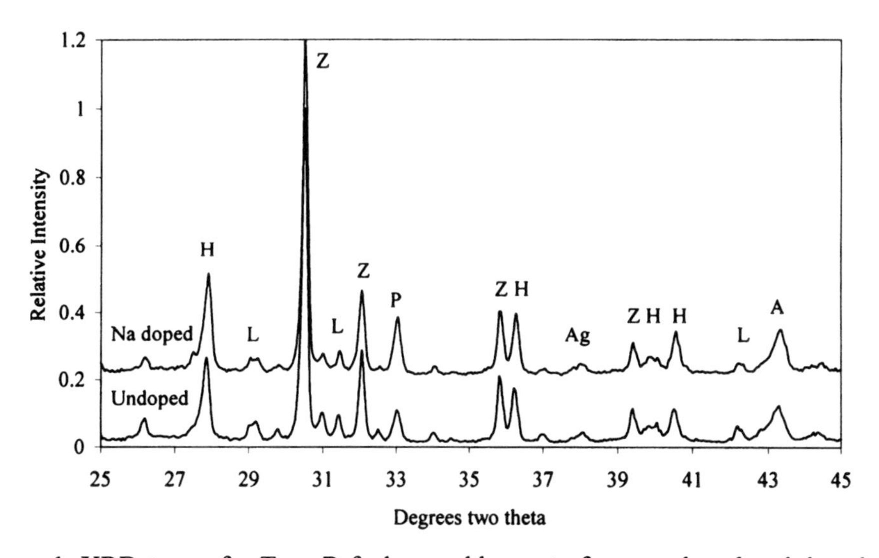

Figure 1. XRD traces for Type B fuel assembly waste form, undoped and doped with Na_2O showing increase in perovskite in doped sample. Zirconolite [Z]; Perovskite [P]; Hollandite [H]; Loveringite [L]; Austenite [A]; silver [Ag].

showing the formation of zirconolite, hollandite, loveringite and perovskite. Further, fuel assembly C contained spinel due to the higher levels of iron and chromium. In addition to the oxide phases, all waste forms contained metallic phases, principally silver, which was used as a simulant for rhodium and palladium, and an austenite solid solution. Quantitative assessment of the XRD traces is necessarily speculative, however, there was clear variability in the level of perovskite. This was most noticeable as an increase in perovskite with sodium as the impurity as shown in Figure 1; the traces are normalised to the zirconolite-2M (221) peak. There was also an apparent decrease in perovskite with phosphate as the impurity. No unequivocal evidence could be found for the formation of previously reported phases such as monazite and pollucite.

A detailed microstructural evaluation of these samples by SEM/EDS is beyond the scope of this paper. Figure 2, however, showing the microstructures of the Type C waste form HIPed at 1200 °C with no impurities and the full complement of impurities, reveals that the impurities promote appreciable grain

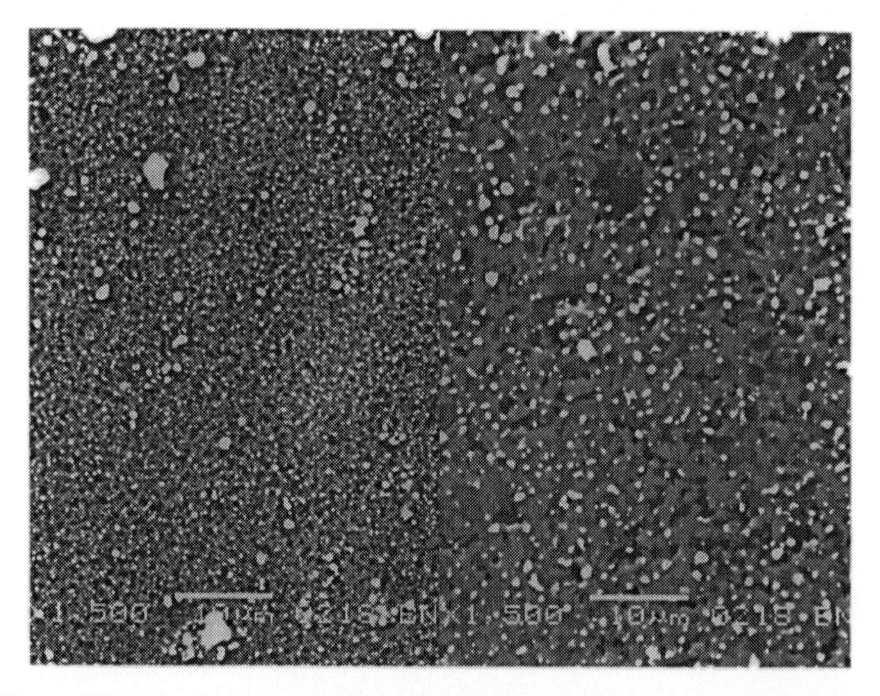

Figure 2. Back scattered electron micrographs of Type C fuel assembly waste form; undoped, left; full complement of impurities, right.

growth. Although not quantified, secondary electron imaging gave the impression of a higher level of porosity in this sample.

Previous work on impurities in Synroc C also noted an increase in Ca and Sr leach rates with phosphate additions and attributed this to the formation of less durable Ca/Sr phosphate phases despite the concomitant observation of monazite.[4] Formation of monazite cannot be confirmed here by XRD because of peak overlaps. Clearly it would be desirable to ensure that all phosphate was precipitated as the more durable monazite. It is postulated that both Ca/Sr phosphates and monazite can precipitate when the ammonium phosphate is added to the waste forms during preparation and that these phases may persist throughout processing with the Ca/Sr phosphates contributing to elevated leach rates. It would therefore be informative to prepare samples in which the phosphate content had been precipitated as monazite at an early stage of specimen history to establish whether Ca/Sr phosphates can be avoided. Clearly the diversion of Ca and Sr to such phases provides a plausible explanation for the decrease in perovskite.

With sodium additions, the increase in intensity of the (200)/(121) perovskite doublet suggests that sodium is accommodated here, probably by conjoint substitution with rare earth elements on the Ca site.

For silica additions, only the Type C waste form, in which the silica content is 0.3 wt%, shows an unequivocal increase in caesium leach rate compared to the range of caesium leach rates observed for undoped waste forms.[2] The silica contents of Types A, B and G are in the range 0.15 - 0.2 wt% which has previously been enough to promote enhanced caesium leaching.[3] Unfortunately, these data were produced by a seven day, 90 °C test which will give lower caesium leach rates than the one day, 100 °C test utilised in this study and thus precludes a meaningful comparison. Further work on microstructural analysis is required to establish whether any silica has been immobilised in the loveringite.

It is clear that the conjoint effect of the additions is significant. This is most apparent in Type C, in which only the barium leach rate is unaffected, and strongly suggests the formation of less durable silicates or amorphous phases. The coarsening of the microstructure in this sample, relative to the undoped formulation, shows that a liquid phase was present during densification of the waste form.

CONCLUSIONS AND FUTURE WORK

The work reported details the effects of sodium, silica and phosphate impurities on a novel titanate ceramic formulation. Increases in leach rates generally mirror patterns observed with these impurities in Synroc C. Phosphate additions cause an increase in calcium and strontium leach rates; sodium additions do not affect leach rates and XRD traces suggest that sodium is immobilised in perovskite. Silica additions below approximately 0.2 % increase caesium leach rates slightly, however, significant increases are seen only when the silica content is 0.3 %. The largest effect on leach rates is seen when all three impurities are present. Further characterisation work is required to establish links between leach rates and microstructure.

REFERENCES

[1]E.R. Maddrell and M.L. Carter, "Titanate Ceramics for the Immobilisation of High Level Waste from Advanced Purex Reprocessing Technology," in *Environmental Issues and Waste Management Technologies in the Ceramic and Nuclear Industries VI*, (St Louis, MO, May 2000) Edited by D.R. Spearing and R.L. Putnam, American Ceramic Society, Westerville, OH, in press.

[2]E.R. Maddrell, "Generalised Titanate Ceramic Waste Form for Advanced Purex Reprocessing," J. Am. Ceram. Soc., **84**[5], (2001)

[3]A.E. Ringwood, S.E. Kesson, K.D. Reeve, D.M. Levins and E.J. Ramm, "Synroc," in *Radioactive Waste Forms for the Future,* edited by W. Lütze and R.C. Ewing (Elsevier, Amsterdam, 1988) pp. 233-334.

[4]W.J. Buykx, D.M. Levins, R.St.C. Smart, K.L. Smith, G.T. Stevens, K.G. Watson and T.J. White, "Processing Impurities as Phase Assemblage Modifiers in Titanate Nuclear Waste Ceramics," J. Am. Ceram. Soc., **73**[2] pp. 217-25 (1990).

[5]W.L. Gong, R.C. Ewing, L.M. Wang and H.S. Xie, "Crichtonite Structure Type as a Host Phase in Crystalline Waste Form Ceramics," pp. 807-15 in *Scientific Basis for Nuclear Waste Management XVIII* (Kyoto, Japan, October 1994), edited by T. Murakami and R.C. Ewing, Materials Research Society Symposium Proceedings **353,** Pittsburgh PA 1995.

[6]W.J. Buykx, K. Hawkins, D.M. Levins, H. Mitamura, R.St.C. Smart, G.T. Stevens, K.G. Watson, D. Weedon and T.J. White, "Titanate Ceramics for the Immobilisation of Sodium-Bearing High-Level Nuclear Waste," J. Am. Ceram. Soc., **71**[8] pp. 678-88 (1988).

[7]E.R. Vance and G.J. Thorogood, "Immobilisation of Sodium in Perovskite," J. Am. Ceram. Soc., **74**[4] pp. 854-55 (1991).

An Investigation of Sintering Distortion in Full-Size Pyrochlore Rich Titanate Wasteform Pellets Due to Rapid Heating to 1350°C in Air

P. A. Walls, J. Ferenczy, S. Moricca, P. Bendeich & T. Eddowes

ANSTO Materials Division

Lucas Heights Laboratories

New Illawarra Rd., Menai

NSW 2234 AUSTRALIA

ABSTRACT

Full-size (Ø65mm x 25mm thick) pyrochlore rich titanate ceramic pellets designed to immobilise PuO_2 and UO_2 were sintered in air at 1350°C for 4 hours. CeO_2 was substituted for UO_2 and PuO_2 in this work. Heating rates between 2 and 15°C/min were used to determine the effect of heating rate on the integrity of the sintered pellet. Distortion of pellets was observed when heating rates in excess of 5°C/min were used. Ultra-violet dye penetrant tests revealed high levels of porosity in the more rapidly heated pellets. Finite element analysis modelling indicated that use of rapid heating rates (10 – 15°C/min), set up thermal gradients between the surface and centre of the pellet, this in turn introduced compressive stresses at the surface and tensile stresses at the core of the pellet. These tensile stresses were the most probable cause of elliptical cracks/pores formed in the centres in these rapidly heated pellets. The FEA results were supported by evidence from the dilatometer, which showed that rapidly heated pellets did not appear to expand as much during heating in the nominal 350°C – 1350°C range as pellets heated slowly.

To the extent authorized under the laws of the United States of America, all copyright interests in this publication are the property of The American Ceramic Society. Any duplication, reproduction, or republication of this publication or any part thereof, without the express written consent of The American Ceramic Society or fee paid to the Copyright Clearance Center, is prohibited.

INTRODUCTION

The aim of this work was to examine the effects of heating rate to the sintering temperature, on the densification and physical properties of pyrochlore-rich titanate ceramic wasteform pellets designed to immobilise PuO_2 and UO_2.

EXPERIMENTAL PROCEDURE

The pellet fabrication process involved micro-blending the component oxides by attrition milling, calcination of the mixed powders, granulation of the powder, and uniaxial compaction of pellets at 21 MPa in a double acting press. The density distribution in a green pellet was checked by cutting samples from the pellet, weighing them and determining their volume with a mercury pycnometer.

A series of heat treatments were then performed in which a 2°C/min heating rate to 350°C was used, and between 350°C and the hold temperature, 1350°C, heating rates between 2 and 15°C/min.

A schematic diagram of the heating patterns is shown in Figure 1.

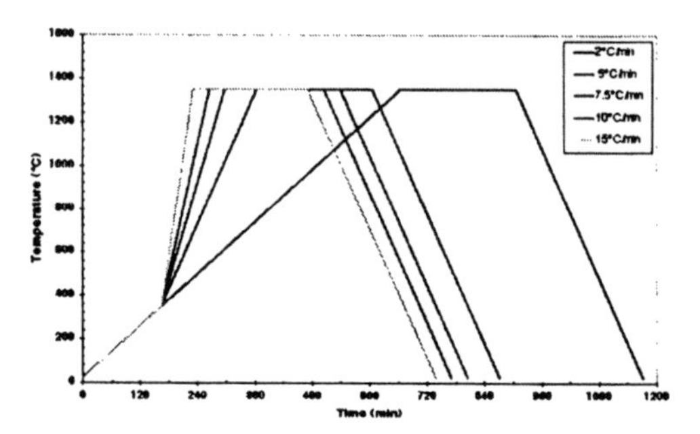

Figure 1. Heating patterns for pellets heated to 1350°C and held at this temperature for 4 hrs.

Green pellets were weighed and measured before being sintered in a scanning laser dilatometer. The dilatometer was used to continuously monitor the diameter of pellets during heat treatment. The laser beam was positioned to scan across the mid-thickness diameter of the pellet 150 times a second with an average dimension reading being made over 4 seconds (i.e. 600 scans). Temperature and dimensional data were also logged throughout the heat treatment.

The density distribution across the sintered pellets was measured with the water displacement method for determining Archimedes density. Scanning electron microscopy and XRD analyses were performed on fracture surfaces taken from the centres of the pellets.

After heat treatment, sintering curves were examined in detail. Significant distortion of the pellets, especially those heated at the higher rates (e.g. 10 and 15 °C/min) was readily evident. Cross-sectional profiles were then measured across a diameter of the pellet with a contact dial gauge.

Ultra-violet dye penetrant was applied to polished sections of pellets to determine porosity distribution. High intensity UV illumination was used together with a digital camera to capture images of the porosity distribution.

RESULTS & DISCUSSION

The density distribution across one of the green pellets was found to differ by 2-3 %. This indicated that the degree of distortion was only due to the heating rate used. Figure 2 shows the cross-sectional profiles of the pellets. In general; the faster the heating rate the greater the degree of distortion observed. Greatest distortion was noted on the lower surface of the pellet. This additional upward "bowing" may be related to the reduced heat transfer from the lower surface of the pellet, which rests on a zirconia grit support bed on top of an alumina setter plate. It should also be noted that the distribution of pores in the pellets was not uniform, as shown in the UV penetrant results, (Figure 3). Large ellipsoidal pores or cracks can be seen in the SEM fractograph of the pellet heated at 15 °C/min.

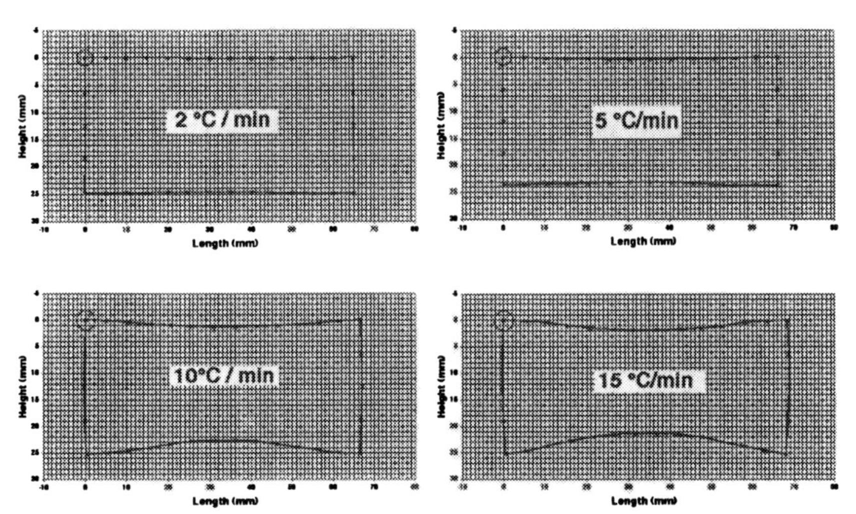

Figure 2. Profiles of pellets heated at different rates to 1350°C with a hold at temperature of 4 hours.

 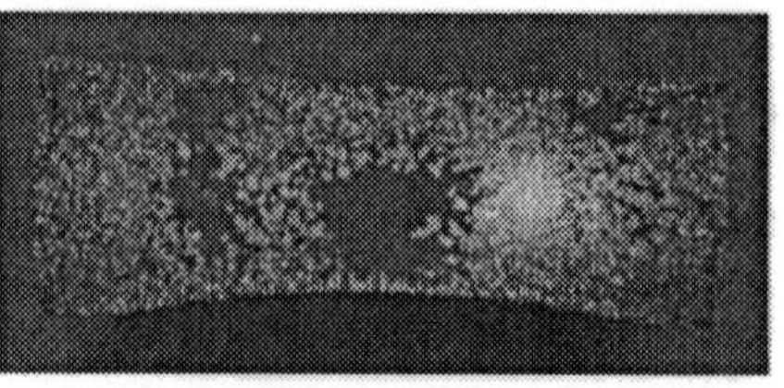

a) 2°C/min b) 15°C/min

Figure 3. UV penetrant distribution indicating level of porosity in sintered pellets (diameter (width) of samples ~Ø65mm).

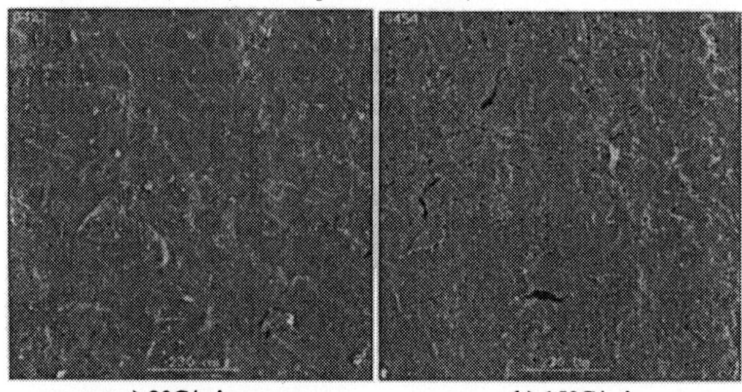

a) 2°C/min b) 15°C/min

Figure 4. SEM fractographs of the centre of pellets sintered to 1350 °C and heated at different rates.

The sintering curve for the pellet sintered at 2°C/min to 1350°C, exhibited a linear increase in diameter with temperature; however, when a heating rate of 5°C/min or greater was used, a significant departure from linearity was noted (Figure 5). This deviation was more pronounced the higher the heating rate. The most probable cause of this "under-expansion" is related to the existence of temperature gradients across the pellet. If the centre of the pellet were cooler than the surface (assumed to be at the furnace temperature), the pellet would be impeded from expanding to its expected extent. If this was the case, the faster the heating rate the more pronounced the thermal gradient across the pellet.

Thermo-mechanical parameters used in the Finite Element Analysis (FEA) model were estimated from knowledge of the starting oxide composition. These parameters included Young's modulus, specific heat capacity and thermal conductivity. The thermal expansion coefficient for the pellet was estimated from the thermal dilation profile.

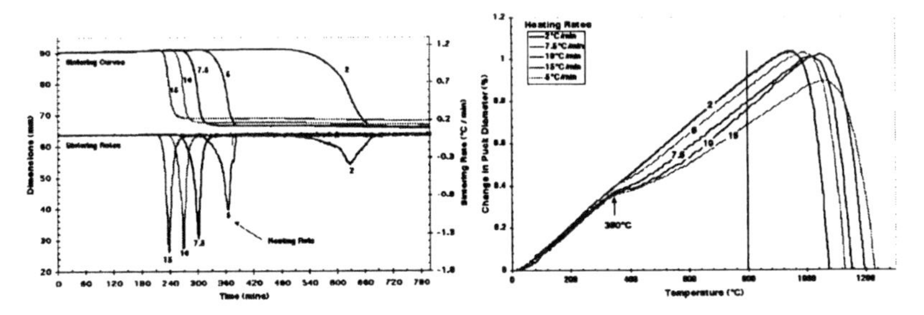

Figure 5. Sintering curves for pellets   Figure 6. Sintering curves for pellets

a) Temp. distribution

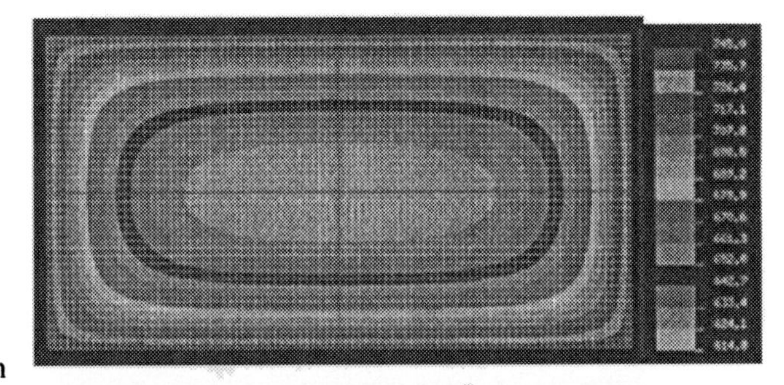

b) Stress distribution

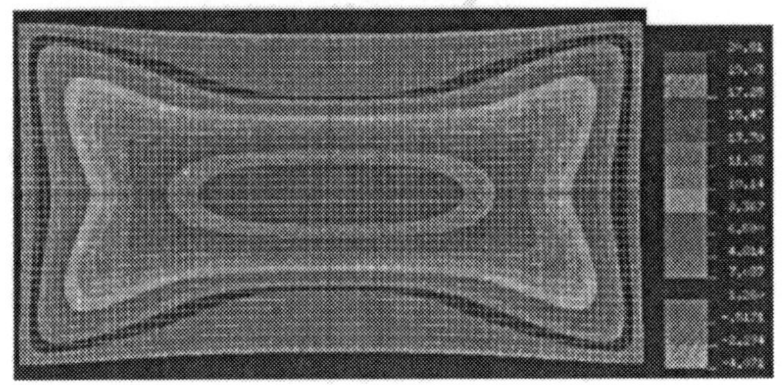

(614 to 745 °C, grey to red)   (-4 to 20 MPa, grey to red)

Figure 7 FEA model of pellet with surface at 745°C, heating rate is 16°C/min (grey at the bottom and red at the top of the color-code (grey-black) on the right)

These parameters were used in the FEA model, in which the temperature of the surface of the pellet was heated at 2°C/min to 350°C and then 16°C/min to 900°C. In this model, the surface of the pellet was assumed to be heated uniformly from all directions. In reality the lower surface of the pellet rests on a powder bed, which is in turn situated on top of an alumina setter plate, so the rate of heat transfer from the lower surface toward the centre of the pellet would be expected to be lower than shown. The full sintering curves for this series of pellets are shown in Figure 5. There was negligible difference in the peak sintering rate (1.5mm / min) when heating at or above 10°C/min, indicating this to be the maximum rate for these pellets under the fabrication conditions used. Figure 8 was plotted from the FEA model. It should be compared with the laser dilatometer traces shown in Figure 6. The similarity of the curves is readily visible, even the initial dip in the curve between 20 – 100°C, although this is only a preliminary FEA investigation, it is highly indicative that too rapid a heating rate is responsible for the observed deformation in the pellets.

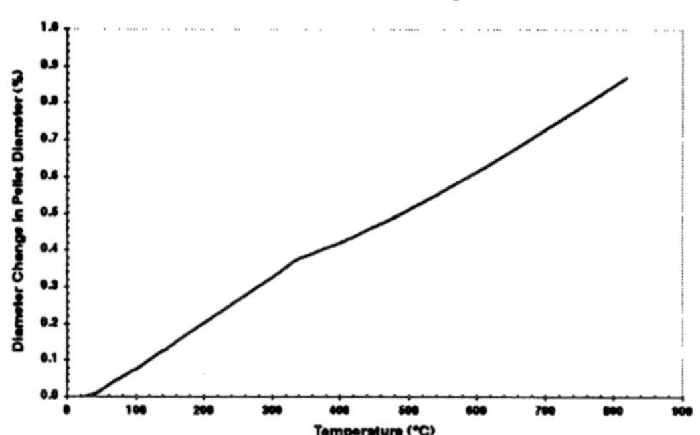

Figure 8. Thermal dilation calculated by the FEA Model.

CONCLUSION

Highest density pellets were obtained when heating at 2°C/min, but densities above 91%TD could still be obtained by heating rates as high as 7.5°C/min. The heating rate used between 350 and 1350°C determined the size and distribution of porosity in the pellet. Marked distortion of pellets occurs with heating rates significantly in excess of 5°C/min. The onset of sintering temperature was also affected by the heating rate. The minimum sintering onset temperature recorded was for the pellet heated at 5°C/min to 1350°C. Initial FEA modelling agreed well with findings from the laser dilatometer.

REFERENCES

1. P. A. Walls et al. " Final Report on Task 3.3 Binder Burnout & Sintering to LLNL for Contract B345772" ANSTO Report No. R00m037 (2000)

This work was supported by funding on Lawrence Livermore National Labs Contract No. B345772.

KEYWORD AND AUTHOR INDEX

Lightning Source UK Ltd.
Milton Keynes UK
UKOW04n0956131115

262605UK00001B/51/P